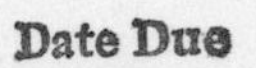

APR 13 1991			

PROPAGATION OF WAVES

PROPAGATION OF WAVES

by

P. DAVID
Ingénieur Général Honoraire des Télécommunications

and

J. VOGE
Ingénieur en Chef,
Centre National d'Études de Télécommunications,
Issy-les-Moulineaux, France

TRANSLATED BY

J. B. ARTHUR

PERGAMON PRESS
OXFORD · LONDON · EDINBURGH · NEW YORK
TORONTO · SYDNEY · PARIS · BRAUNSCHWEIG

Pergamon Press Ltd., Headington Hill Hall, Oxford
4 & 5 Fitzroy Square, London W.1
Pergamon Press (Scotland) Ltd., 2 & 3 Teviot Place, Edinburgh 1
Pergamon Press Inc., 44-01 21st Street, Long Island City, New York 11101
Pergamon of Canada Ltd., 207 Queen's Quay West, Toronto 1
Pergamon Press (Aust.) Pty. Ltd., 19a Boundary Street,
Rushcutters Bay, N.S.W. 2011, Australia
Pergamon Press S.A.R.L., 24 rue des Écoles, Paris 5e
Vieweg & Sohn GmbH., Burgplatz 1, Braunschweig

First English edition 1969

This is a translation of *Propagation des Ondes*
published in 1966 by Eyrolles Éditeur, Paris

L brary of Congress Catalog Card No. 68-18521

Printed in Hungary

08 012114 4

CONTENTS

INTRODUCTION

In this book we assume a knowledge of the principles of radiation and propagation in a homogeneous, isotropic, lossless dielectric—i.e. one which is "transparent" to electromagnetic waves—with no obstacles.

We now go on to consider the real case of wave propagation around the Earth, as influenced by its curvature and its surface irregularities and by passage through atmospheric layers which may be refracting, absorbing or ionized.

This is an extremely difficult problem which radio technologists cannot claim to have solved completely, in spite of 50 years of theoretical and experimental research. Indeed, the calculation of the diffraction of waves around a perfect, homogeneous sphere with known electrical properties is difficult enough; in addition, the nature and relief of the surface add complications which are practically impossible to treat by calculation. The lower layers of the atmosphere introduce refraction, the extent of which varies according to the temperature, pressure, humidity, wind, etc. The upper, ionized layers refract, reflect and absorb waves according to complicated laws which are functions of place, time, season and year, with rather unpredictable irregularities. In addition, the radiocommunication frequency range extends over such a large interval that phenomena which are important at one extremity are negligible at the other, and vice versa.

It is thus impossible to find a completely satisfactory solution to the general problem of wave propagation. One is obliged to reduce the difficulties by introducing various simplifying hypotheses in turn, considering several types of propagation and several frequency ranges separately; and where necessary, from the user's point of view, one has to choose and strike a balance between these partial and incomplete solutions.

We feel that we ought to warn the reader that all his patience and concentration will be required.

In any case, it is absolutely impossible within the scope of this volume to reproduce the derivation of the principal formulae, for which reference should be made to the original texts. What we can do is to give first of all a general sketch of the behaviour of waves in the various media and at their interfaces (assumed planar for simplification): it will thus be possible to understand the different basic phenomena: absorption, refraction, reflection, interference, etc. Application to the case of the terrestrial sphere is then presented as a natural generalization; in the absence of the actual calculations, we do at least give the results in the form of curves and graphs which are easy to use. After discussing the diffraction of the "ground" wave around the Earth, we summarize the role and properties of the troposphere, and then those of the ionosphere from a general physical point of view. Finally, in the light of all these considerations we examine separately each frequency band and discuss the rules and formulae which are most appropriate in each case. Examples and maps are given to illustrate the use of the various methods in the determination of ranges or favourable wavelengths. We also discuss problems encountered in the new field of space communications.

Thus we hope to provide the reader with the means of either solving for himself, with this book in his hands, a large number of problems or else using the forecast bulletins of the various specialist services without any trouble.

It has not been possible to provide a comprehensive bibliography as this would have reached prohibitive dimensions; we have, however, felt compelled to add a few references to the main publications of recent years on each topic; many of these references are to the documents of the Comité Consultatif International des Radiocommunications (C.C.I.R.) and the Union Radio-Scientifique Internationale (U.R.S.I.) which summarize the work and the deliberations of specialists from all over the world.

CHAPTER 1

GENERALITIES

1.1. Basic formulae

The calculation of radio links is usually done in stages. One starts from "basic formulae" which express what happens in "free space", i.e. a vacuum or an isotropic, homogeneous dielectric, such as dry air. One then adds, in the form of correction terms—sometimes very large ones—the effects of the ground, the obstacles, the real atmosphere (particularly the troposphere and ionosphere), etc.

These basic formulae result directly from Maxwell's equations. We shall reproduce them here, (a) to avoid continual references to other books and (b) to recast them in various forms which will be convenient to have at hand.

There are various types of problems that one can tackle. For example, the transmitter and receiver might have to be considered separately; this is the case in broadcasting, where a single transmitter serves a large number of independent receivers for which it is not responsible; the *electromagnetic field* in the region served must then be taken as the intermediate step in the calculation; it will be determined from the characteristics of the transmitter (power, height, gain and directivity of the aerial) and the propagation trajectory, the effect of this field on the receiving aerial is then calculated and, hence, the voltage or power supplied.

In other cases—for example, Post Office radio links—the entire link is between one transmitter and one receiver, both under control of the same service; here it is advantageous to calculate its *overall* efficiency (the ratio of the received power to the transmitted power), or some analogous parameter, directly; and in order to improve it, one treats the transmitting and receiving aerials symmetrically, both being equally important.

So as to be able to treat these various problems indifferently, we shall give the basic formulae, first for transmission, then for reception and finally for the whole combination.

1.1.1. Transmission formulae

It is known that in free space, an electromagnetic field consists of two perpendicular components—an electric field E and a magnetic field M, connected by the relationship:

$$M\,(\mathrm{A/m}) = \frac{E}{120\pi} = \frac{E\,(\mathrm{V/m})}{377\,(\mathrm{ohms})} \tag{1.1}$$

where the denominator may be thought of as the "impedance" of the medium.

The power flux per unit area is given by Poynting's formula:

$$\Phi\left(\frac{\mathrm{W}}{\mathrm{m}^2}\right) = E\times M = \frac{E^2\,(\mathrm{V/m})}{377\,(\mathrm{ohms})}\,. \tag{1.2}$$

This field can be created by various kinds of radiating elements.

Let us assume firstly an "isotropic radiator", i.e. one which radiates uniformly in all directions. If the emitted power is W_t, it is distributed uniformly over spheres of increasing radius and since there are no losses, the flux density at distance d is

$$\Phi = \frac{W_t}{4\pi d^2} \tag{1.3}$$

in units which correspond to those chosen for W_t and d, for instance watts per square metre.

From formula (1.2), this power corresponds to a field

$$E\left(\frac{\mathrm{V}}{\mathrm{m}}\right) = \sqrt{120\pi\times\Phi} = \sqrt{30}\times\frac{\sqrt{W_t\,(\mathrm{W})}}{d\,(\mathrm{m})}\,. \tag{1.4}$$

It is often practical to change the units, adopting millivolts per metre for E, kilowatts for W_t and kilometres for d; we then have the form:

$$E\left(\frac{\text{mV}}{\text{m}}\right) = 173\,\frac{\sqrt{W_t\,(\text{kW})}}{d\,(\text{km})}. \tag{1.4a}$$

(Isotropic radiator)

But the isotropic radiator is only a theoretical concept; it does not exist. The radiating element of an antenna is the *small doublet*, i.e. an element of current of constant intensity I and of length l very small compared with the transmitted wavelength λ. The electric field in its equatorial plane at distance d is

$$E_0 = \frac{2\pi l \cdot I \cdot c}{\lambda \cdot d} \tag{1.5}$$

in coherent e.m.c.g.s. units (where c is the velocity of light), or, in the same units as above,

$$E_0\left(\frac{\text{mV}}{\text{m}}\right) = 60\pi\,\frac{I\,(\text{A}) \times l\,(\text{m})}{\lambda\,(\text{m}) \times d\,(\text{km})}. \tag{1.5a}$$

The product $l \cdot I$ is sometimes called the *moment* of the doublet, and the factor

$$E_0 \times d = \frac{60\pi l I}{\lambda}$$

is called its "cymomotive force".

The three terms l, I, λ determine the radiated power W_t. It can be shown that its value is

$$W_t\,(\text{W}) = 80\pi^2\left(\frac{l}{\lambda}\right)^2 I^2\,(\text{A}), \tag{1.6}$$

as if the radiation introduced an apparent resistance, known as the "radiation resistance":

$$R_r = 80\pi^2\left(\frac{l}{\lambda}\right)^2 \text{ ohms} \tag{1.7}$$

(where l and λ are in homogeneous units, e.g. metres).

Formula (1.5a) can therefore be put in the form:

$$E_0\left(\frac{\text{mV}}{\text{m}}\right) = \frac{212\sqrt{W_t\,(\text{kW})}}{d\,(\text{km})}. \tag{1.8}$$

(small doublet)

Comparison with (1.4a) immediately shows that for the same radiated power, the "small doublet" produces a slightly greater field than the isotropic radiator, by a factor: 212/173 = 1·22, or 20 log 1·22 = 1·75 dB (corresponding to a power gain of 1·5).

The above expressly assumes that the "doublet" is small compared with the wavelength; this restriction may not be valid, and, in particular, one often meets the *half-wave dipole* which has the advantage of being "tuned", i.e. its impedance reduces to a pure resistance. This modifies slightly the field distribution in the plane of the dipole in two respects:

1. The electrical length l to be used in the field formula (called the *effective height* or "radiation height") becomes

$$l = \frac{2}{\pi} \times \frac{\lambda}{2} = \frac{\lambda}{\pi}.$$

2. The radiation resistance becomes: $R_r = 73{\cdot}1$ ohms.

This results in a further increase in the radiated field, which becomes

$$E' = 60\pi \frac{I \cdot \lambda}{\pi \lambda d} = 60 \frac{I}{d}, \tag{1.9}$$

or, as a function of the radiated power $W_t = 73{\cdot}1 \times I^2$ (watts),

$$E'\left(\frac{\text{mV}}{\text{m}}\right) = \frac{222\sqrt{W_t\,(\text{kW})}}{d\,(\text{km})}. \tag{1.10}$$

(half-wave dipole)

The gain with respect to the small doublet is thus 222/212 = 1·045, or 0·38 dB (a power gain of 1·09).

Finally, there is the type of aerial formed by grouping a number of radiating elements in such a way and with such a phase difference that the radiated energy in a given direction is enhanced (at the expense of other directions, of course). The *power gain G* of such a directional aerial is then defined as the ratio between the apparent power in this preferred direction and the power which would be obtained from one of the above simple aerials as a reference.

The simple aerial must be specified: compared with the "isotropic radiator", the power gain of the "small doublet" is 1·5; that of the half-wave dipole is $1{\cdot}09 \times 1{\cdot}5 = 1{\cdot}64$.

The field will then be obtained by multiplying G by the power W_t in the appropriate formula (1.4a, 1.8 or 1.10), or by multiplying the field by $\sqrt{G}$ or increasing it by 10 log G decibels, as desired.

For example, if the gain G_d is given with respect to the small doublet we have

$$E_0\left(\frac{\text{mV}}{\text{m}}\right) = \frac{212\sqrt{G_d \times W_t\,(\text{kW})}}{d\,(\text{km})}. \tag{1.11}$$

The gain depends, above all, on the *area* S of the aerial. In the best possible case, it can be shown that the theoretical maximum, with respect to the isotropic radiator, is

$$G_i = \frac{4\pi S}{\lambda^2}, \quad \text{or in decibels:} \quad 10 \log\left(\frac{4\pi S}{\lambda^2}\right). \tag{1.12}$$

In actual fact, imperfections in the aerial (unequal distribution of intensities, edge effects, etc.) significantly lower this limit and it is not wise to rely on more than

$$G_i = (6\text{–}7)\frac{S}{\lambda^2}, \tag{1.12a}$$

or, with respect to the small doublet, multiplying by $(173/212)^2$,

$$G_d = (4\text{–}5)\frac{S}{\lambda^2} \tag{1.12b}$$

(where S and λ^2 are in the same units).

Moreover, the area S is, in practice, limited by considerations of size, so that it is rare to exceed:

G_d = 10–100 (10–20 dB) for decametre waves,

G_d = 1000–10,000 (30–40 dB) for centimetre waves.

However, this limit is not absolute. In centimetre wave stations engaged in radio-astronomy or communications with space craft, enormous aerials are acceptable and the gain G_d can be as high as 10^8 or 10^9.

Note. In the above formulae, we have introduced the radiated power W_t corresponding only to the "radiation resistance" of the aerial. But the aerial may exhibit other losses and its resistance R' may be greater than the radiation resistance R_r; the power supplied by the aerial, W_a, should then be increased in the same proportion:

$$\frac{W_a}{W_t} = \frac{R'}{R_r}, \tag{1.13}$$

or, to put it another way, there will be in the aerial a loss of

$$L_t = 10 \log \left(\frac{R'}{R}\right) \text{ decibels.} \tag{1.13a}$$

1.1.2. Reception formulae

The receiving system may be analysed in two parts:

1. An *aerial* which captures energy from the field and supplies (with an internal impedance Z_i) a given voltage or power to the rest of the receiver.

2. The *receiver* proper, whose "input circuit" presents an impedance Z_u to the aerial.

It may happen that Z_i and Z_u are pure resistances R_i, R_u or may be reduced to pure resistances by the addition of suitable tuning reactances; in this case, it is known that the transferred power is a maximum when $R_i = R_u$ and that this power (sometimes called "available" power) is a quarter of that which would have been dissipated in the aerial if short-circuited ($R_u = 0$). One often endeavours to realize this condition and this "available power" characterizes the aerial.

Let us calculate it for a few simple cases:

Suppose first that the aerial is *linear*, i.e. its transverse dimensions are negligible compared with its dimension in the direction of the electric field E; it can then be characterized by an "effective length" (or height) l by which the field E must be multiplied to obtain the e.m.f. $\mathcal{E}$:

$$\mathcal{E}\ (\text{V}) = l\ (\text{m}) \times E\ (\text{V/m}). \tag{1.14}$$

If the internal impedance reduces to the resistance R_i and the aerial is short-circuited ($R_u = 0$), the power dissipated in the aerial will be

$$W_0 = \frac{\mathcal{E}^2}{R_i}. \tag{1.15}$$

But, of course, this power has to be used in an external receiver, of resistance R_u; the maximum possible "available power", obtained when $R_u = R_i$, will be:

$$W\,(\mathrm{W}) = \frac{\mathcal{E}^2\,(\mathrm{V})}{4R_i\,(\mathrm{ohm})}. \tag{1.16}$$

Consider, in particular, the "small doublet" discussed in the preceding section. If its internal resistance reduces to its radiation resistance (1.7), the available power becomes

$$W\,(\mathrm{W}) = \frac{\lambda^2\,(\mathrm{m}) \times E^2\,(\mathrm{V/m})}{4 \times 80\pi^2}. \tag{1.17}$$

In the case of the *half-wave dipole*, by virtue of the principle of reciprocity, we have, by analogy with emission:

$$l = \lambda/\pi \quad \text{and} \quad R_r = 73{\cdot}1 \text{ ohms}, \tag{1.18}$$

and consequently the available power is

$$W = 1{\cdot}09 \times \frac{\lambda^2 \times E^2}{4 \times 80\pi^2}. \tag{1.19}$$

For a *directional aerial*, defined by its *geometrical area S*, the reciprocity principle again shows that the "power gain" calculated for transmission is valid for reception: the above values will therefore be multiplied by this power gain. For example, if the gain is given with respect to the small doublet (G_d) the available power will be

$$W = \frac{\lambda^2 E^2}{4 \times 80\pi^2} \times G_d \tag{1.20}$$

and if it is given with respect to the isotropic aerial ($G_i = 1{\cdot}5G_d$),

$$W = \frac{\lambda^2 E^2}{6 \times 80\pi^2} \times G_i. \tag{1.21}$$

Now, it is natural and instructive to compare these powers with the power passing through an area σ intersecting the field. The power density of the field E, given by formula (1.2), is

$$\phi = \frac{E^2\,(\mathrm{V/m})}{120\pi},$$

and the total flux passing through the area σ is

$$W_0 = \frac{E^2}{120\pi}\times\sigma. \tag{1.22}$$

Comparing this with (1.17), (1.19) and (1.21), respectively we see that the areas corresponding to the available powers are

$$\left.\begin{aligned}
\sigma &= \frac{3}{8\pi}\lambda^2 = 0{\cdot}12\lambda^2 && \text{for the small doublet (a)}\\
\sigma &= 1{\cdot}09\times\frac{3}{8\pi}\lambda^2 = 0{\cdot}13\lambda^2 && \text{for the half-wave dipole (b)}\\
\sigma &= \frac{\lambda^2}{4\pi}G_i && \text{for the directional aerial of area } S \text{ (c)}
\end{aligned}\right\} \tag{1.23}$$

These areas are called *equivalent areas* or *capture cross-sections*.†

It will be observed that if the directional area gave the maximum theoretical gain (1.12), formula (c) would give $\sigma = S$; in fact, non-uniformity of the field strength over the area S and inevitable imperfections limit the gain to about half of this maximum [formula (1.12a)], so that the "equivalent cross-section" of a directional aerial is only of the order

$$\sigma = (0{\cdot}5\text{–}0{\cdot}6)S. \tag{1.24}$$

These formulae are useful for evaluating quickly the sensitivity of the receivers associated with a given aerial.

† These definitions are sometimes liable to some variations and corrections. For example, in Report No. 227 of the C.C.I.R. (Appendix), a case is envisaged where the radiation resistance (denoted r) of the aerial differs from the value (r') which it would have in free space; the available power, and therefore the "equivalent" cross-section, are then multiplied by r'/r.

(This same correction would be applied, of course, if r was the total resistance of the aerial, including losses.)

A different definition is also given in the case of a "radar target" (see sections 4.7, 4.7.1 and 8.5.4).

1.1.3. Formulae for the combined link

If we now consider the transmitter and receiver together as a whole, it is natural to combine the foregoing formulae in order to calculate the "efficiency" (or its inverse, the "attenuation") of the link.

However, there are numerous possible variants and risks of confusion.

Consider first the case of ideal, lossless *small doublets*, in optimum orientation. The field is given by (1.5) and (1.6) and the available power at the receiver by (1.17); combining them, we find

$$w\,(\mathrm{W}) = \frac{\lambda^2\,(\mathrm{m})}{320\pi^2} \times \frac{(60\pi)^2}{d^2\,(\mathrm{m})} \times \frac{W_t\,(\mathrm{W})}{80\pi^2} = \left[0{\cdot}0143\left(\frac{\lambda}{d}\right)^2\right] \times W_t\,(\mathrm{W}). \tag{1.25}$$

We can say either that the efficiency is

$$\varrho = \frac{w}{W_t} = 0{\cdot}0143\left(\frac{\lambda}{d}\right)^2, \tag{1.26}$$

or that the transmission attenuation is

$$L\,(\mathrm{dB}) = 10\log\frac{1}{0{\cdot}0143} \times \left(\frac{d}{\lambda}\right)^2 = 10\log\left(\frac{d}{\lambda}\right)^2 + 18{\cdot}44. \tag{1.27}$$

As d is always much greater than λ, the "efficiency" is pitifully small, the attenuation enormous: radiocommunication is evidently a poor way to transmit energy. Notice that these figures are independent of the length of the doublets (as long as they remain below $\lambda/2$).

If, instead of small doublets, we were to consider *isotropic aerials*, we should lose, as we saw above, 1·7 dB at the transmitting end and, therefore, the same amount at the receiving end, and the attenuation (defined as the *reference* attenuation, as it is purely theoretical) would become

$$\begin{aligned} L_{\mathrm{ref}}\,(\mathrm{dB}) &= 10\log\left(\frac{d}{\lambda}\right)^2 + 22 \\ &= 20\log d\,(\mathrm{km}) + 20\log f\,(\mathrm{Mc/s}) + 32{\cdot}45. \end{aligned} \tag{1.28}$$

On the other hand, with *half-wave dipoles*, we should gain a factor of 1·045 (0·38 dB) at each end, whence the attenuation would be

$$L_{\text{half-wave}}\,(\text{dB}) = 10 \log \left(\frac{d}{\lambda}\right)^2 + 17{\cdot}7. \qquad (1.29)$$

Finally, for *directional aerials* with gains G_t and G_r (defined with respect to a reference aerial), these gains would multiply the received power and hence the efficiency; or, in other words, the transmission attenuation ((1.27), (1.28) or (1.29), depending on the reference aerial) would be decreased by their values in decibels.†

These definitions are simple; they can be made more complicated by introducing other factors. For example, we can take account of the fact that aerials are not usually "ideal" and that their true resistance R' is greater than their radiation resistance R_r. This means adding the attenuation calculated from formulae (1.13, 1.13a) at the transmitting end and an attenuation $L_r = 10 \log R'/R_r$ at the receiving end. The total attenuation then becomes

$$L_{\text{total}} = L + L_t + L_r. \qquad (1.30)$$

We can also distinguish the radiation resistance and gain in free space, and their values near the ground, and the name "*propagation* attenuation" is given to the value which the transmission attenuation would have if the gain and the radiation resistance of the actual aerials were the same as in free space. (We shall return later to discuss the effect of the ground.)

There are many more possible variants: we can introduce frequency instead of wavelength, "cymomotive force" $(E \cdot d)$ instead of field E, the areas of the aerials in place of their gain; we can deduce the range from the ratio of the powers w, W_t, and so on.

Thus, knowing the emitted power W_t and the necessary reception power w, formula (1.25) gives the range in free space between

† This is true in free space but may be optimistic in certain types of transmission, either because the field is distorted at the receiver, or because the mode of propagation (tropospheric diffusion) becomes less favourable when the beam is too narrow; this increase in propagation attenuation produces the same effect as if the overall gain in directivity was less than the product $G_t \cdot G_r$.

small doublets

$$d\,(\mathrm{m}) = \lambda\sqrt{\frac{W\times 0{\cdot}0143}{w}} = 0{\cdot}122\lambda\,(\mathrm{m})\sqrt{\frac{W\,(\mathrm{W})}{w\,(\mathrm{W})}}. \quad (1.31)$$

If *directional aerials* are used (defined by their power gains G_t, G_r, with respect to the doublet) the range is multiplied by $\sqrt{G_t \cdot G_r}$; if the aerials are defined by their areas S_t, S_r, and the maximum practical gain given by formula (1.12b) is taken, the range becomes approximately,

$$d = \frac{0{\cdot}6}{\lambda}\sqrt{S_t \cdot S_r \times \frac{W}{w}} \quad (1.32)$$

(in homogeneous units, for example: d and λ in metres, S_t, S_r in square metres, W and w in watts).

These formulae show that considerable ranges are indeed possible (in the absence of interference), even with low radiated power, provided one uses very sensitive receivers and highly directional aerials and, finally, reduces the pass-band. Table 1.1 gives feasible orders of magnitude for three typical cases.

TABLE 1.1

		Between aircraft at high altitudes	Television or multiplex telephony	Ground receiving station for space craft
Wavelength λ (m)		2	0·1	0·1
Transmitter	power W (W)	1	1	1
	area S_t (m²)	—	4	—
	gain G_t	1	2×10^3	1
Receiver	bandwidth B (c/s)	2×10^4	6×10^6	10
	noise factor (dB)	10	13	—
	noise temperature (°K)	—	—	50
	signal/noise ratio (dB)	30	50	6
	received power w (W)	$0{\cdot}75\times10^{-12}$	$4{\cdot}8\times10^{-8}$	$2{\cdot}8\times10^{-20}$
	area S_r (m²)	—	4	5000
	gain G_r	1	2000	$2{\cdot}5\times10^6$
Obtained range d (km)		280	115	115×10^6

1.2. Statistical study of variable fields

Given a fixed intensity in a fixed aerial, the field is perfectly stable in the free space around it. But, as we shall see later, in a large number of practical cases, propagation introduces elements of uncertainty and variables of such complexity that the instantaneous value of the field can no longer be either calculated or predicted; the same is also true of certain irregular perturbation phenomena (atmospheric interference). We then have to be content with statistical data, drawn from sufficiently extensive observations, which enable us to estimate the order of magnitude of the variations; although we cannot state that the field will have such-and-such a value at such-and-such an instant—and, consequently, whether or not a link will work—we can indicate a certain *probability* for these results; this concept of probability, moreover, is becoming more and more common throughout modern physics.

Let us recall briefly the parameters and formulae used in this type of work.

1.2.1. Amplitude probability distribution

The first point to consider is the distribution law for amplitudes, i.e. for the fraction of the time during which a given value is exceeded.

It can be represented by a curve (Fig. 1.1) having as ordinates, for example, the levels of the variable, or their logarithms, and

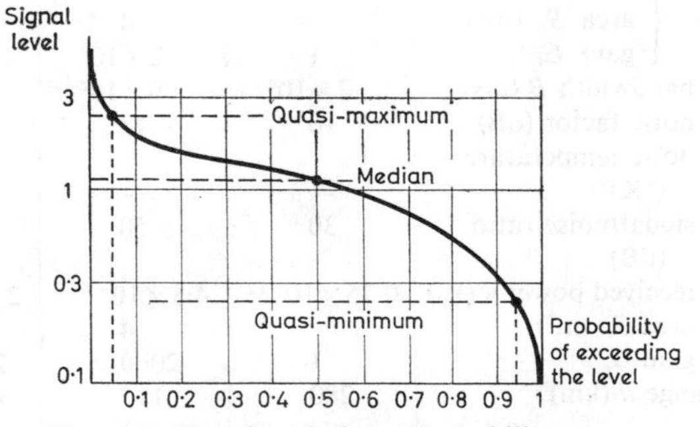

Fig. 1.1. Signal variations. Probability curve

as abscissae the probabilities that the ordinate will be exceeded (or its inverse).

Certain points can be marked on this curve:

(a) the *median* value, corresponding to the probability $\frac{1}{2}$, i.e. the value which has as much chance of being exceeded as not: let this value be E_m,†

(b) the "quartiles", with probabilities $\frac{1}{4}$ and $\frac{3}{4}$; the "deciles", with probabilities 0·1, 0·2, . . ., 0·9,

(c) sometimes a "quasi-minimum" q_m which is the level considered necessary in order to have an acceptable service and which corresponds to a probability of 0·9, or 0·95, or even 0·99, etc., as required; and similarly, there is the "Quasi-Maximum" Q_M which is the upper limit acceptable for perturbation phenomena, e.g. for 5%, 1%, etc., of the time.

It is important to be familiar with this curve, in order to calculate the effect of given changes or of correcting apparatus (gain controls, diversity, correlators); in particular, it is profitable to be able to approximate by various simple, theoretical formulae, of which the following are the principal ones.

Gauss' law

This is the distribution due to a large number of unordered, random causes: the values of the variable y are distributed on either side of the mean μ with relative deviations $x = (y-\mu)/\sigma$ such that the probability of a deviation lying between x and $(x+\mathrm{d}x)$ is

$$p(x) = \frac{1}{\sqrt{2\pi}} \mathrm{e}^{-x^2/2} \cdot \mathrm{d}x. \tag{1.33}$$

This is the well-known "bell-shaped curve", Fig. 1.2, in which the maximum ordinate is about 0·4 for $x = 0$.

σ is the error (sometimes called "typical" or "standard") with probability:

$$p_0 = \frac{1}{\sqrt{2\pi}} \mathrm{e}^{-1/2} = 0{\cdot}243. \tag{1.34}$$

† Not to be confused with the *mean* value E_μ.

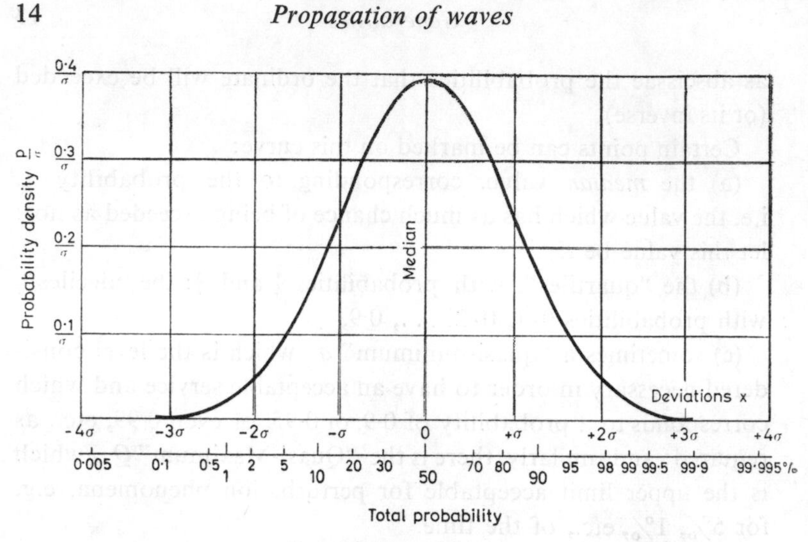

FIG. 1.2. Gauss' law

We can also define the "probability density" as

$$\phi(x) = \frac{p(x)}{\sigma},$$

and the "cumulative distribution" as

$$\Phi(X) = \int_{-\infty}^{X} \varphi(x)\,dx, \tag{1.35}$$

and the "total probability" of having a deviation x, less than a given limit X:

$$P(X) = \frac{1}{\sqrt{2\pi}} \int_{-\infty}^{X} e^{-x^2/2} \cdot dx. \tag{1.36}$$

Normal logarithmic law

This is the above law applied not to the variable itself, but to its *logarithm* (or to its value in decibels). For example, if we are considering a field E, with median value E_m, we put

$$x = \log(E/E_m), \tag{1.37}$$

and deduce that the total probability that the field will be less than the value E will be

$$P\left(\frac{E}{E_m}\right) = \frac{1}{\sqrt{2\pi}} \int_{-\infty}^{E} e^{-1/2(\log E/E_m)} . d\left(\log \frac{E}{E_m}\right). \quad (1.38)$$

The total probability that it will be greater than or equal to E would be given by the same integral between the limits E and $+\infty$.

Rayleigh's law

This is obtained, for example, by studying the variation in length of a vector OR obtained by the geometric addition of a large number of small vectors OA of approximately equal amplitudes and random phases.

As before, if the variable is E and its median value E_m, the probability of a level lying between E and $(E+dE)$ is

$$p\left(\frac{E}{E_m}\right) = 1{\cdot}386 \frac{E}{E_m} \times e^{-0{\cdot}693(E/E_m)^2} . d\left(\frac{E}{E_m}\right). \quad (1.39)$$

The *total* probability that the field is greater than E is then

$$P\left(\frac{E}{E_m}\right) = e^{-0{\cdot}693(E/E_m)^2}. \quad (1.40)$$

We can also find the "variance"

$$\sigma = \frac{E_m}{1{\cdot}1774}$$

by writing

$$p\left(\frac{E}{\sigma}\right) = \frac{E}{\sigma^2} \times e^{-1/2(E/\sigma)^2} \times dE. \quad (1.41)$$

This law is represented by Fig. 1.3.

In order to see how well an experimentally found distribution law can be approximated by one of the above formulae, it is particularly convenient to plot it on a special graph in which the law in question would be represented by a straight line. Thus the "normal logarithmic" distribution will be a straight line in the coordinate system in which abscissae are plotted on the scale of (1.38) and ordinates on a log E/E_m scale (Fig. 1.4), whilst on the same graph Rayleigh's law will be represented by curve 1;

on the other hand, if we use the scale of (1.40) for abscissae, it is Rayleigh's law which becomes linear; and yet the two scales are similar, which can lead to confusion. Figure 1.5 shows how they compare.

The points and curves 2 and 3 in Fig. 1.4 are experimental data which we shall be discussing later; it can be seen that they lie between the Rayleigh and "normal logarithmic" laws. According to the C.C.I.R., rapid variations in a field (a few minutes) follow Rayleigh's law and slow variations (15–60 min) the

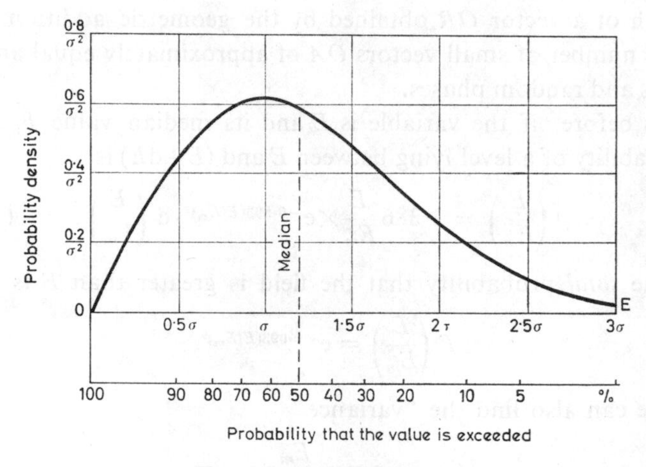

FIG. 1.3. Rayleigh's law

"normal logarithmic" law; however, it is an unfortunate fact that some authors make it a point of honour to find a fit with one of the laws and do not hesitate to twist their results to achieve it. We can see no point in this; the phenomena of propagation are far more complex than the hypotheses of Gauss and Rayleigh.

1.2.2. Speed of variation

The statistical distribution of amplitudes is not sufficient to characterize a variable field; if, for example, it is known that the signal will be below the desired value for 5% of the time, it is not irrelevant whether this fraction consists of a tenth of a

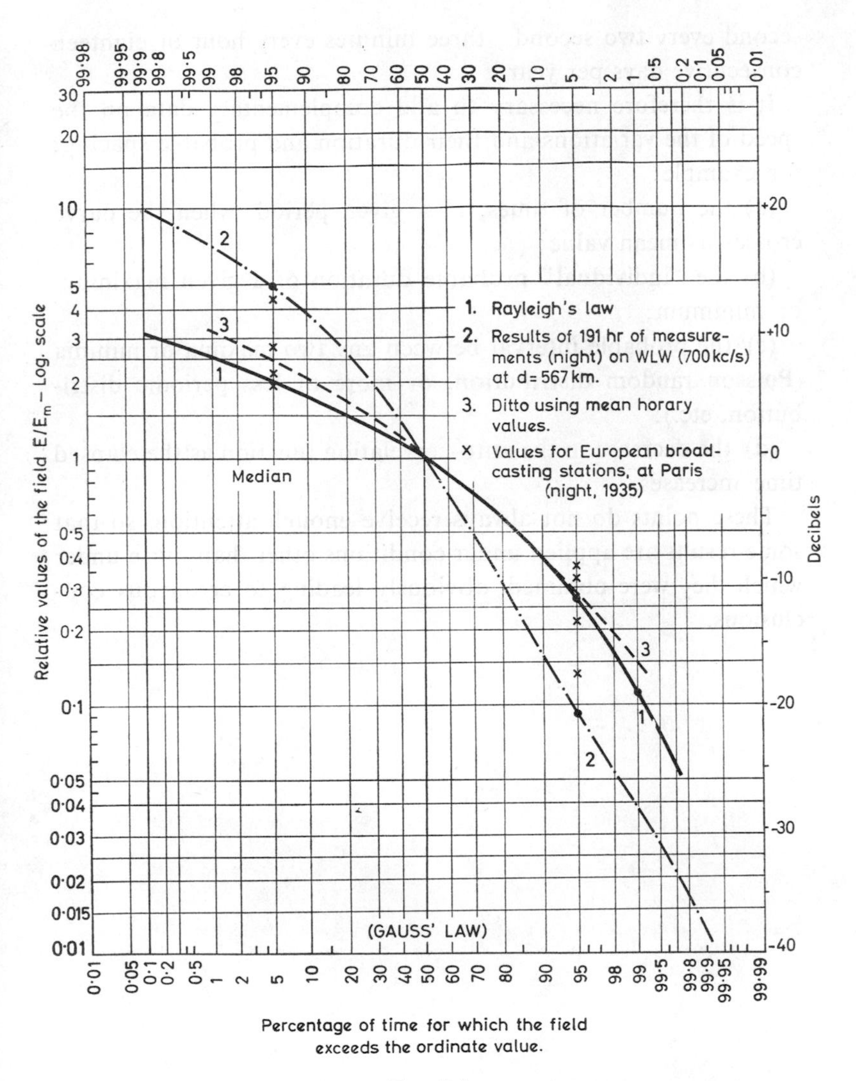

FIG. 1.4.

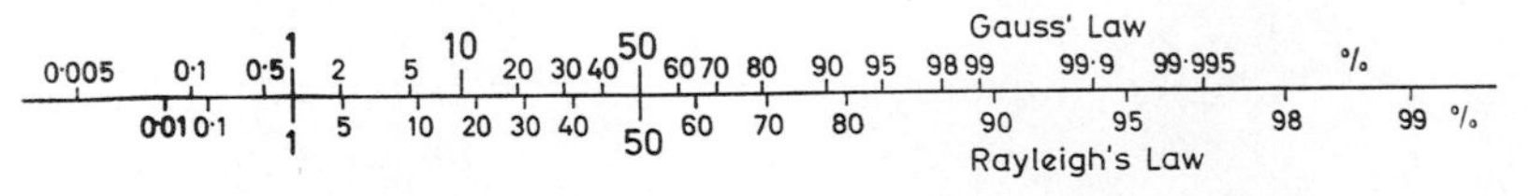

FIG. 1.5. Comparison of the Gauss and Rayleigh scales

second every two seconds, three minutes every hour or eighteen consecutive days per year.

It is therefore necessary to add complementary data on the speed of the variations and their duration and probable spacing; for example:

(a) the number of times, in a given period, when the curve crosses its mean value;

(b) the "individual" probable duration of a given maximum or minimum;

(c) the probable interval between any two maxima or minima (Poisson random distribution, or more or less periodic distribution, etc.);

(d) the decrease in the auto-correlation function as the elapsed time increases.

These points do not always receive enough attention, so that some results are applied under conditions other than those under which they were obtained, obviously leading to erroneous conclusions.

CHAPTER 2

CURRENTS AND PROPAGATION IN VARIOUS MEDIA

2.1. Properties of various types of currents

The study of wave propagation can be deduced from the study of the various types of currents possible in the media involved: air which can be dense and humid or rarified and ionized; a surface which is more or less conducting (land, sea, etc.); various obstacles.

We shall therefore commence by summarizing these various currents, their values and the relationships between them.

In every case, an electron of charge $(-q)$, placed in a sinusoidal field $\mathcal{E} = \bar{E}\cdot e^{j\omega t}$, is acted on by a force:

$$F = -q\bar{\mathcal{E}} = -q\cdot\bar{E}\cdot e^{j\omega t}.$$

But its consequent motion and, therefore, the current vary according to the properties of the medium.

Displacement current

In a dielectric, electrons are acted on by a restoring force proportional to their displacement x, say $F_1 = -Kx$.

The total force is therefore $(F-F_1)$ and the electron acquires an acceleration:

$$\gamma = \frac{F-F_1}{m} \quad (m \text{ being its mass}), \tag{2.1}$$

that is

$$\frac{d^2x}{dt^2} = -q\frac{\bar{\mathcal{E}}}{m} - \frac{K}{m}x$$

or:

$$\frac{d^2x}{dt^2}+\frac{K}{m}x = -q\frac{\bar{\mathcal{E}}}{m}, \tag{2.2}$$

the equation of sinusoidal motion when the displacement is in phase with the field; if $K/m \gg \omega^2$ the amplitude is

$$x = -\frac{q}{K}\bar{E}\cdot e^{j\omega t}$$

and therefore the velocity

$$v = \frac{dx}{dt} = -j\omega\frac{(-q)}{K}\bar{E}\cdot e^{j\omega t},$$

and if there are N electrons per unit volume, unit cross-section is traversed in unit time by Nv electrons carrying a charge $Nv(-q)$; the intensity per unit cross-section is therefore:

$$i_d = j\left(N\frac{q^2}{K}\right)\omega\bar{\mathcal{E}} = j\varepsilon\omega\cdot\bar{\mathcal{E}} \tag{2.3}$$

(like a "capacity" current, in quadrature with $\bar{\mathcal{E}}$).

The coefficient ε is the dielectric constant of the medium; the dielectric constant of a vacuum multiplied by 4π is taken as the unit in the e.s. c.g.s. system; in the rationalised m.k.s. system it is:

$$\varepsilon_0 = \frac{1}{36\pi}\cdot 10^{-9} = 8{\cdot}854\times 10^{-12}\ \text{F/m}, \tag{2.4}$$

But this is an awkward number to deal with and we usually arrange to incorporate it in the coefficients in the formulae and only work with the ratio $\varepsilon_r = \varepsilon/\varepsilon_0$ of the dielectric constant of the medium considered *relative* to that *in vacuo*. This is what we shall do here.

Conduction current

In a conductor (for example, a metal) the electrons are "free", i.e. the restoring force F_1 does not exist.

The differential equation of motion is, therefore, simply:

$$\gamma = \frac{d^2x}{dt^2} = \frac{F}{m} = -\frac{q}{m}\bar{\mathcal{E}} = -\frac{q}{m}\bar{E}\cdot e^{j\omega t}, \tag{2.5}$$

but the "mean free path" of these electrons is very small, in other words, the motion of an electron is abruptly halted after a time θ very short compared with the period $2\pi/\omega$. The field has not, therefore, changed appreciably during this interval, nor has the force, and v is obtained from γ simply by multiplying by θ:

$$\bar{v} = \bar{\gamma}\theta = -\frac{q}{m}\,\theta\bar{E}\cdot e^{j\omega t},$$

and the above reasoning therefore gives for the intensity:

$$\bar{i}_c = (-Nq)\,\bar{v} = \left(\frac{Nq^2\theta}{m}\right)\bar{\mathcal{E}} = \sigma\bar{\mathcal{E}}. \tag{2.6}$$

The current this time is therefore *in phase* with the field and the coefficient $\sigma = Nq^2\theta/m$ is the *conductivity.*

Electron current in a rarefied ionized medium

An ionized medium, like a conductor, contains free electrons. But, if it is sufficiently rarefied, the "mean free path" of these electrons can be much longer than the period $2\pi/\omega$, so that the motion produced by the force F is no longer perturbed by collisions. v is therefore obtained from γ by normal integration, which gives:

$$\bar{v} = \int \bar{\gamma}\,dt = \frac{-q}{m}\frac{1}{j\omega}\bar{\mathcal{E}}, \tag{2.7}$$

and, thus, the intensity:

$$\bar{i}_e = -N\overline{qv} = -j\left(\frac{Nq^2}{m\omega^2}\right)\omega\bar{\mathcal{E}} = -j\varepsilon'\omega\bar{\mathcal{E}}. \tag{2.8}$$

This current is therefore in *antiphase* with the displacement current (2.3) of the dielectric and subtracts from it, behaving as if the dielectric constant (of the non-ionized medium) were decreased (due to the presence of N electrons arising from the ionized molecules) and reduced to

$$\varepsilon'' = \varepsilon - \varepsilon' = \varepsilon - \frac{Nq^2}{m\omega^2} = \varepsilon - \frac{Nq^2}{4\pi^2 m f^2} \tag{2.9}$$

(This decrease can proceed until ε'' becomes zero at the critical frequency

$$f_c = \sqrt{\frac{Nq^2}{4\pi^2 m\varepsilon}} \tag{2.9a}$$

and we shall see the consequences of this later.)

2.2. Complex media. Ionosphere. Terrain

The three types of current can easily occur together and be superimposed on each other in one and the same medium if electrons of the three categories exist therein.

For example, if a medium (air) is ionized, but not sufficiently rarefied for the number of collisions to drop to zero, there might be N free electrons and ν collisions per unit time; the complete calculation then shows† that the relation (2.8) should be replaced by

$$i_e = -j\left[\frac{Nq^2}{m(\omega^2+\nu^2)}\right]\omega\bar{\mathcal{E}}, \tag{2.10}$$

i.e. ε' is decreased, and there also appears a conductivity:

$$\sigma' = \frac{Nq^2}{m}\times\frac{\nu}{\nu^2+\omega^2}. \tag{2.11}$$

It will be seen later that this conductivity causes considerable absorption in the ionosphere when ω decreases (long waves).

Another medium of particular interest is the *ground* which is a semiconductor having a dielectric constant ε as well as a conductivity σ.

Naturally, these two parameters depend to a great extent on the nature of the terrain and particularly on its humidity (and slightly on frequency).

A large number of measurements have been made over the

† See papers on the ionosphere, for example, Jouaust, *Note préliminaire LNR*, nos. 20, 33, 45, etc.

whole range of radio frequencies. Table 2.1 shows the *mean* results† of several authors.

TABLE 2.1. *Terrain constants*

Wavelength	> 3 m		10 cm		3 cm		1 cm	
	ε_r with respect to vacuum	σ[a] mho/m	ε_r	σ	ε_r	σ	ε_r	σ
Sea water	80	1–5	69	6·5	65	16	22	50
Fresh water	80	0·001–0·1						
Humid soil, clay	30	0·01–0·02	24	0·6				
Fertile cultivated soil	15	0·005						
Grass, meadow, race courses, sports grounds			3–6	0·05–0·11				
Rocky ground	7	0·001						
Urban areas, large towns	5	0·001						
Dry soil	4	0·01						
Very dry soil, deserts	4	0·001–0·0001	2	0·03	about 3	0·007–0·1		

[a] σ is sometimes given in e.m. c.g.s. units, which are equivalent to 10^{11} mho/m; the conductivity of the sea is then $(1-5)\times 10^{-11}$ e.m.u.

In such a complex medium, we frequently have to considre the *total* current, the sum of the displacement and conduction intensities:

$$i_t = i_d + i_e = (j\omega\varepsilon)\mathcal{E} + \sigma\mathcal{E} = j\omega\left(\varepsilon - j\frac{\sigma}{\omega}\right)\mathcal{E}. \qquad (2.12)$$

Expressing ε in terms of the relative ε_r with respect to the

† Obviously, we can only talk about mean values—the reader must not look for any illusory precision in these figures. Even for well-defined terrains there may be some uncertainty: the conductivity of the sea depends on the degree of salinity and differs appreciably between the Mediterranean and the White Sea; the dielectric constant and conductivity of a given terrain depend largely on its humidity, and therefore on the rainfall and the season; a forest behaves very differently depending on whether or not there are leaves on the trees. The classification of soils as "fertile", "rocky", "dry" or "very dry" is obviously very loose.

vacuum value ε_0 [eqn. (2.4)] and noting that:

$$\omega = \frac{2\pi c}{\lambda} \quad (c = \text{velocity of propagation} = 3\times 10^8 \text{ m/sec}),$$

this becomes:

$$i = i_d + i_e = j\omega\varepsilon_0 \times \left(\varepsilon_r - j\frac{\sigma}{\omega\varepsilon_0}\right)\mathcal{E} = j\omega\varepsilon_0(\varepsilon_r - j\cdot 60\sigma\lambda)\mathcal{E} \quad (2.12a)$$

(σ being still in mhos per metre and λ in metres).

There is thus a critical wavelength λ_c for which $\varepsilon_r = \sigma\lambda_c$, i.e. the conduction current is *equal* to the displacement current. It is about 0·33 m in sea water and 10–100 m over land, depending on the humidity (Fig. 2.1).

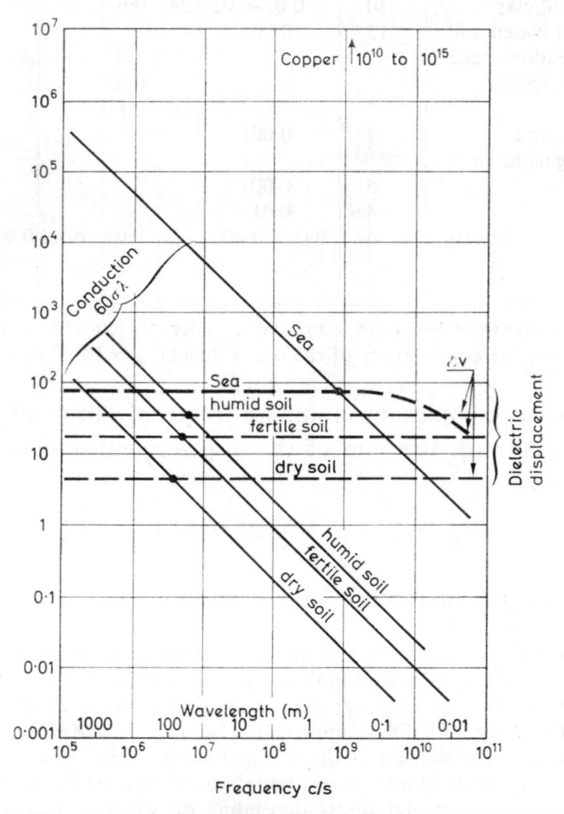

Fig. 2.1. Comparison of conduction and displacement currents for different terrains

At shorter wavelengths, the displacement current is predominant, i.e. the medium is practically a *dielectric;* it can be seen that this is the case for *all soils* at decimetre and shorter wavelengths.

On the other hand, for wavelengths *longer* than λ_c, the conduction current is preponderant, i.e. the medium is practically a *conductor;* it can be seen that this is the case for *all soils* at long wavelengths (kilometre waves).

The transition from one category to the other occurs somewhere in the medium to short wave range, depending on the type of terrain.

These observations are important and we shall come back to them several times.

2.3. Equations of propagation

It is well known that propagation through a homogeneous medium can be deduced from Maxwell's equations: calling the magnetic field $\mathscr{H}$ and the permeability μ,† and assuming that propagation is along the Ox-axis, we have

$$\left\{\begin{aligned} \frac{\partial \mathscr{E}}{\partial x} &= -j\omega\mu\overline{\mathscr{H}}, &\quad (2.13)\\ \frac{\partial \overline{\mathscr{H}}}{\partial x} &= -(j\omega\varepsilon+\sigma)\bar{\mathscr{E}}. &\quad (2.14)\end{aligned}\right.$$

Note that these are similar to the equations for propagation along a line with constants R, L, C, G:

$$\left\{\begin{aligned} \frac{\partial \mathscr{V}}{\partial x} &= -(R+jL\omega)\mathscr{J}, &\quad (2.15)\\ \frac{\partial \mathscr{J}}{\partial x} &= -(G+jC\omega)\mathscr{V} &\quad (2.16)\end{aligned}\right.$$

except that the real term analogous to R (which represents the losses) is missing from eqn. (2.13); this is because the losses intro-

† Note that, in the rationalized m.k.s. system, the permeability of free space is $\mu_0 = 4\pi\times10^{-7} = 1{\cdot}257\times10^{-6}$ henry/m.

3*

duced by magnetism are, in effect, neglected in the elementary theory. They are known to exist, however. We can therefore represent them by introducing into eqn. (2.13) a real coefficient τ so as to obtain:

$$\left\{\begin{aligned} \frac{\partial \overline{\mathcal{E}}}{\partial x} &= -(\tau + j\mu\omega)\overline{\mathcal{H}}, &\quad (2.13a)\\ \frac{\partial \overline{\mathcal{H}}}{\partial x} &= -(\sigma + j\varepsilon\omega)\overline{\mathcal{E}}, &\quad (2.14a)\end{aligned}\right.$$

which is a perfectly symmetrical system analogous to (2.15) and (2.16).

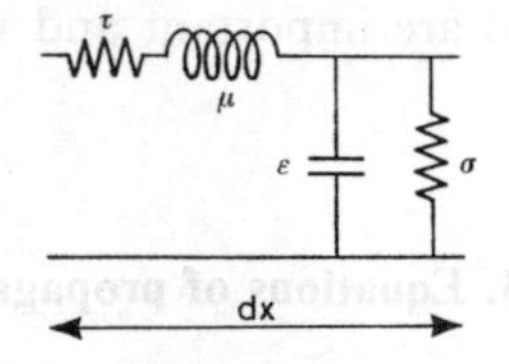

FIG. 2.2. Analogy between propagation in space and along a line

If we want to neglect the losses and revert to (2.13), we merely have to make $\tau = 0$; this is usually the case; but as we shall see there is a notable exception (the case of absorbing substances) in which τ plays a big part.

The formal analogy between the systems (2.13a, 2.14a) and (2.15, 2.16) leads to the prediction that wave propagation in the medium with parameters μ, τ, ε, σ is analogous to propagation along the transmission line shown schematically in Fig. 2.2.

We now proceed to verify this.

It is known that in the steady state the solutions of the differential equations (2.13a, 2.14a) are obtained immediately in the form:

$$\mathcal{E} = E_0 \cdot e^{-\gamma x} \cdot e^{j\omega t}, \quad \mathcal{H} = H_0 \cdot e^{-\gamma x} \cdot e^{j\omega t}, \tag{2.17}$$

$\gamma = \alpha + j\beta$ being given by the relation:

$$\gamma^2 = (\alpha + j\beta)^2 = (\tau + j\mu\omega)(\sigma + j\varepsilon\omega), \tag{2.18}$$

which separates into

$$\left\{\begin{aligned} \alpha^2 - \beta^2 &= \sigma\tau - \varepsilon\mu\omega^2, \\ 2\alpha\beta &= \omega(\tau\varepsilon + \sigma\mu), \end{aligned}\right. \tag{2.19}$$

analogous to the classical relations of the "propagation constant" in transmission lines; to stress the analogy, we can define the "characteristic impedance" of the medium as the quotient which has, in fact, the dimensions of an impedance):

$$\frac{\mathcal{E}}{\mathcal{H}} = z_c = \sqrt{\frac{\tau + j\mu\omega}{\sigma + j\varepsilon\omega}}. \tag{2.20}$$

A general solution to the set of equations (2.19) can be obtained without difficulty, but we shall confine ourselves here to a few simple particular cases.

2.4. Lossless medium: $\tau = \sigma = 0$

We find

$$\alpha = 0, \tag{2.21}$$

i.e. there is no attenuation during propagation;

$$\beta = \omega\sqrt{\varepsilon\mu}, \tag{2.22}$$

representing a velocity of propagation:

$$v = \frac{\omega}{\beta} = \frac{1}{\sqrt{\varepsilon\mu}} \quad \text{and} \quad z_c = \sqrt{\frac{\mu}{\varepsilon}}. \tag{2.23}$$

The "refractive index" $n = \sqrt{\varepsilon\mu}$. (2.23a)

If we substitute in these last two equations the free space values,

$$\mu_0 = 4\pi \times 10^{-7} \quad \text{and} \quad \varepsilon_0 = \frac{1}{36\pi} \times 10^{-9},$$

we find, as expected,

$$c = 3 \times 10^8 \text{ m/sec},$$

$$z = 377 \text{ ohms}.$$

The values for atmospheric air are very similar; but the differences, although very small (a few parts in 10,000), play an important part in tropospheric propagation, as we shall see in Chapter 5.

We have verified the analogy between these properties and those of transmission lines (Fig. 2.2); this is a low-pass circuit but since the elements are infinitely small, the cut-off frequency is infinite.

The reader may be wondering how to interpret the case of the *ionized* medium where ε is reduced by $(-\varepsilon') = (-Nq^2/m\omega^2)$ and may tend to zero, which, from (2.22), would give a velocity of propagation greater than the velocity of light and tending ultimately to infinity. It must be remembered that this velocity is only a steady state parameter, the *phase* velocity; the velocity of transmission of the signals is the *group* velocity which varies inversely with the phase velocity† and tends to zero with ε; the correct physical interpretation, then, is that as the ionization increases, the signals are propagated more and more slowly and ultimately are not propagated at all. We shall return to this point at some length in connection with the ionosphere.

2.5. Low-loss medium

When either τ or σ is non-zero, γ contains a non-zero real part α, i.e. attenuation of the transmission occurs.

The cases which are of interest in radiocommunication are, in general, those in which the attenuation is small enough to allow the possibility of long range transmission (as in the case of lines); (2.19) can then be reduced to:

$$\left.\begin{aligned} -\beta^2 &= -\varepsilon\mu\omega^2 \quad \text{as above,} \\ \alpha &= \frac{\omega}{2\beta}(\tau\varepsilon+\sigma\mu) = \frac{1}{2\sqrt{\varepsilon\mu}}\cdot(\tau\varepsilon+\sigma\mu); \end{aligned}\right\} \tag{2.24}$$

in practice, τ is zero in the majority of cases, and we are left

† It is frequently stated that the product of the phase and group velocities is constant and equal to c^2. In the ionosphere, this is only an approximation, due to the continual variation of the properties (see Cotte, Note prélim. Lab. Nat. Radio., no. 16, 1946).

with:

$$\alpha = \frac{1}{2}\sigma\sqrt{\frac{\mu}{\varepsilon}}; \tag{2.25}$$

the exponential attenuation factor α is thus proportional to the conductivity σ. It is independent of frequency if σ is. However, we have seen [eqn. (2.11)] that in an ionized medium such as the upper atmosphere the apparent conductivity increases with wavelength; therefore, from (2.25), the attenuation also increases. This has, in fact, been proved experimentally: the absorption in the lower layers of the ionosphere is, in a given interval, more pronounced for long waves; arising from this is the concept of a "minimum usable frequency" to which we shall return later.

There is, however, one case in which we are forced to increase the electrical and magnetic losses simultaneously, namely that of absorbent ("antiradar") substances; indeed, in order to avoid reflections at their surface, the loss angles

$$\delta_\mu = \arctan\frac{\tau}{\mu\omega} \quad \text{and} \quad \delta_\varepsilon = \arctan\frac{\sigma}{\varepsilon\omega}$$

must be of the same order (see Chapter 3). Under these conditions, we obtain

$$\alpha = \frac{\omega^2\varepsilon\mu}{2\beta}\left(\frac{\tau}{\mu\omega}+\frac{\sigma}{\varepsilon\omega}\right) = \frac{\pi}{\lambda}(\tan\delta_\mu+\tan\delta_\varepsilon). \tag{2.26}$$

And since no one knows how to make a material in which $\tan\delta_\mu$ is greater than unity, we are obliged to limit the conductivity so as to keep $\tan\delta_\varepsilon$ to the same value; formula (2.26) then shows that in order to obtain a really effective attenuation a thickness of several wavelengths is necessary; although we are dealing with the wavelength in the medium (which is less than the wavelength in air), this represents a considerable thickness even at decimetre wavelengths and is quite prohibitive at metric wavelengths.

2.6. High-loss medium

It may also be of interest to calculate the attenuation in media with high losses, such as certain soils or metals; we know at the outset that the attenuation will be too great for transmission over any appreciable distance, but we may want to know what thickness is necessary to reduce the field by a given fraction.

The simultaneous equations (2.19) can obviously be solved without difficulty, but let us take for simplicity the limiting case in which τ is zero and σ much greater than $\varepsilon\omega$; eqns. (2.19) reduce to

$$\alpha^2 - \beta^2 = -\varepsilon\mu\omega^2,$$
$$2\alpha\beta = \sigma\mu\omega,$$

and the difference $(\alpha^2 - \beta^2)$ is therefore small compared with the product $2\alpha\beta$, that is to say, α and β are of the same order; we can therefore say

$$\alpha = \sqrt{\frac{\sigma\mu\omega}{2}};$$

the attenuation thus increases as the square root of the frequency, or, in other words, the penetration depth (for a given attenuation, say $1/e$) is proportional to the square root of the wavelength. This effect is well known in submarine reception, or as the "skin effect" in copper conductors: very thin screens are sufficient to block short waves; screens of a few millimetres of copper, or a layer of sea water a few metres thick are sufficient to block even long waves (see § 8.6).

2.7. Summary

Due to the complexity of the problem and the number of cases which can arise, it is useful to summarize the above in a table listing the hypotheses and conclusions.

TABLE 2.2. *Propagation in various media*

Medium	Type of current	Propagation: Velocity	Propagation: Attenuation
(1) *Pure dielectric (dielectric constant ε)* Vacuum, dry air at normal pressure, etc. (all electrons acted on by an elastic restoring force)	"Displacement" $i_d = j\cdot\varepsilon\omega\bar{E}$	$v = \frac{1}{\sqrt{\varepsilon\mu}}$	Zero
(2) *Good conductor (conductivity σ)* Copper, sea at long wavelengths (free electrons, with very short "mean free path" θ)	"Conduction" $i_c = \frac{Nq^2\theta}{m}\bar{E} = \sigma\bar{E}$	Reduced	$\alpha = \sqrt{\frac{\sigma\mu\omega}{2}}$ High. Penetration less than a few millimetres in copper and a few metres in sea
(3) *Ionized, very rarefied* (N free electrons with long mean free path)	"Electronic" $i_e = -j\frac{Nq^2}{m\omega^2}\omega\bar{E}$ *Total:* $i_d+i_e = j\omega\left(\varepsilon-\frac{Nq^2}{m\omega^2}\right)\bar{E}$	*Phase vel.:* increased *Group vel.:* reduced	Zero
(4) *Ionized, less rarefied* (ν collisions)	$i_c=-j\frac{Nq^2}{m(\omega^2+\nu^2)}\bar{E}$	*Phase vel.:* increased *Group vel.:* reduced	$\alpha = \frac{1}{2}\sigma\sqrt{\frac{\mu}{\varepsilon}}$ with $\sigma = \frac{Nq^2}{m}\times\frac{\nu}{\nu^2+\omega^2}$ increasing with λ

(Table 2.2 cont.)

Medium	Type of current	Propagation	
		Velocity	Attenuation
(5) "*Average*" *terrain* (Large dielectric constant and conductivity)	(1) and (2) superimposed	Slightly reduced	$\alpha = \frac{1}{2}\sigma\sqrt{\frac{\mu}{\varepsilon}}$
(6) *Absorbent substances* (Electric and magnetic losses high and of same order) $\frac{\tau}{\omega\mu} \approx \frac{\sigma}{\varepsilon\omega}$	(1) and (2) superimposed	Slightly reduced	$\alpha = \frac{\pi}{\lambda}\left(\frac{\tau}{\mu\omega}+\frac{\sigma}{\varepsilon\omega}\right)$

CHAPTER 3

PASSAGE FROM ONE MEDIUM TO ANOTHER. REFLECTION. REFRACTION

LET us now examine the case of a plane wave passing from one medium to another through their interface which we shall in the first instance assume to be infinite, plane and perfectly smooth.

The two media are defined by their parameters $\tau_1\mu_1\sigma_1\varepsilon_1$ and $\tau_2\mu_2\sigma_2\varepsilon_2$, that is to say, using the notation (2.20), their impedances are:

$$z_1 = \sqrt{\frac{\tau_1+j\mu_1\omega}{\sigma_1+j\varepsilon_1\omega}}, \quad z_2 = \sqrt{\frac{\tau_2+j\mu_2\omega}{\sigma_2+j\varepsilon_2\omega}}. \tag{3.1}$$

3.1. Normal incidence

Let us take first the simple case of *normal incidence* (as in ionospheric probes or "antiradar" absorbent screens).

The incident fields $\mathcal{E}_1$ and $\mathcal{H}_1$, perpendicular to the direction of propagation, are then both parallel to the plane of separation (Fig. 3.1).

They give rise to *reflected* fields $\mathcal{E}_1'$ and $\mathcal{H}_1'$, and *refracted* fields $\mathcal{E}_2$ and $\mathcal{H}_2$ (all reckoned positive in the same direction) and the conditions for continuity at the interface are simply:

$$\begin{aligned} \mathcal{E}_1+\mathcal{E}_1' &= \mathcal{E}_2, \\ \mathcal{H}_1+\mathcal{H}_1' &= \mathcal{H}_2. \end{aligned} \tag{3.2}$$

If we introduce the above-mentioned "impedances" z_1 and z_2 and replace $\mathcal{E}_1$ by $z_1\mathcal{H}_1$, $\mathcal{E}_1'$ by $-z_1\mathcal{H}_1'$† and $\mathcal{E}_2$ by $z_2\mathcal{H}_z$, the

† Since the fields are reckoned positive in the same direction and the reflected ray is propagated in the opposite direction to the incident ray, the sign of z_1 must be changed.

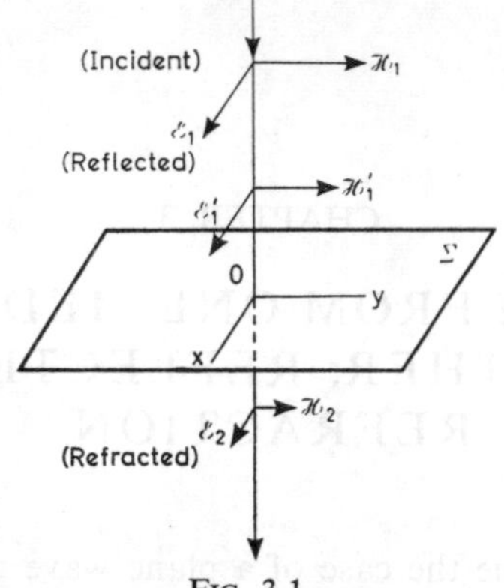

FIG. 3.1.

first equation becomes

$$z_1\mathcal{H}_1 - z_1\mathcal{H}_1' = z_2\mathcal{H}_2$$

or, putting $K = z_2/z_1$,

$$\mathcal{H}_1 - \mathcal{H}_1' = K\mathcal{H}_2.$$

If we now combine this with the second equation, we find

$$\left\{\begin{aligned} \mathcal{H}_2 &= \frac{2}{1+K}\mathcal{H}_1 \\ \mathcal{H}_1' &= \frac{1-K}{2}\mathcal{H}_2 = \frac{1-K}{1+K}\mathcal{H}_1 \end{aligned}\right. \tag{3.3}$$

from which

$$\left\{\begin{aligned} \mathcal{E}_1' &= -z_1\mathcal{H}_1' = -z_1\frac{1-K}{1+K}\mathcal{H}_1 = \frac{K-1}{K+1}\mathcal{E}_1 \\ \mathcal{E}_2 &= z_2\mathcal{H}_2 = z_2\frac{2}{1+K}\mathcal{H}_1 = \frac{2K}{1+K}\mathcal{E}_1. \end{aligned}\right. \tag{3.4}$$

We can now write down the *reflection* and *transmission coefficients* of the two components of the field:

$$\left\{\begin{aligned} R_{\mathcal{E}} &= \frac{\mathcal{E}_1'}{\mathcal{E}_1} = \frac{K-1}{K+1}, & T_{\mathcal{E}} &= \frac{\mathcal{E}_2}{\mathcal{E}_1} = \frac{2K}{K+1}, \\ R_{\mathcal{H}} &= \frac{\mathcal{H}_1'}{\mathcal{H}_1} = \frac{1-K}{1+K} = -R_{\mathcal{E}}. & T_{\mathcal{H}} &= \frac{\mathcal{H}_2}{\mathcal{H}_1} = \frac{2}{1+K}. \end{aligned}\right. \tag{3.5}$$

The reader will recognize the analogy between these formulae and those for the voltage and current along a line of impedance z_1 terminated by an impedance z_2; if we again put $K = z_2/z_1$, the reflection coefficients $R_{\mathscr{E}}$ and $R_{\mathscr{H}}$ are identical with the reflection coefficients for the voltage and current; we should thus have a *standing wave ratio:*

$$\tau = \frac{1+|R|}{1-|R|}. \tag{3.6}$$

It is of interest to take a quick look at this ratio, or, rather, at the amplitude and sign of the reflected field.

Let us assume that the first medium is *air;* then $\tau_1 = \sigma_1 = 0$ and $\mu_1 = \mu_0$ *(in vacuo).*

If the *second medium is a good conductor*, the term σ_2 is predominant, z_2 is very much smaller than z_1, K is very small and $R_{\mathscr{E}}$ tends to -1.

Total reflection occurs with change of sign for the electric vector [in the immediate vicinity of the surface, the total electric field $(\mathscr{E}_1+\mathscr{E}_1')$ is therefore zero as the two terms cancel out].

The transmission coefficient $T_{\mathscr{E}}$ is zero and there is no electric field in the metal; the surface of separation acts as a mirror ($T_{\mathscr{H}}$ is not zero at the actual surface but the magnetic field decreases rapidly with depth of penetration).

If the *second medium is a dielectric*, $\sigma_2 = 0$, $\mu_1 = \mu_0$ and the coefficient K is given by $\sqrt{\varepsilon_1/\varepsilon_2}$. If $\varepsilon_2 = \varepsilon_1 = 1$, we have $K = 1$, giving continuity between the two media. There is no reflection and the surface of separation is transparent.

If $\varepsilon_2 > \varepsilon_1$ (the general case for soil, especially humid soil, or for a dielectric hood (or "radome") protecting a radar) or if $\varepsilon_2 < \varepsilon_1$ (e.g. in penetrating a rarefied ionized layer, § 1.1), a reflected wave appears, its intensity depending on the disparity between ε_2 and ε_1. In particular, if ε_2 drops to zero and becomes negative (ionosphere at low frequencies), K becomes imaginary and so the moduli of the reflection coefficients become unity: we have total reflection again (this is the principle of vertical ionospheric probing).

Finally, it may be of interest to examine to what extent one can realize an *absorbent layer* which, when placed on a metallic surface, would inhibit reflection of waves by that surface: the problem of the "electrically deaf room" or of the "antiradar" varnish.

The above formulae show that in order to avoid reflection on entering this layer, its impedance must be equal to that of the air, i.e.

$$\sqrt{\frac{\tau_2+j\mu_2\omega}{\sigma_2+j\varepsilon_2\omega}} = \sqrt{\frac{\mu_0}{\varepsilon_0}} \qquad \text{(real)},$$

a condition which can be satisfied with $\tau_2 = \sigma_2 = 0$ if $\mu_2/\varepsilon_2 = \mu_0/\varepsilon_0$; but there would then be no losses in passing through the layer (which is not absorbent but transparent); the wave would therefore be reflected at the metal surface which it is intended to protect and would come back out, the layer serving no purpose whatever. What is needed, on the other hand, is a large absorption in a small thickness, which requires τ_2 and σ_2 to be large, with the condition:

$$\frac{\tau_2}{\sigma_2} = \frac{\mu_2}{\varepsilon_2} = \frac{\mu_0}{\varepsilon_0}.$$

Unfortunately, there is no known material which exhibits such properties. One is forced to resort to mixtures of rubber, lampblack and very fine iron filings; one can also achieve better results by adjusting the thickness of the layer so as to obtain phase opposition between the wave reflected at the outer surface and that reflected by the metal surface underneath, or even by putting down several layers in succession with differing composition, simulating a continuous variation of the parameters τ, σ, μ, ε. But these artifices, as well as the undesirable variation of the parameters with frequency, make it very difficult to obtain a really effective absorption over a large band of frequencies.

3.2. Oblique incidence

Let us now generalize to the case of *oblique incidence*. We now have to take into account the orientation of the fields:

1. Electric vector *perpendicular* to the plane of incidence (the polarization is then conventionally said to be "horizontal");

2. Electric vector in the plane of incidence ("vertical" polarization).

If the electric vector occupies an intermediate position, the two projections are considered separately so as to conform to the above limiting cases.

3.2.1. Horizontal polarization (perpendicular to plane of incidence)

If i is the angle of incidence and the direction of the electric vector is chosen as the y-axis, we have at the origin O (Fig. 3.2):

$$\left\{\begin{array}{ll} \mathcal{E}_x = 0, & \mathcal{H}_x = H_0 \cos i \cdot e^{j\omega t}, \\ \mathcal{E}_y = E_0 \cdot e^{j\omega t}, & \mathcal{H}_y = 0, \\ \mathcal{E}_z = 0, & \mathcal{H}_z = H_0 \sin i \cdot e^{j\omega t}. \end{array}\right. \tag{3.7}$$

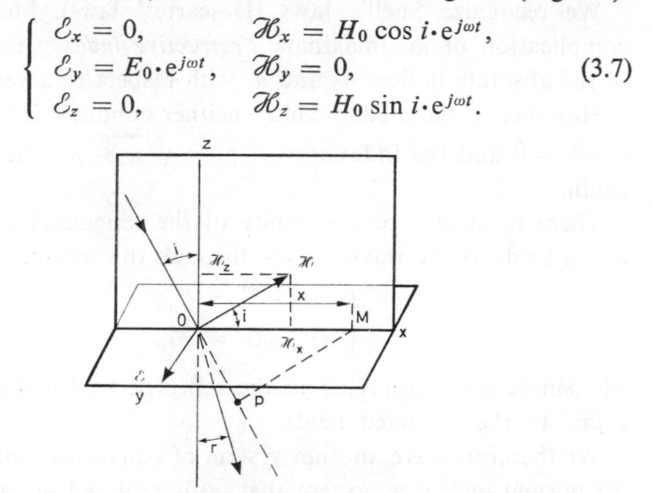

Fig. 3.2. Reflection of a horizontally polarized wave

The wave reaches a point M, at a distance x in the plane of separation, when the wavefront has travelled a distance $OP = x \sin i$, i.e. with an amplitude multiplied by

$$e^{-\gamma_1(x \sin i)}.$$

If we consider next the reflected wave $\mathcal{E}_1'\mathcal{H}_1'$, we find an analogous relation for the angle of reflection i'; and, finally, if we consider the refracted wave $\mathcal{E}_2\mathcal{H}_2$, we have an analogous relation for the angle of refraction r and the propagation constant γ_2 of the second medium.

As there must be continuity during progression of the wave, the ratios of the reflected and refracted fields to the incident field must remain constant, i.e. the amplitude variation must be the same for all three and therefore:

$$\gamma_1 \sin i = \gamma_1 \sin i' = \gamma_2 \sin r,$$

from which, obviously;

$$\left\{\begin{array}{ll} i = i', & \text{the angle of reflection is equal to the angle of incidence,} \\ \dfrac{\sin i}{\sin r} = \dfrac{\gamma_2}{\gamma_1} = & \underset{\text{(a complex constant)}}{\text{the refractive index}} = n = \dfrac{n_2}{n_1}. \end{array}\right. \tag{3.8}$$

We recognize Snell's laws (Descartes' laws), but with the complication of an imaginary "*refractive index*", the quotient of the absolute indices n_2 and n_1 with respect to a vacuum.

However, if the media exhibit neither conductivity nor losses, $\sigma = \tau = 0$ and the index $n = \gamma_2/\gamma_1 = \sqrt{\mu_2\varepsilon_2/\mu_1\varepsilon_1}$ becomes real again.

There must also be continuity of the tangential components of the fields as the wave passes through the surface $\sum$,† giving

$$\left\{\begin{array}{l} \mathcal{E}_y + \mathcal{E}'_y = \mathcal{E}''_y, \\ \mathcal{H}_x + \mathcal{H}'_x = \mathcal{H}''_x \end{array}\right. \tag{3.9}$$

(the single prime applying to the reflected field and the double prime to the refracted field).

We therefore have another system of equations similar to that for normal incidence, except that $\mathcal{H}$ is replaced by $\mathcal{H}_x$.

We can therefore find analogous expressions for the reflection coefficient by replacing the quotient $z_1 = \mathcal{E}/\mathcal{H}$ by $z'_1 = \mathcal{E}/H_0 \cos i = z_1/\cos i$ in the first medium and $z_2 = \mathcal{E}''_x/\mathcal{H}''$ by $z'_2 = \mathcal{E}/H_2 \cos r = z_2/\cos r$ in the second, whence the new coefficient (instead of the K of normal incidence) is

$$K_H = \frac{z_2}{\cos r} \times \frac{\cos i}{z_1} = K \frac{\cos i}{\cos r}, \tag{3.10}$$

which we shall discuss later.

† On the other hand, the normal components (in this case H_2) suffer a discontinuity which maintains the rotational vector constant. See a text-book on theoretical electricity, for example Bloch.

3.2.2. Vertical polarization (in the plane of incidence)

An analogous calculation, except that the fields $\mathscr{E}$ and $\mathscr{H}$ are interchanged and, in consequence, the apparent impedances are not divided but multiplied by cos *i* and cos *r*, gives

$$K_V = K\frac{\cos r}{\cos i}. \tag{3.11}$$

Applying this substitution, the formulae (3.5) give the reflection coefficients $R_{\mathscr{E}}$ and $R_{\mathscr{H}}$ relative to the *tangential* component $\mathscr{E}_x$ of the electric field and the tangential component (and therefore the *whole*) of the magnetic field $\mathscr{H} = \mathscr{H}_y$.

It is also of interest to know the reflection coefficient for the *normal* (vertical) component of the electric field $\mathscr{E}_z$ and the *total* electric field $\mathscr{E}$; but these coefficients can be deduced immediately from the preceding ones; in fact, the ratio $\mathscr{E}/\mathscr{H}$ depends only on the medium and is therefore the same after reflection; the reflection coefficient for $\mathscr{H}$ (or $\mathscr{H}_y$) therefore applies also to $\mathscr{E}$.

Since the same also applies to $\mathscr{E}_x$ (except for sign), it must therefore be the same for $\mathscr{E}_z$ as well. In short, the reflection coefficient is

$$\begin{cases} R_{\mathscr{H}} = \dfrac{1-K}{1+K} \quad \text{for} \quad \mathscr{H}_y, \mathscr{H}, \mathscr{E}_z \text{ and } \mathscr{E}, \\ \text{and} \\ -R_{\mathscr{H}} = \dfrac{K-1}{1+K} \quad \text{for} \quad \mathscr{E}_x. \end{cases} \tag{3.11a}$$

3.2.3. Discussion of the oblique incidence reflection coefficients

Discussion of the values of the reflected field, i.e. of the *reflection coefficients*, is fundamental in propagation problems.

If we therefore repeat the calculation in § 3.1 to find $R_{\mathscr{E}}$ and $R_{\mathscr{H}}$ and replace K by the above values K_H or K_V (3.10, 11 and

11a), we obtain

$$\begin{cases} (R_{\mathscr{E}})_H = -(R_{\mathscr{H}})_H = \dfrac{K\dfrac{\cos i}{\cos r}-1}{K\dfrac{\cos i}{\cos r}+1} = \dfrac{K\cos i-\cos r}{K\cos i+\cos r}, \\ \\ (R_{\mathscr{E}})_V = +(R_{\mathscr{H}})_V = \dfrac{1-K\dfrac{\cos r}{\cos i}}{1+K\dfrac{\cos r}{\cos i}} = \dfrac{\cos i-K\cos r}{\cos i+K\cos r}. \end{cases} \quad (3.12)$$

Let us now confine ourselves to the case where the media considered are the air, the ground and the ionosphere: for these, $\tau = 0$ and μ is the same. Thus the ratios $K = z_2/z_1$ and $\sin i/\sin r = \gamma_2/\gamma_1 = \bar{n}$ [eqn. (3.8)] depend only on the "complex dielectric constants" $\eta = \varepsilon - j\sigma/\omega$ and we have

$$K = \frac{z_2}{z_1} = \sqrt{\frac{\tau_2+j\mu_2\omega}{\tau_1+j\mu_1\omega}\times\frac{\sigma_1+j\varepsilon_1\omega}{\sigma_2+j\varepsilon_2\omega}} = \sqrt{\frac{\sigma_1+j\varepsilon_1\omega}{\sigma_2+j\varepsilon_2\omega}}$$

$$= \sqrt{\frac{\eta_1}{\eta_2}} = \frac{1}{\bar{n}} \quad (3.13)$$

(if the first medium is air, $\sigma_1 = 0$ and this reduces to $1/K = \bar{n} = \sqrt{\varepsilon_2/\varepsilon_1 - j\sigma_2/\varepsilon_1\omega}$) from which

$$\begin{cases} (R_{\mathscr{E}})_H = -(R_{\mathscr{H}})_H = \dfrac{\cos i-\bar{n}\cos r}{\cos i+\bar{n}\cos r}, \\ \\ (R_{\mathscr{E}})_V = -(R_{\mathscr{H}})_V = \dfrac{\bar{n}\cos i-\cos r}{\bar{n}\cos i+\cos r}. \end{cases} \quad (3.14)$$

At grazing incidences, it is often convenient to consider, in place of the angle of incidence i, its complement the angle "of arrival" $\varphi = \pi/2 - i$.

If we, then, replace $\cos i$ by $\sin\varphi$ and $\cos r$ by

$$\sqrt{1-\sin^2 r} = \sqrt{1-\frac{\sin^2 i}{n^2}} = \sqrt{1-\frac{\cos^2\varphi}{n^2}}$$

we obtain

$$\begin{cases} (R_{\mathcal{E}})_H = \dfrac{\sin\varphi-\sqrt{\bar{n}^2-\cos^2\varphi}}{\sin\varphi+\sqrt{\bar{n}^2-\cos^2\varphi}} \\[2ex] (R_{\mathcal{E}})_V = \dfrac{\bar{n}^2\sin\varphi-\sqrt{\bar{n}^2-\cos^2\varphi}}{\bar{n}^2\sin\varphi+\sqrt{\bar{n}^2-\cos^2\varphi}} \end{cases} \tag{3.15}$$

[$\bar{n}$ is defined by (3.13)].

So that we shall not have to go on indefinitely using these two different expressions, we can put them into the same form by the following artifice (Norton):

We introduce a "polarization constant C" defined by:

for horizontal polarization

$$\mathcal{C}_H = (\eta-\cos^2\varphi) = C_H e^{-j(\pi/2-b_H)}$$

$$\text{with } \begin{cases} C_H = |\eta-\cos^2\varphi| \\ b_H = \pi-b', \end{cases}$$

for vertical polarization

$$\mathcal{C}_V = \frac{\eta-\cos^2\varphi}{\eta^2} = C_V e^{-j(\pi/2-b_V)}$$

$$\text{with } \begin{cases} C_V = \left|\dfrac{\eta-\cos^2\varphi}{\eta^2}\right| \\ b_V = 2b''-b', \end{cases}$$

$$\text{and } \begin{cases} \eta = n^2 = \dfrac{\varepsilon_2}{\varepsilon_1} - j\dfrac{\sigma_2}{\varepsilon_1\omega} = \varepsilon_r - j60\sigma_2\lambda, \\[2ex] \tan b' = \dfrac{\varepsilon_r-\cos^2\varphi}{60\sigma_2\lambda} \qquad \tan b'' = \dfrac{\varepsilon_r}{60\sigma_2\lambda}. \end{cases} \tag{3.16}$$

It is only necessary then to rewrite (3.15) in the form:

$$(R_{\mathcal{E}})_H = \frac{\dfrac{\sin\varphi}{\sqrt{n^2-\cos^2\varphi}}-1}{\dfrac{\sin\varphi}{\sqrt{n^2-\cos_2\varphi}}+1} \quad \text{and}$$

$$(R_{\mathcal{E}})_V = \frac{\dfrac{n^2\sin\varphi}{\sqrt{n^2-\cos^2\varphi}}-1}{\dfrac{n^2\sin\varphi}{\sqrt{n^2-\cos^2\varphi}}+1}, \tag{3.15a}$$

to see that, using the notation (3.16), these two expressions can both be written:

$$R_{\mathscr{E}} = \frac{\dfrac{\sin\varphi}{\sqrt{\mathscr{C}}} - 1}{\dfrac{\sin\varphi}{\sqrt{\mathscr{C}}} + 1} = \varrho \cdot e^{j\psi}, \tag{3.17}$$

provided $\mathscr{C}$ takes the value $\mathscr{C}_H$ or $\mathscr{C}_V$ appropriate to the polarization considered.

This formula will be of use to us later.

Applying it, we can easily find the values of the reflection coefficient when a wave propagating in air reaches the surface of another medium.

Consider first the case of a *highly conductive terrain*: *sea* at *long waves*; the factor 60 $\sigma_2\lambda$ is very large, therefore $\bar{n}^2$ or η is large; therefore $\mathscr{C}_H$ is large, and $\mathscr{C}_V$ small. Therefore $(R_{\mathscr{E}})_H$ is approximately equal to -1: the tangential (horizontal) reflected electric field is in antiphase with the incident field.

On the other hand, $(R_{\mathscr{E}})_V$ is nearly $+1$ (unless $\sin\varphi$ is extremely small),† i.e. the vertical reflected field is of the same order and sign as the incident field. This type of reflection is sometimes said to be "metallic".

Let us next take the case of a terrain which is a poor conductor but has a *high dielectric constant*: fresh water, or perhaps sea at short waves, 60 $\sigma_2\lambda$ being less than ε. In this case, n and η are still greater than unity, but not much greater, and become real; we reach essentially the same conclusion for $(R_{\mathscr{E}})_h$ (horizontal reflected field in antiphase with the incident field); however, for $(R_{\mathscr{E}})_v$, the conclusion depends on the angle φ; if it is large, $\sin\varphi \simeq 1$ and $(R_{\mathscr{E}})_v$ is in the region of 1; but if angle φ is decreased until $\sin\varphi$ is less than $\sqrt{\mathscr{C}_v}$, the coefficient $(R_{\mathscr{E}})_v$ decreases, be-

† It is not necessary to consider $\varphi = 0$ in Fresnel's formula which assumes a plane wave coming from distant transmitter; the case of $\varphi = 0$ means that the transmitter is at ground level, a different case which will be treated later.

comes zero,[†] then changes sign and tends towards (-1); when this happens the vertical reflected field is also in antiphase with the incident field.

This case, sometimes called "vitreous" reflection, is, as we shall see, essential to the study of ultrashort wave propagation.

Between these two cases, of course, lies the common case of a *"medium" terrain at "medium" waves* for which $60\ \sigma_2\lambda$ is of the order of ε_2; here, the reflection coefficients $R_{\mathscr{E}}$ vary in a complicated manner with large, progressive phase changes. Their values have been calculated by various authors[‡] and Fig. 3.3 shows some typical results. (The reader must guard against attaching too much precision to them because of the practical uncertainty in the terrain constants; here we have assumed:

$$\text{sea} \quad \varepsilon = 80, \qquad \sigma = 4 \text{ mho/m}$$

$$\text{land} \quad \varepsilon = 10, \qquad \sigma = 0{\cdot}001 \text{ mho/m.)}$$

Finally, we can also have the case of a wave *incident on an ionospheric layer:* as we have seen [eqn. (1.9)], the dielectric constant ε_r decreases and tends to zero (the conductivity σ_2 being assumed to be zero). Thus the reflection coefficient increases with angle φ and its modulus becomes unity when $(n^2-\cos^2\varphi)$ becomes negative, i.e. from eqn. (1.9), when:

$$\left(1-\frac{Nq^2}{m\omega^2}\right)-\cos^2\varphi < 0,$$

$$\frac{Nq^2}{m\omega^2} > \sin^2\varphi, \tag{3.18}$$

† This occurs at the "Brewster" angle, given by

$$n^2 \sin\varphi = \sqrt{n^2-\cos^2\varphi}\,, \quad \text{i.e.} \quad \sin\varphi = \cos i = \frac{1}{\sqrt{n^2+1}}.$$

It is easily verified that in this case:

$$\cos i = \sin r, \quad \text{i.e.} \quad (i+r) = \frac{\pi}{2}.$$

‡ Notably Burrows *(Bell Syst. Tech. J.*, Jan. 1937, pp. 54–61), and McPetrie *(Proc. Wir. Sect. El. Eng.*, March 1938, pp. 47–52): their curves giving the amplitude and phase of R have been frequently reproduced, notably by Terman.

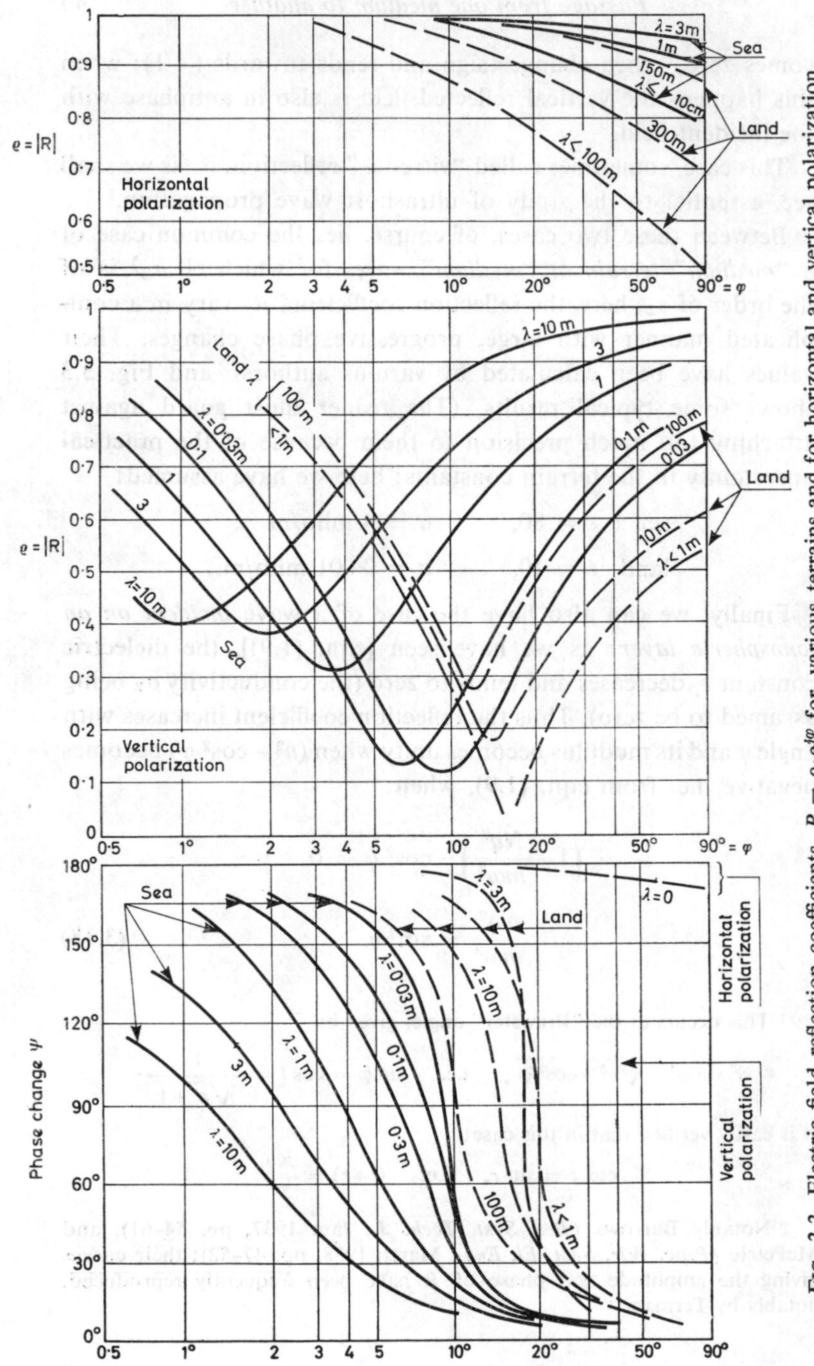

FIG. 3.3. Electric field reflection coefficients $R = \varrho \cdot e^{j\psi}$ for various terrains and for horizontal and vertical polarization

or

$$\omega < \sqrt{\frac{Nq^2}{m}} \times \frac{1}{\sin \varphi} = \omega_c \frac{1}{\sin \varphi}, \tag{3.18}$$

a very important formula giving the condition for total reflection by the ionosphere, to which we shall refer later.

3.2.4. Resultant Electric Field in the Neighbourhood of the Surface of Separation

In the immediate vicinity of the surface of separation, the effective field obviously arises from the *addition* of the incident field (as it would be if the surface of separation did not exist) and the reflected field. It is often important to state precisely the change which results from this addition; this is particularly so for *ground level* reception, i.e. at a height which is low compared with the wavelength (we shall see later the case where, as the height increases, an additional path difference between the incident and reflected waves is introduced).

When the reflection coefficient is in the neighbourhood of +1, the reflected field is nearly equal to the incident field and in phase with it: *the resultant amplitude is therefore nearly doubled.* This is the case with long waves over high-conductivity terrain for a vertically polarized field and we thus find a factor of 2 when we consider an "earthed" dipole.

On the other hand, when the reflection coefficient is in the neighbourhood of −1, the reflected field opposes the incident field and *their resultant is very small:* this is the case for horizontal polarization over high-conductivity terrain and for *both* types of polarization over low-conductivity terrain (we recognize here the well-known property that there is no tangential electric component at the point of contact with a perfect conductor). This conclusion seriously affects all propagation problems: at long wavelengths it precludes the use of horizontal polarization; at short wavelengths it makes it essential to increase the height of the stations so as to introduce a phase change in the reflected field. We shall be returning to this point later.

3.2.5. The case of a finite reflecting surface. Fresnel zones

So far we have assumed a plane wave falling on a perfectly plane, polished, infinite surface (i.e. the irregularities are of dimensions negligible compared with the wavelength).

This is, of course, an ideal case. Let us see when modifications are required for a real case.

Let us assume firstly that the reflecting surface Σ is of finite dimensions but still perfectly polished (Fig. 3.4).

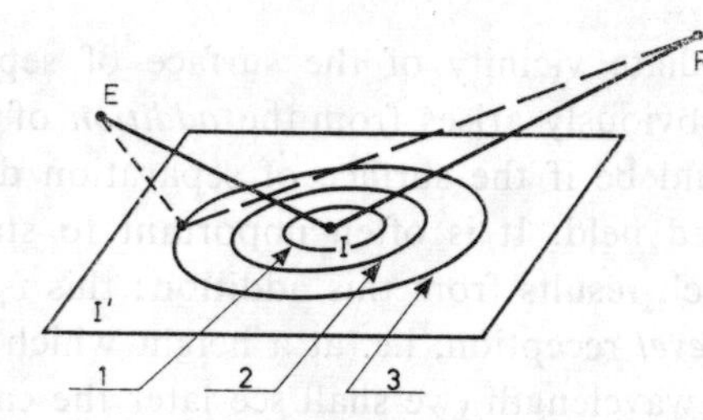

Fig. 3.4. Fresnel zones

Let E be the source, R a receiver, I the point in the plane of incidence normal to Σ for which $i = r$, i.e. the path length EIR is a minimum. This means that in a small cone around I, the path length remains constant, i.e. radiation emitted from E, after reflection at Σ, arrives at R in phase: radiation in this cone will therefore behave like the plane wave of the preceding paragraph, i.e. as if it came from the image E' of E with respect to the surface.

However, if the point of incidence I' is an appreciable distance from I, the path $EI'R$ is longer and the wave arriving at R is out of phase with the previous wave. We therefore no longer obtain exact addition; as long as the point I' lies within a certain zone such that the path difference ($EI'R$–EIR) is less than half a wavelength, some reinforcement nevertheless occurs. But, beyond this is a second zone giving a path difference lying between $\lambda/2$ and λ; waves reflected in this zone will be more or less in phase opposition to those of the first zone and will attenuate them. Then there is a third zone with path differences lying between λ and $3\lambda/2$, where the reflected waves are once more in phase with those from the first zone and reinforce them; and so on. These zones are called "*Fresnel zones*".

The total reflected field therefore becomes a sum of terms of opposite signs and decreasing amplitude (since the conditions become progressively less favourable from one zone to the next); calculation shows that the *first* zone provides the principal contribution and that if all the rest, which partially cancel each other out, were suppressed, the total reflected field would hardly be affected.

This first zone thus gives an approximate idea of the *active part* of the reflecting surface Σ; as long as Σ extends beyond this zone, it can be considered infinite, to a first approximation; on the other hand, if it is smaller, the reflection decreases appreciably.

The dimensions of this zone are of some interest to us and they have been calculated by various authors;† the formulae obtained are, unfortunately, complicated. We shall confine ourselves here to giving a few examples in order to fix orders of magnitude; Table 3.1 relates to two stations at a distance d = 10,000 m, heights h_1 = 100 m and h_2 = 1000 m or 25 m, wavelengths λ = 0·1 m and 1 m.

TABLE 3.1. *Dimensions of the first Fresnel zone (in metres)*

	h_2 = 1000 m		h_2 = 25 m	
	λ = 0·1 m	λ = 1 m	λ = 0·1 m	λ = 1 m
Major axis	166 m	524 m	2000 m	5520 m
Minor axis	18 m	58 m	26 m	88 m

It can be seen that the zones are relatively extensive, especially in the propagation direction (it is easy to observe this effect optically by observing the image of a light source on a rippling sheet of water; the reflection appears to be long and narrow).

† See, for example, § 5.3.1 and Kerr, *Propagation of Short Radio Waves*, §§ 5–4.

3.2.6. THE CASE OF AN IRREGULAR SURFACE. RAYLEIGH'S CRITERION. DIFFUSION

Let us now assume that the surface exhibits irregularities with dimensions large compared with the wavelength.

Clearly, the phenomenon is going to become much more complicated; specular reflection will disappear, giving way to a series of small elementary reflections, more or less random, in all directions, in other words, to *diffusion*.

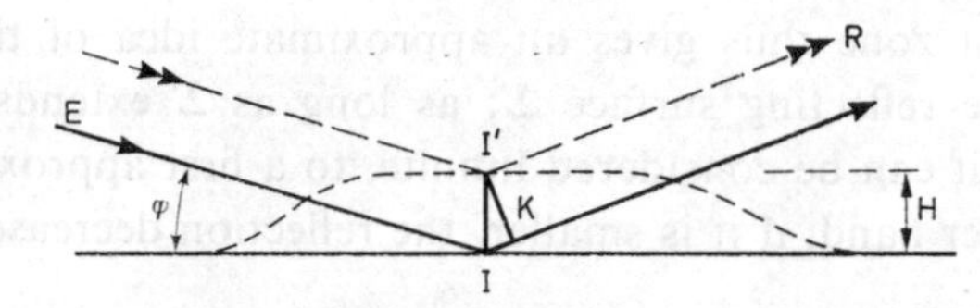

FIG. 3.5. Rayleigh's criterion

We cannot be at all precise until we make some assumption about the shape, magnitude and ultimately the "periodicity" of the surface irregularities.

As a first approximation we assume that the surface consists of flat humps of uniform height H (Fig. 3.5). Assuming that reflection takes place at the peak I' of these humps in the same way as it does at I on the plane of the base, the path EIR will merely be shortened by an amount:

$$2IK = 2H \sin \varphi,$$

where φ is the angle which the incoming wave makes with the surface.

The resultant phase change will be negligible below a certain value; on the other hand, if it is large, the reflection coefficient will be reduced. The following criteria are found to apply:

for $H \sin \phi < \frac{\lambda}{120}$, negligible reduction ($<$ 10%);†

for $H \sin \phi < \frac{\lambda}{16}$, coefficient R reduced to $R/2$;

† Norton, C.C.I.R., London, 1953, doc. 11F, p. 7.

for $H \sin \phi < \frac{\lambda}{4}$, coefficient R reduced to $R/10$.†

Obviously, the fraction of energy which is not returned with angle $r = i$ "diffuses" in other directions, sideways and even backwards.

This approximate, but simple, criterion allows a primary evaluation of the effects of irregularities in the terrain to be made; it should be noted that the smaller the angle φ, the less important this effect, and that it is therefore negligible at grazing incidence where the reflection coefficient R is unaffected.

Of course, many other forms of irregularities can exist; moreover, the two media concerned can have slightly different complex constants so that refraction at the surface of separation can be just as important as reflection (or even more so). Such cases are common. For example:

(a) *waves* on the surface of the sea; they can be a serious nuisance in navigational radars by producing "clutter" on the screen; this effect has been satisfactorily analysed by various authors;‡

(b) the superstructure of a ship, the body of an aircraft: we shall say more on this subject when we discuss obstacles and radar targets (§ 4.8.6);

(c) atmospheric "turbulence", which can be likened to spherical "bubbles" with a dielectric constant which differs slightly from that of the ambient atmosphere; we shall also be returning to this feature in connection with tropospheric propagation (Chapter 5);

(d) analogous phenomena in the lower part of the ionized layer of the upper atmosphere (Chapter 6);

† Kerr, *Propagation of Short Waves*, 1951, §§ 5–10; *Proc. I.E.E.*, part B, November 1955, pp. 827–30.

‡ On the subject of the effect of waves, and more generally reflection at the surface of the sea, refer to:

Davies, *Proc. I.E.E.* part III, March 1954, p. 118 (summary only).

Lagrone, *I.R.E. Trans.*, AP3, 2, April 1955, pp. 48–52.

Court, *Proc. I.E.E.*, part B, Nov. 1955, pp. 827–30.

Ament and Katzin, *I.R.E. Trans.*, CS4, no. 1, March 1956, pp. 118–22.

Sofaer, *Proc. I.E.E.*, July 1958, pp. 383–94.

Du Castel, Misme, Voge, Spizzichino, *Ann. Télécomm.* Jan.–Feb. 1959, pp. 33–40.

(e) finally, of course, all relief features of the ground, trees, forests, houses, buildings, power lines, etc., which are so random that practical values of the reflection coefficient can only be determined by experiment.†

3.3. Experimental verifications

The above theoretical formulae, based on simple hypotheses and rigorous calculations, are a good representation of reality as long as we are dealing with a plane surface of separation and the constants of the two media are known.

In particular, for long waves over sea, the multiplication of the vertical field by the factor 2 and the virtual elimination of the horizontal field have been verified ever since the early days of radio by all possible means. It has also been proved that, over land, conditions are little different and, in particular, that the horizontal component of the field is still too weak to be usable in the ordinary way (except in the exceptional case of the very long "Beverage" aerial).

At short waves and especially at very short waves, the formulae appear to be less reliable, due to the fact that the surface of the ground is no longer ideally planar: apart from the case of a desert of perfectly level sand or a perfectly calm sea, the surface exhibits irregularities with dimensions of the order of the wavelength: on land we have hills, vegetation, trees, buildings, etc.; on sea, waves of various heights.

Since, in this domain, the reflection coefficient can, as we have seen, play on important part in the determination of the range, it is interesting to verify its value directly under practical conditions. Various series of experiments have been carried out—mainly in England and America in the early days of radar—and the results are quite clear.

Over *calm sea*, the calculated reflection coefficients are found

† Reflection coefficients of irregular ground:
Bullington, *Proc. I.R.E.*, Aug. 1954, pp. 1258–62.
Sherwood, *Proc. I.R.E.*, July 1955, pp. 877–8.

for *vertical* polarization down to wavelengths of 10 and 3 cm (Fig. 3.6, righthand curves; the points represent the experimental results). For *horizontal* polarization, the agreement is not so good and there is considerable dispersion; the observed coefficients are lower than the calculated values (Fig. 3.6, left-hand).†

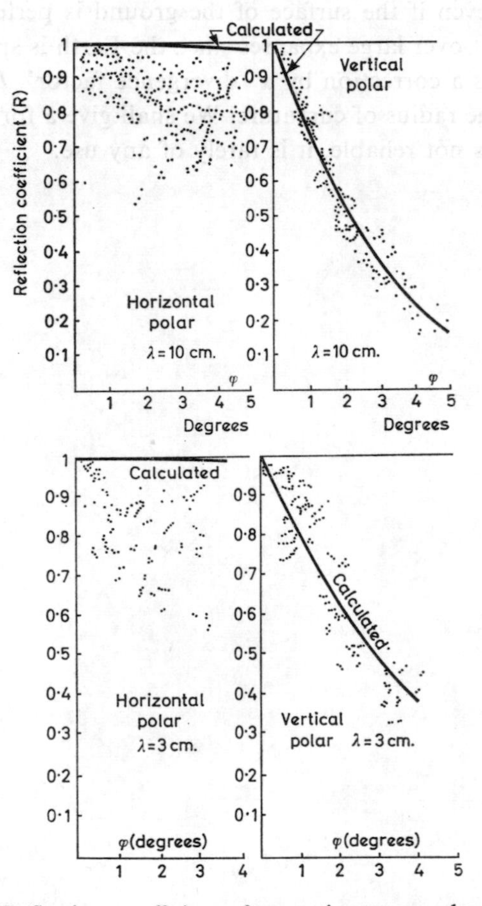

FIG. 3.6. Reflection coefficients for centimetre wavelengths over *calm sea*

† From Kerr, *Microwave Propagation*, § 5.6. There have been, however, some experiments in which the reflection is better for horizontal polarization than vertical polarization (see Lamont and Watson, *Nature*, 158, 28 Dec. 1946, p. 983). See also the bibliography of the preceding sections.

Over *rough sea* and *land*, the disagreement and dispersion are worse still. The reflection coefficient only approaches unity over exceptionally flat areas, aerodrome runways, for example. Over cultivated land, fields, forests, towns, etc., it is rarely greater than 0·2 at decimetre wavelengths.

Lastly, even if the surface of the ground is perfectly regular, it is not flat over large expanses since the Earth is spherical. This necessitates a correction by a "divergence factor" D to take account of the radius of curvature. We shall give a formula in § 4.3 but, as it is not reliable, it is rarely of any use.

CHAPTER 4

THE ROLE OF THE TERRAIN

4.1. General

We saw in Chapter 1 the "basic formulae" which are valid for a perfect dielectric, far from any obstacle.

We shall now dicuss to what extent the proximity of the Earth modifies these conclusions. The simplest case is that of stations high above the ground.

We shall then jump to the other extreme: the case of stations in contact with the ground.

Finally, we shall consider the more complex intermediate case of stations at any height.

All this is assuming that the Earth is "smooth"; we shall therefore finish by examining the effect of obstacles which may stand out from its surface.

4.2. The real case of very high stations

In practice, there is one situation in which, to a reasonable approximation, the ideal theoretical conditions are met, namely, the case of *very high* stations in a sufficiently *homogeneous* and *transparent* atmosphere. Let us define these restrictions:

By "*very high*" we mean that the heights h_1, h_2 of the two stations are (a) very much greater than the wavelength and (b) great enough compared with the distance d for the "*direct*" ray joining the stations to be much shorter than any "*indirect*" ray reflected from the ground. (We shall later find formulae which define these inequalities.)

By a "*homogeneous, transparent atmosphere*" we mean one

whose refractive index is reasonably constant, with no movement up or down, no temperature inversions ("subsidences") and no liquid or solid "precipitations" (thick fog, clouds, rain, snow, hail, etc.). (We shall return to these points in §§ 5 *et seq.*)

These conditions may all be present simultaneously for links between *spacecraft*, *aircraft at short range* or *mountain peaks*.

The formulae of §§ 1.1.1–1.1.3 show how favourable this situation is, i.e. how the sensitivity of modern receivers makes vast ranges feasible with low power transmitters.

For example, it is easy to transmit $W_t = 10$ W $= 0{\cdot}01$ kW; and there is no problem in receiving telegraphy or telephony with a field of $E_0 = 0{\cdot}1$ mV/m; formula 1.8 then indicates a range of $d = 212$ km $= 132$ miles.

By using directional aerials, i.e. by multiplying the field by the gains G_1, G_2 (much greater than 1), or by making even more sensitive receivers (which is not difficult), one would obtain any range compatible with our hypotheses, even with transmitted powers W_t of the order of a hundredth of a watt. In fact, under these conditions the question of "transmitter power" does not arise in practice. This is used to advantage by all directional centimetre wave "relay" or "radio telegraphy" systems.†

The margin of safety obtained is often so great that it is hardly worth while discussing to what extent the air behaves effectively as a perfect dielectric. We shall return to this point in the paragraphs devoted to the *troposphere;* we merely remark now that even at *metric* wavelengths we occasionally come across zones of "opacity" due to refraction in vertical air currents, and that at wavelengths *shorter* than 5 cm considerable absorption may be encountered due to "precipitations" (rain, fog, etc.), water vapour and even atmospheric oxygen.

† See, for example, *Onde électrique*, special issue, April–May 1953, and extensive measurements by Millar and Byam over 42 miles *(Proc. I.R.E.*, June 1950, pp. 618–26).

4.3. Stations near the ground but in direct line of sight. Interference zone

The above case ("very high stations") would be represented schematically in Fig. 4.1 by points beyond T_1 and R_1.

As the stations become less elevated, they first enter a region

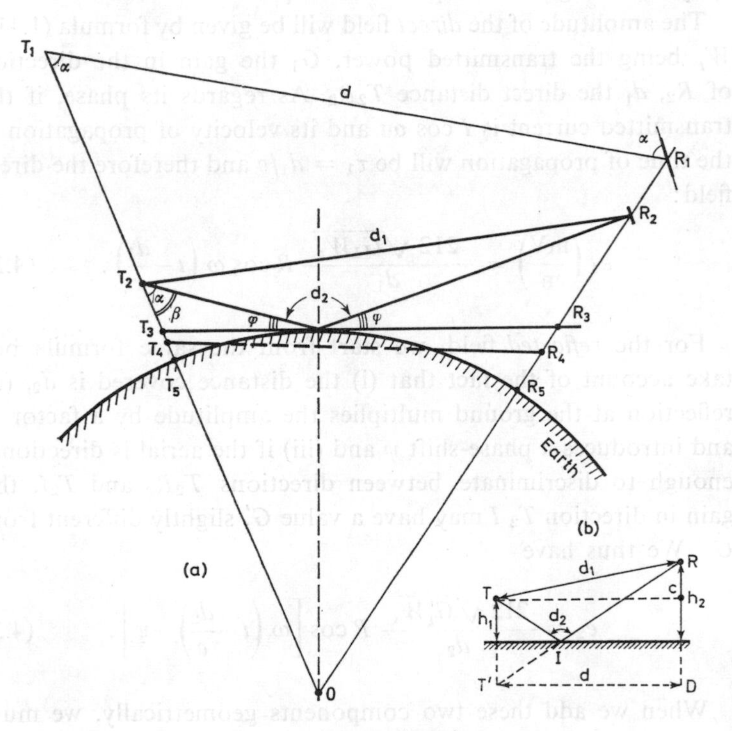

FIG. 4.1. Elevated stations

(represented by T_2R_2) where they are still *in direct line of sight*, i.e. their distance d is still less than the limit:†

$$d_1\,(\text{km}) = 4{\cdot}1\,(\sqrt{h_1}+\sqrt{h_2}). \qquad (4.1)$$

(m) (m)

† This formula takes account of the refraction of the atmosphere as will be explained later. It can also be written:

$$d_1\,(\text{nautical miles}) = 1{\cdot}225\,(\sqrt{h_1}+\sqrt{h_2})\,(\text{feet}).$$

As long as the conditions $h_1 \gg \lambda$, $h_2 \gg \lambda$ are satisfied, we can, to a first approximation, consider the radiation received by receiver R_2 as the sum of two intensities, the "direct", propagated along T_2R_2 as in free space, and the "reflected", along T_2IR_2. It is therefore sufficient to add the two components geometrically, knowing their amplitude and phase.

The amplitude of the *direct* field will be given by formula (1.11), W_t being the transmitted power, G_1 the gain in the direction of R_2, d_1 the direct distance T_2R_2. As regards its phase, if the transmitted current is $I \cos \omega t$ and its velocity of propagation c, the time of propagation will be $\tau_1 = d_1/c$ and therefore the direct field:

$$e_1 \left(\frac{\text{mV}}{\text{m}}\right) = \frac{212\sqrt{G_1 W_t}}{d_1} R \cos \omega \left(t - \frac{d_1}{c}\right). \tag{4.2}$$

For the *reflected* field, we start from the same formula but take account of the fact that (i) the distance covered is d_2, (ii) reflection at the ground multiplies the amplitude by a factor R and introduces a phase-shift ψ and (iii) if the aerial is directional enough to discriminate between directions T_2R_2 and T_2I, the gain in direction $T_2 I$ may have a value G_1' slightly different from G_1. We thus have

$$e_2 = \frac{212\sqrt{G_1' W_t}}{d_2} R \cos \left[\omega \left(t - \frac{d_2}{c}\right) - \psi\right]. \tag{4.3}$$

When we add these two components geometrically, we must bear in mind that they have practically the same orientation (except for sign). Actually, Fig. 4.1 is not to scale; the heights h_1, h_2 are, in practice, much less than the distance d and even smaller compared with the ground ray a. The geometrical sum of e_1 and e_2 therefore reduces approximately to the algebraic sum:

$$\frac{212\sqrt{G_1 W_t}}{d_1} \left\{\cos \omega \left(t - \frac{d_1}{c}\right) + \frac{d_1}{d_2}\sqrt{\frac{G_1'}{G_1}} R \cos \left[\omega \left(t - \frac{d_2}{c}\right) - \psi\right]\right\}. \tag{4.4}$$

Putting

$$\begin{cases} \left(\dfrac{d_1}{d_2}\sqrt{\dfrac{G_1'}{G_1}}\times R\right) = R', \\ \omega\dfrac{(d_1-d_2)}{c}-\psi = -\dfrac{\omega\,\Delta}{c}-\psi = -\theta, \end{cases} \tag{4.5}$$

the trigonometric term may be written

$$\cos\omega\left(t-\frac{d_1}{c}\right)[1+R'\cos\theta]+\sin\omega\left(t-\frac{d_1}{c}\right)\times R'\sin\theta,$$

the amplitude being

$$\sqrt{(1+R'\cos\theta)^2+(R'\sin\theta)^2} = \sqrt{(1-R')^2+4R'\cos^2\frac{\theta}{2}}.$$

The equation therefore represents the amplitude:

$$E = |e| = \frac{212\sqrt{G_1W_t}}{d_1}\times\sqrt{(1-R')^2+4R'\cos^2\frac{\theta}{2}}. \tag{4.6}$$

It can be seen that if any of the quantities h_1, h_2, d vary, $\Delta = (d_2-d_1)$ and ψ do likewise and so θ varies and $\cos\theta/2$ passes alternately through the values 0 and 1; thus the two vectors e_1 and e_2 are alternately in phase and in antiphase and their resultant e varies periodically between the maximum $(1+R')$ and the minimum $(1-R')$; this classical phenomenon of *interference* has given its name to the zone we are discussing.

The general trend of the variation is thus easy to forecast; the exact determination of the positions and values of the maxima and minima is rather more complicated. We shall consider two cases.

4.3.1. Distance small. Earth's curvature negligible

Let us suppose firstly that the distance d is *small enough for the curvature of the earth to be negligible*, i.e. its surface is coincident with the tangent plane through I; this is valid when the distance d is less than a third of the optical range (4.1).

The position of the point I is then obtained immediately by considering the "image" T' of the transmitter in the tangent plane (considered as an optical mirror) (Fig. 4.2).

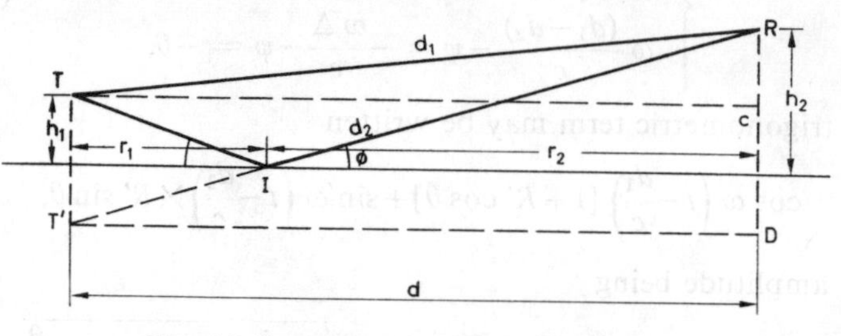

FIG. 4.2. Interference on the tangent plane

We see easily

$$\left\{\begin{array}{l} \dfrac{r_1}{d} = \dfrac{h_1}{h_1+h_2} = \dfrac{1}{1+u}, \\ \text{putting} \quad u = \dfrac{h_2}{h_1}, \\ \varphi = \text{arc tan}\dfrac{h_1}{r_1}, \end{array}\right. \tag{4.7}$$

$$d_1 = \sqrt{d^2+(h_2-h_1)^2}$$

$$d_2 = \sqrt{d^2+(h_2+h_1)^2}$$

from which, as h_1 and h_2 are very much less than d:

$$\left\{\begin{array}{l} \Delta = d_2-d_1 \approx \dfrac{2h_1h_2}{d}, \\ \text{and the phase change} \quad \dfrac{\omega\,\Delta}{c} = \dfrac{4\pi h_1 h_2}{\lambda d}, \end{array}\right. \tag{4.8}$$

and therefore:

$$\frac{\theta}{2} = \frac{2\pi h_1 h_2}{\lambda d} + \frac{\psi}{2}.$$

The field is a maximum [multiplied by $(1+R')$] when $\theta/2 = k\pi$ and a minimum [multiplied by $(1-R')$] when

$$\frac{\theta}{2} = (2k+1)\frac{\pi}{2}.$$

One very interesting special case is that of reflection from a *calm sea*, or very flat ground, especially for horizontal polarization. Angle φ being small, it can be seen from Fig. 3.3 that in this case (for very short waves, of course) we have approximately:

$$R \approx 1, \qquad \psi \approx 180°.$$

As d_2 is about the same as d_1 and G_1' close to G_1, we also have $R' \approx 1$ (typically 0·9 to 0·99). Consequently, the resultant field varies between maxima of nearly double the "free" field e_1 and very small minima of the order of a tenth or a hundredth of e_1; the maxima occur when $2\pi h_1 h_2/\lambda d = (2k+1)\pi/2$, the minima when $2\pi h_1 h_2/\lambda d = k\pi$. This can be seen in the middle of the "interference zone", Fig. 4.3.

At shorter distances, φ increases and therefore R decreases and may even vanish in the case of vertical polarization; when this happens, the fluctuation disappears and we are left with $e = e_1$ (point B, Fig. 4.3).

At greater distances, the arc $2\pi h_1 h_2/\lambda d$ ultimately becomes less than $\pi/2$ and tends to zero; in other words, cos $\theta/2$ tends to sin $2\pi h_1 h_2/\lambda d = 2\pi h_1 h_2/\lambda d$ and expression (4.6) then tends to:

$$E = \frac{212\sqrt{G_1 W_t}}{d_1} \times \frac{4\pi h_1 h_2}{\lambda d}. \tag{4.9}$$

As d_1 is practically equal to d, it can be seen that the decrease in this range follows a regular $1/d^2$ law.†

If the distance is constant and the height of one of the stations varies—h_2, for example—it can be seen that a vertical diagram consisting of a series of "leaves" or "lobes" is obtained (see Fig.

† As this rapid decrease is usually undesirable, one wonders whether it would be possible to avoid it by removing the cause, i.e. by preventing the counter-reflection from the ground around the point I; for example, by choosing the siting of the stations, if possible, so that the "first Fresnel zone" (§ 3.2.5) around the point I is as absorbent as possible; or, if the locations of the stations are given, by placing an absorbent obstacle in the first Fresnel zone, provided it is accessible. Appreciable results have been obtained by obstructing half of this zone with a 42 ft high metal grill (Lorant, *Wir. World*, Feb. 1953, pp. 87–8).

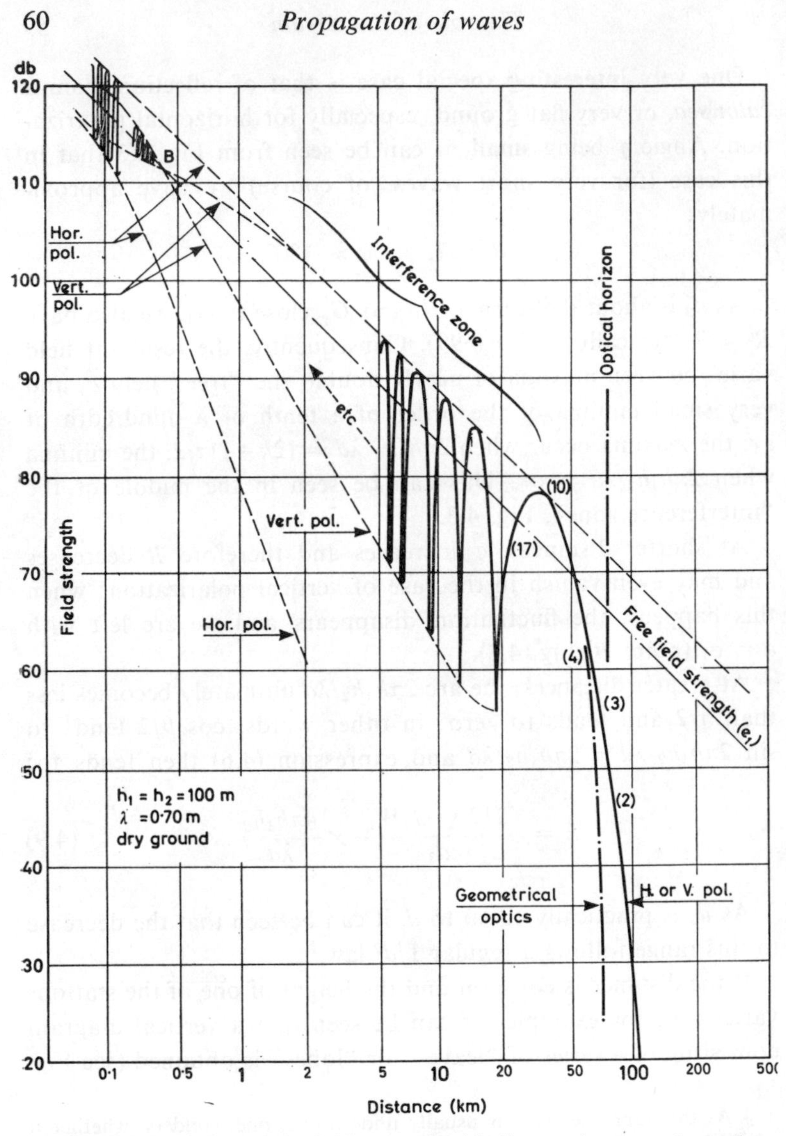

FIG. 4.3. Variation of the field strength with distance between elevated stations

4.4) with approximately equal maxima: below the lowest maximum (the first when approaching from a distance or from the ground), the field strength decreases as h_2 (hence the point of having the stations at a certain height).

Although the numerical application of these formulae presents no difficulty, here are two examples to fix orders of magnitude:

1. $h_1 = 50$ m, $h_2 = 1500$ m,
   $G_1 = 100$, $W_t = 1$ W,
   $d = 100$ km, $\lambda = 1$ m,

horizontal polarization over sea.

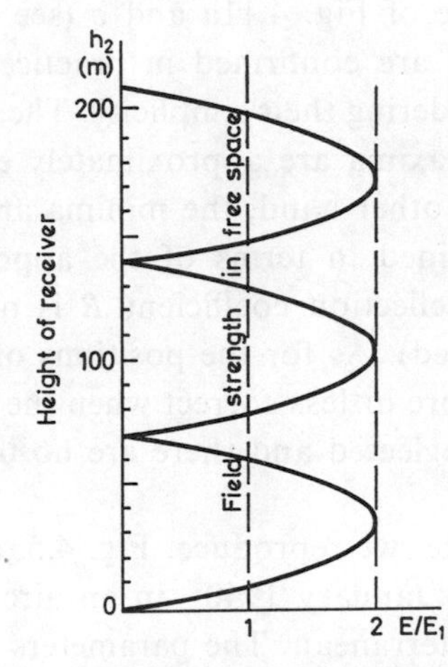

FIG. 4.4. Variation of field strength with receiver height (at fixed distance) in the interference zone
$h_1 = 360$ m; $d = 10$ km; $\lambda = 5$ m; horizontal polarization over sea

We can calculate immediately:

$$\Delta = \frac{2h_1h_2}{d} = \frac{2\times 50\times 1500}{100{,}000} = 1{\cdot}5 \text{ m which is } \frac{3\lambda}{2} \text{ exactly,}$$

$$u = \frac{h_2}{h_1} = 30, \quad r_1 = d\times\frac{1}{1+u} = \frac{100}{31} = 3{\cdot}2 \text{ km,}$$

$$\varphi = \arctan\frac{h_1}{r_1} = \arctan\frac{50}{3200} = 0{\cdot}0155 = 0{\cdot}9°.$$

Figure 3.3 shows that R is approximately equal to 1 and

$$\psi = 180°, \quad \text{and so} \quad \frac{\psi}{2} = \frac{\pi}{2} \quad \text{and} \quad \frac{\theta}{2} = 2\pi.$$

We are therefore at the second maximum; the "free" field is doubled and:

$$E\left(\frac{\mathrm{mV}}{\mathrm{m}}\right) = \frac{212\sqrt{100\times 0{\cdot}001}}{100}\times 2\cos 2\pi = 0{\cdot}67.$$

2. $h_1 = 360$ m, $h_2 = 1060$ m, $\lambda = 5$ m,

distance variable, horizontal polarization over sea. For Δ we find the upper curve of Fig. 4.11a and b (see below).

These predictions are confirmed in practice to quite a satisfactory extent, considering their simplicity. The periodic variation is very clear; the maxima are approximately equal to twice the "free" field; on the other hand, the minima are not zero, which can be easily explained in terms of the approximations made (in particular, the reflection coefficient R is not quite equal to unity, as was assumed). As for the positions of the maxima and minima, they are more or less correct when the Earth's curvature can be effectively neglected and there are no intervening irregularities of terrain.

By way of example, we reproduce, Fig. 4.5, a record made by the French Navy in January 1940† in an aircraft in horizontal flight over the Mediterranean. The parameters were $\lambda = 5{\cdot}00$ m, $h_1 = 360$ m, $h_2 = 1075$ m, d from 30 to 10 km.

It can be seen that the positions of the recorded maxima and minima coincide well, over a wide range, with the positions M_6, m_6, M_7, m_7, etc., calculated from the above example (see Fig. 4.11b).

It was also possible to verify the variation with d fixed and h_2 variable (Fig. 4.4) by using a captive balloon. Other similar observations have been made subsequently in Britain and America, at shorter wavelengths (10 and 3 cm), but always over sea‡ or particularly regular ground (aerodrome concrete runways, sports fields, desert sand, etc.).

† Technical note CEP.TSF. Commandant Demanche and the author. Figure 4.5 shows three sections of the recording one above the other.

‡ See, for example, Megaw, *Radiolocation Convention*, London, 1946, and Kerr, *Propagation of Short Radio Waves*, 1951, Fig. 5.15.

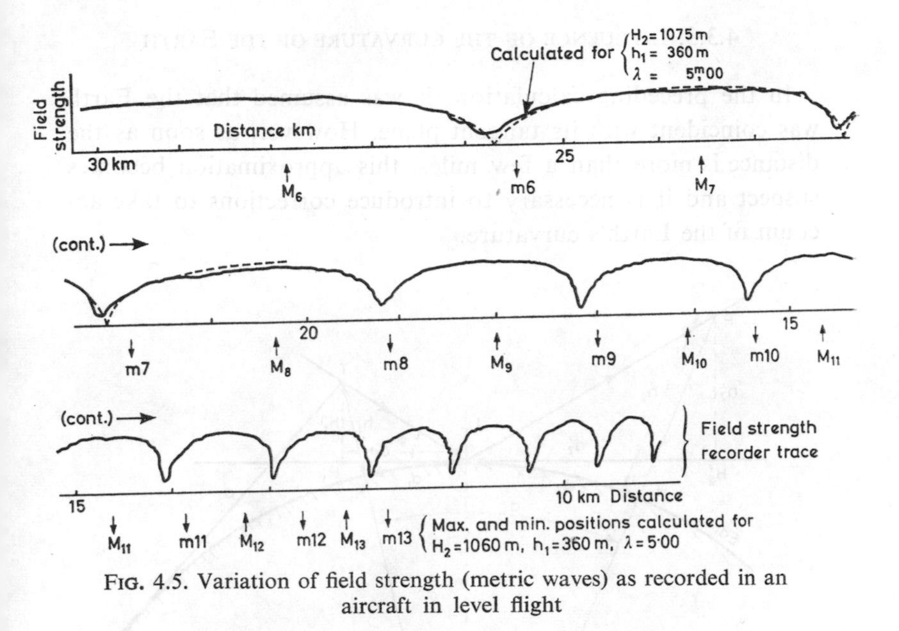

FIG. 4.5. Variation of field strength (metric waves) as recorded in an aircraft in level flight

4.3.2. EFFECT OF SURFACE IRREGULARITIES

By contrast, if the experiment is repeated over any other kind of surface—even over rough sea, at centimetre wavelengths—the periodic fluctuation predicted by formulae (4.4) and (4.6) is no longer observed, only irregular variations, with an occasional decrease of the field strength, approximately as $1/d^2$, at the end.†

This is explained by the fact, already seen in § 3.2.5, that on land, irregularities of the ground, vegetation, perhaps buildings, greatly reduce the value of the reflection coefficient R; and furthermore, hills and valleys displace the point of reflection I and consequently introduce a random, irregular path difference. It is not surprising, therefore, that the periodic variation of the factor (4.6) becomes small and is masked by others.

† Various examples and curious experiments will be found in Kerr, *Propagation of Short Radio Waves*, §§ 5.9–5.12, such as that of a soil, raked to simulate waves of length 60–120 cm and height 5–16 cm.

4.3.3. Influence of the Curvature of the Earth

In the preceding calculation, it was assumed that the Earth was coincident with its tangent plane. However, as soon as the distance is more than a few miles, this approximation becomes suspect and it is necessary to introduce corrections to take account of the Earth's curvature.

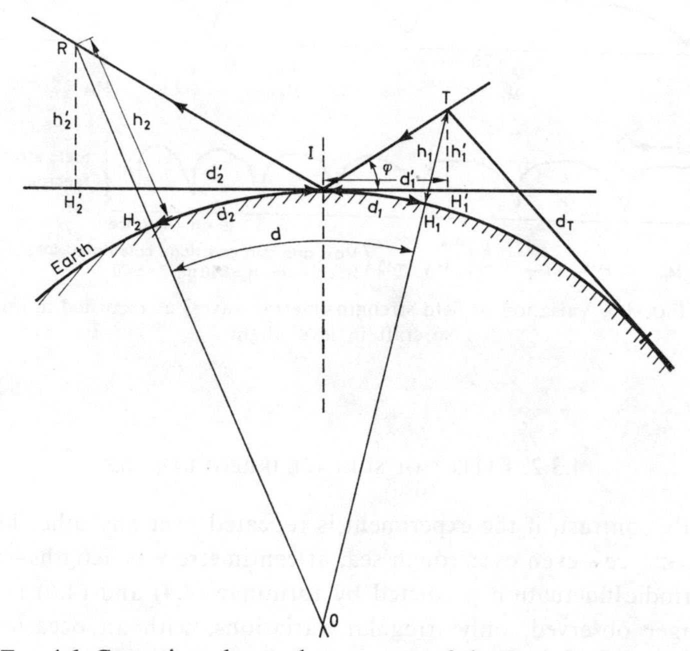

Fig. 4.6. Corrections due to the curvature of the Earth in the interference zone

Let T and R (Fig. 4.6) be the stations, d the distance at ground level, h_1 and h_2 the heights ($h_1 < h_2$).

The difficulty arises from the fact that the above simple construction for finding the reflection point I is not applicable; in fact, we cannot take the image of one of the stations in the tangent plane since the position of the latter depends in turn on the point I.

However, let us assume that the problem has been solved (Fig. 4.6); we see that three corrections are necessary:

1. The heights h_1', h_2' with respect to the tangent plane are less than the heights h_1, h_2 above ground; this decreases the angle of incidence φ.

2. The distances d_1', d_2' between the point of reflection I and the feet of the perpendiculars to the tangent plane are different from the distances d_1, d_2 along the ground; they are greater, especially on the side of the higher station, $d_2' > d_2$; this gives a further decrease in angle φ.

3. Finally, since reflection is taking place from a convex, spherical dome, the reflected beam is divergent, i.e. the reflection coefficient R is multiplied by a *divergence factor* D less than unity (as already pointed out in § 2.3.5).

These corrections (the position of the point I, values of the distances d_1', d_2' and of the heights h_1', h_2' to be used in the formula of § 4.3.1 to find the path difference Δ, the angle of arrival φ and the divergence factor D) have been calculated by various authors with various degrees of ingenuity. We give here the results found by Burrows and Atwood;† they characterize the position of the stations by two dimensionless parameters:

(a) the ratio of the heights

$$u = \frac{h_2}{h_1} \tag{4.10}$$

($u > 1$, assuming that $h_2 > h_1$);

(b) the ratio of the distance d to the optical range d_T of the lower station, i.e.

$$v = \frac{d}{d_T} = \frac{d}{\sqrt{2ah_1}} \tag{4.11}$$

where a is the radius of the earth, increased, if necessary, to take account of refraction, as indicated in § 5.2.

These two parameters can be deduced immediately from the given h_1, h_2, d and a.

† *Radio Wave Propagation*, pp. 385 *et seq.* We have redrawn the graphs in a different form for convenience.

A third parameter S is derived from them, namely the root of the equation

$$S^3 - \frac{3}{2}S^2 - \frac{S}{2}\left(\frac{1+u}{v^2} - 1\right) + \frac{1}{2v^2} = 0, \tag{4.12}$$

obtained from the graph, Fig. 4.7.

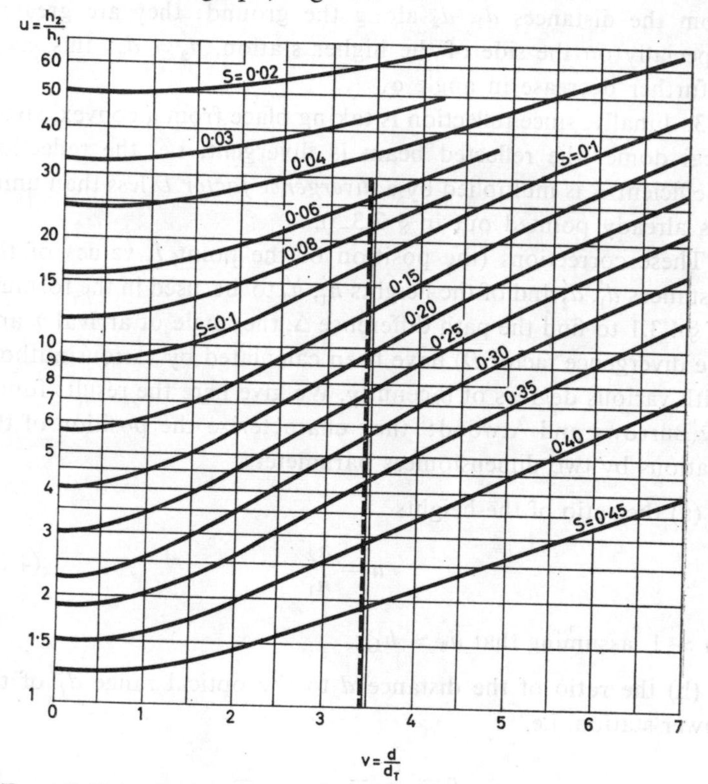

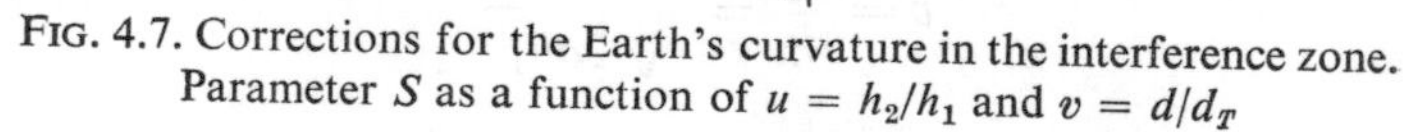

FIG. 4.7. Corrections for the Earth's curvature in the interference zone. Parameter S as a function of $u = h_2/h_1$ and $v = d/d_T$

Knowing these three parameters, one can deduce:†

† We can derive the formulae for a plane surface by making a tend to infinity. v then tends to zero and S to $1/(1+u)$; φ therefore tends to

$$\arc\tan\sqrt{\frac{h_1}{2a}}\times\frac{1}{Sv} = \arc\tan\sqrt{\frac{h_1}{2a}}\times(1+u)\times\frac{\sqrt{2ah_1}}{d}$$

$$= \arc\tan\frac{h_1(1+u)}{d} = \arc\tan\frac{h_1}{d_1}$$

(a) the *divergence coefficient* D from the relationship

$$D = \frac{1}{\sqrt{1+4\dfrac{(Sv)^2(1-S)}{1-(Sv)^2}}}, \tag{4.13}$$

(Fig. 4.8);

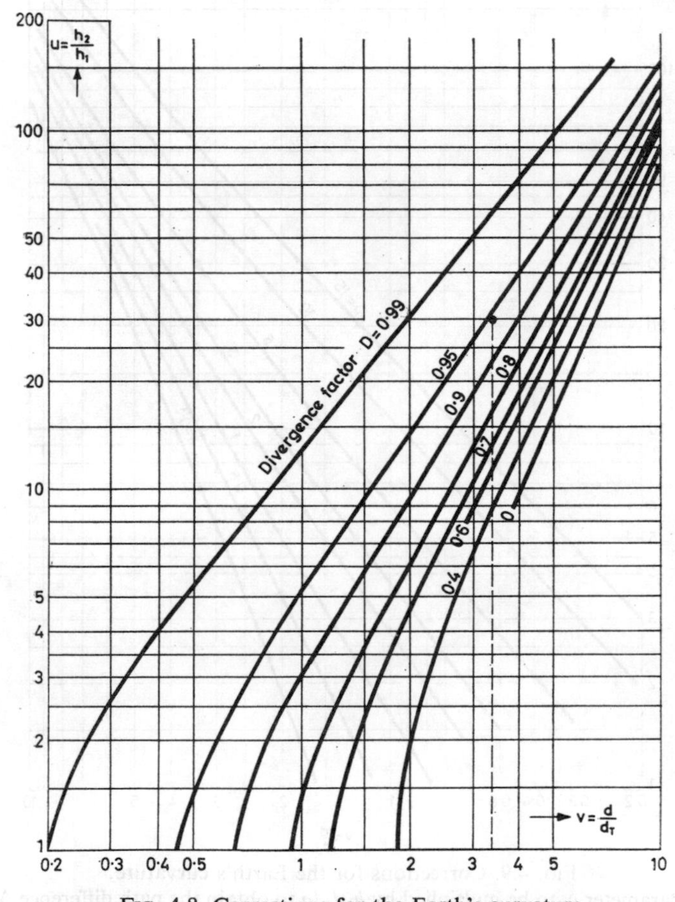

FIG. 4.8. Corrections for the Earth's curvature.
Divergence factor D

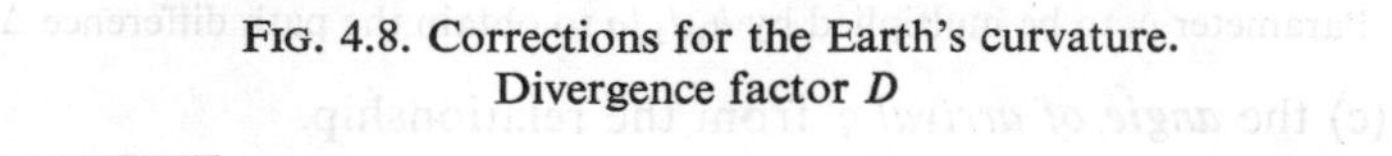

---

and

$$\Delta = \frac{h_1\sqrt{2ah_1}}{a}\left(1-\frac{1}{1+u}\right)\times\frac{h+u}{2}\times\sqrt{2ah_1} = 2h_1^2\cdot\frac{u}{d} = \frac{2h_1h_2}{d}.$$

(b) the *path difference* Δ from the relationship

$$\Delta = \frac{h_1 \times d_T}{a} \times \left[(1-S)\frac{(1-S^2v^2)^2}{Sv} \right] = \frac{h_1 \times d_T}{a} \times \delta \quad (4.14)$$

(Fig. 4.9 gives δ as a function of *u*, *v*);

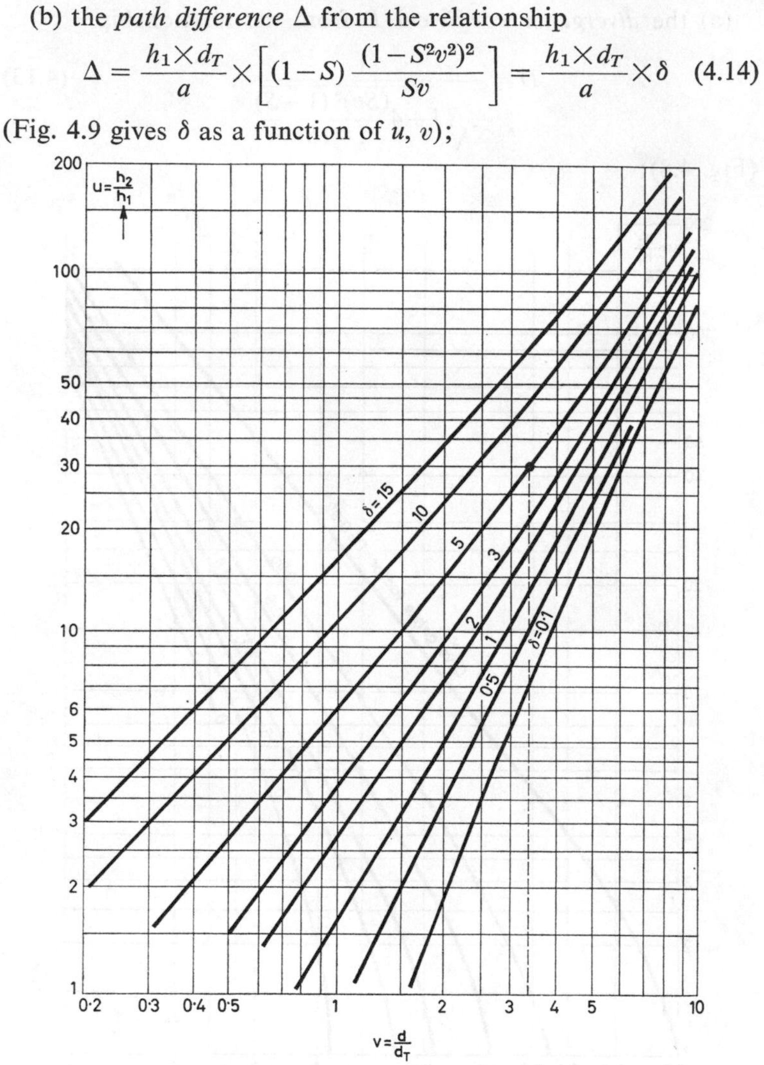

FIG. 4.9. Corrections for the Earth's curvature.
Parameter δ, to be multiplied by $h_1 d_T / a$ to obtain the path difference Δ

(c) the *angle of arrival* φ from the relationship

$$\varphi = \text{arc tan} \sqrt{\frac{h_1}{2a}} \left(\frac{1}{Sv} - Sv \right) \quad (4.15)$$

(Fig. 4.10, which gives φ as a function of *Sv*).

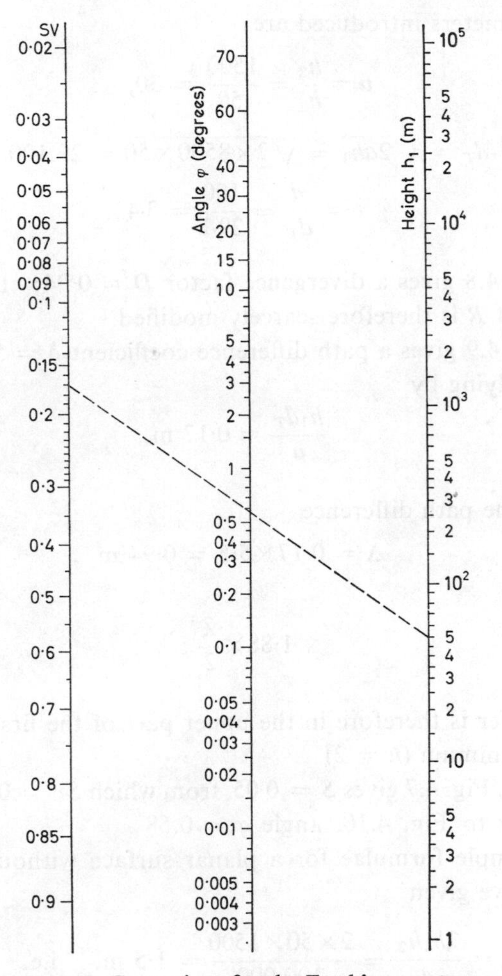

FIG. 4.10. Corrections for the Earth's curvature.
Arrival angle φ as a function of the height h_1(m) and the factor Sv:

$$\left[\varphi = \text{arc tan}\,\frac{\sqrt{h_1}}{4120}\left(Sv - \frac{1}{Sv}\right)\right]$$

By way of example, let us take again the case of

$$h_1 = 50 \text{ m}, \qquad h_2 = 1500 \text{ m},$$

$$d = 100 \text{ km}, \qquad \lambda = 1 \text{ m}.$$

The parameters introduced are

$$u = \frac{h_2}{h_1} = \frac{1500}{50} = 30,$$

$$d_T = \sqrt{2ah_1} = \sqrt{2\times 8500\times 50} = 29{,}100 \text{ m},$$

$$v = \frac{d}{d_T} = \frac{100}{29{\cdot}1} = 3{\cdot}43.$$

Figure 4.8 gives a divergence factor $D = 0{\cdot}94$; the reflection coefficient R is therefore scarcely modified.

Figure 4.9 gives a path difference coefficient $\delta = 5{\cdot}5$.

Multiplying by

$$\frac{h_1 d_T}{a} = 0{\cdot}17 \text{ m},$$

we find the path difference

$$\Delta = 0{\cdot}17\times 5{\cdot}5 = 0{\cdot}94 \text{ m}$$

which is

$$1{\cdot}88\times\frac{\lambda}{2};$$

the receiver is therefore in the upper part of the first lobe, quite near a minimum ($n = 2$).

Finally, Fig. 4.7 gives $S = 0{\cdot}05$, from which $Sv = 0{\cdot}17$, whence, according to Fig. 4.10, angle $\varphi = 0.58°$.

The simple formulae for a planar surface without correction would have given

$$\Delta' = \frac{2h_1h_2}{d} = \frac{2\times 50\times 1500}{100{,}000} = 1{\cdot}5 \text{ m}, \quad \text{i.e.} \quad \frac{3\lambda}{2},$$

and therefore a field *maximum* instead of a near minimum, and

$$\varphi' = \text{arc tan}\,\frac{h_1}{d_1} = \text{arc tan}\,\frac{h_1(1+u)}{d} = \text{arc tan}\,\frac{50\times 31}{100{,}000} = 0{\cdot}9°.$$

There is an appreciable difference and these corrected formulae therefore seem to be much more exact than the simple formulae for a flat surface. It should be noted, however, firstly, that the difference is less when the distance is smaller; secondly, that the apparent accuracy depends largely on the value adopted for a the Earth's radius, corrected for refraction. We have taken $a =$ 8500 km corresponding to "normal" atmospheric refraction; but, as we shall see in the chapter on the Troposphere, the atmosphere is seldom "normal" and consequently, the correction may be illusory. Thus Fig. 4.11 shows the path differences calculated by the two methods, for the example illustrated in Fig. 4.5 ($h_1 = 360$ m, $h_2 = 1060$ m, $\lambda = 5$ m, distance variable). It can

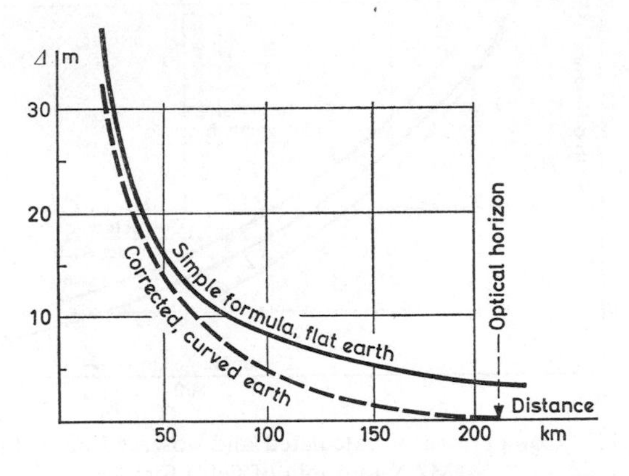

FIG. 4.11a. Influence of the Earth's curvature on the path difference for $h_1 = 360$ m, $h_2 = 1060$ m

be seen (Fig. 4.11a) that the difference is particularly great as one approaches the limit of optical visibility (here 210 km); the corrected formulae in fact give $\Delta = 0$ for $d = 210$ km, whereas the simple formula would still give $\Delta = 3{\cdot}5$m, which is absurd. But if we reproduce on the useful portion of these curves (Fig. 4.11b) the positions of the maxima and minima observed (Fig. 4.5), we see that they agree better with the results of the simple formula than with those of the improved formula (the points indicated by crosses, which would apply if what we have assumed to be the sixth maximum were only the fifth, are no more satisfactory). On that day, therefore (January 1940), atmospheric refraction was more pronounced than "normal" and the effective radius of curvature of the earth was practically infinite.

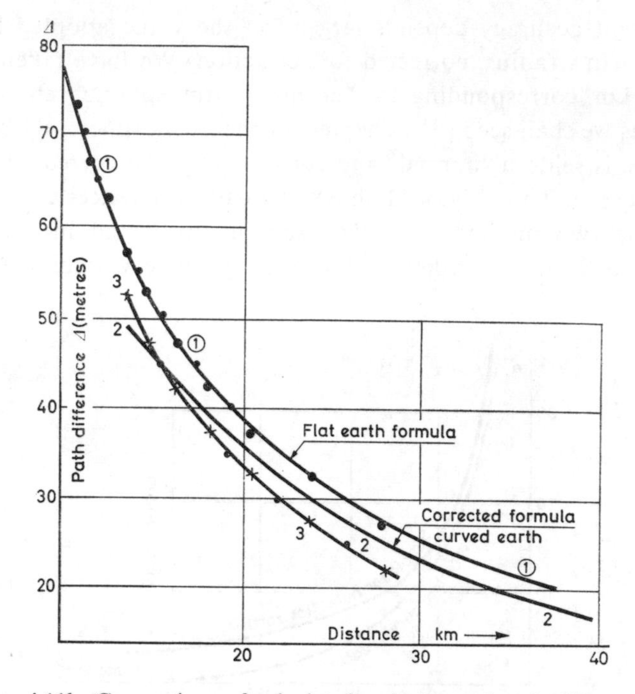

FIG. 4.11b. Comparison of calculated and observed path differences (MAX and min of field) for:

$h_1 = 360$ m; $h_2 = 1060$ m; $\lambda = 5$ m

———— calculated, simple formula, flat Earth

— — — calculated, corrected formula, round Earth, $a = 8500$ km

●.●.● MAX and min observed (assuming that the MAX at 27·5 km is the 6th)

×.×.× MAX and min observed (assuming that the MAX at 27·5 km is the 5th)

4.3.4. MODIFICATION OF THE APPARENT CURVATURE OF THE EARTH BY THE SHAPE OF THE GROUND

It may happen that in the vicinity of the aerials, or in the Fresnel "reflection zone" (Fig. 3.4), there is some feature of the surface whose general shape produces the effect of a curvature very different from that of the terrestrial sphere: a smaller radius in the case of a hill, a radius of opposite sign in the case of a hollow.

We can then adjust part of the preceding calculation by generalizing the "divergence coefficient D" to include one or two different radii of curvature, possibly negative (in which case there is a convergence effect with $D > 1$). Norton† claims to have greatly improved the agreement between calculated and measured values in this way. Certainly, one can explain the reinforcement in the field which is frequently observed in certain areas where the ground slopes concavely towards the transmitter.

Note. In all problems of visibility, heights with respect to the horizon, effect of obstacles, etc., it is usually convenient to plot graphs with an extended "scale of heights" and curves representing the apparent curvature of the Earth (including refraction) on this scale.

4.4. Diffraction (or "shadow") zone

4.4.1. General

When the heights of the stations are reduced, they reach the tangent plane (Fig. 4.1, T_3R_3), and then pass below it (T_4R_4). The preceding formulae, derived from geometrical optics, which would indicate zero field, now cease to be valid. In fact the waves can, to a certain extent, follow the curvature of the Earth by the phenomenon of *diffraction*. But this is an extremely difficult physical optics phenomenon which requires a complete revision of the method of calculation starting from Maxwell's equations in spherical coordinates with the "limiting condition" that the properties of the medium change at the air–ground surface of separation. The differential equations thus obtained may be integrated by expansion into series; but the large range of variation of the parameters (notably f and σ) and the slow rate of convergence of the series make the solution extremely laborious.

The efforts of numerous eminent mathematicians—Sommerfeld, H. Poincaré, Watson, Laporte, Eckersley, Van der Pol and

† Norton, *Trans. I.R.E. Professional Group on Antennas and Propagation*, PGAP. 3, Aug. 1952, p. 152 (166) and C.C.I.R. London, 1953, doc. 11.

Bremmer, Burrows, Norton, etc.—and innumerable international commissions have failed to produce a general formula which can be used by engineers. One therefore has to subdivide the problem into various categories, use for each one graphs prepared in advance and interpolate, sometimes quite arbitrarily, in intermediate categories.

We start with the simplifying assumptions that the heights h_1, h_2 of the stations are fixed and chosen in advance to simplify the solution; the chosen value is nearly always zero, i.e. the stations are assumed to be at ground level. In exceptional cases, we can also take for this value a certain "natural unit", a function of the wavelength.†

Once this part of the problem has been solved, we next study the effect of the heights. As long as they are low (below a certain "critical" value), the effect is purely "local" and may be likened either to a reduction in the range or a supplementary gain multiplying the ground-level field. There are special graphs for taking account of this.

Finally, if the heights are markedly greater than the "critical" value and almost up to the point at which direct visibility is possible, the problem acquires maximum complexity. Only a few particular cases have been treated (for example, that where one station is very high and the other is near the ground); in this zone, known as the *intermediate zone*, one is often reduced to interpolating between the values for the *interference zone* and those for the *shadow zone*.

4.4.2. Stations at ground level

By "ground level" is meant "at a height of less than a wavelength". This is usually the case with long and medium waves, but not with ultrashort waves.

Firstly, it will be recalled that the nearness of the ground considerably modifies the properties of the aerial—the vertical

† Kerr, *Propagation of Short Radio Waves*, Fishback method, chap. 2.

radiation diagram and the effective height—the radiation resistance, etc.

In particular, for a vertical doublet placed on flat ground of infinite conductivity, an "image" is formed in the ground; the field is doubled at equal intensity, but at the expense of doubling the radiation resistance, so that for the same radiated power there is a gain of $\sqrt{2}$ and formula (1.8) becomes

$$E_0\left(\frac{\text{mV}}{\text{m}}\right) = \frac{300\sqrt{W_t\,(\text{kW})}}{d\,(\text{km})}. \tag{4.16}$$

In reality the ground is never of infinite conductivity and the Earth is not flat. The above field therefore has to be multiplied by a supplementary attenuation coefficient A and it decreases with increasing distance at a rate faster than $1/d$, according to a complex law where two effects are superimposed:

1. At *short distances* where the Earth's curvature is unimportant, the attenuation results from an *absorption* of energy by the ground. Calculation shows that the constants of the ground—relative dielectric constant ε_r, conductivity σ (mho/m)—combine with the wavelength λ (m) and the distance d (km) to constitute an overall (complex) parameter which was called by Sommerfeld the *numerical distance*:†

$$p = \frac{\pi \mathscr{C}}{\lambda}\cdot d = \frac{\pi C d}{\lambda}\, e^{-j(\pi/2-b)}, \tag{4.17}$$

$\mathscr{C}$ being the complex constant defined in § 3.2.3 (polarization coefficient), a function of the polarization, but used here with $\varphi = 0$, i.e. $\cos^2\varphi = 1$:

$$\mathscr{C}_V = \frac{\eta-1}{\eta^2}, \quad \mathscr{C}_H = \eta-1.$$

† Some authors (e.g. Burrows) sometimes take twice this value for p. It should be mentioned that Sommerfeld's initial calculation contained several errors which have been the subject of contradictory criticism (see Norton, *Proc. I.R.E.*, Sept. 1937; Eckardt and Kahan, *J. de Phys.*, May 1948; Poincelot, *Ann. Télécomm.*, June 1953; Fannin, *Proc. I.R.E.*, Aug. 1953; Barlow and Cullen, *Proc. I.E.E.*, pt. III, Nov. 1953; and a general historical revue by Boudouris, *Onde Électrique* May 1957). But his work, revised and corrected, is still, nevertheless, the basis for all succeeding efforts; we accord to him here the credit which he deserves.

The supplementary attenuation A_1 which multiplies the field E_0 is a function of this parameter p:

$$A_1 = f(p), \tag{4.18}$$

for which the general expression is very complicated and can be written in various forms; for example, that of Sommerfeld–Norton is

$$A = f(p, b)\,e^{j\varphi} = 1 + j\sqrt{p}\cdot e^{-|p|} \times 2\int_{-j\sqrt{p}}^{\infty} e^{-u^2}\,du, \tag{4.18a}$$

[the integral is the "error function" of the variable $(-j\sqrt{p})$].

An approximate expression by Van der Pol is

$$A = \frac{E}{E_0} = \frac{2+0{\cdot}3p}{2+p+0{\cdot}6p^2} \quad \text{when} \quad \varepsilon_r \ll x. \tag{4.18b}$$

(long waves, conductive terrain)

Figure 4.12a, curve 1, is a reproduction of a classical graph giving the value of this attenuation.†

It can be seen that as d increases, A tends to $1/2p$, i.e. is proportional to $1/d$; thus the total attenuation of the field tends to $1/d^2$; it can also be seen that for highly conductive ground and long waves, $60\,\sigma\lambda \gg \varepsilon_r$: therefore $\mathcal{C}_H \gg \mathcal{C}_V$; the "numerical distance" p is therefore much greater with horizontal polarization, i.e. the field is attenuated more rapidly.

2. At "*large distances*", the most important factor is the curvature of the Earth; it may be followed by diffraction up to a point, but beyond that the attenuation becomes extremely rapid, as a function of the essential parameter:

$$\eta = \beta_0 \frac{d}{a^{2/3}\lambda^{1/3}} = \frac{\beta_0}{a^{2/3}} \times d_c. \tag{4.19}$$

The coefficient β_0 depends slightly on the nature of the terrain and on the polarization [i.e. on the coefficient $\mathcal{C}$ of eqn. (4.17)]; it decreases by about half on going from a very short wavelength over a poor surface to a long wavelength over a high-conductivity

†It will be found, to a higher accuracy, elsewhere, e.g. Norton, *Proc. I.R.E.* Dec. 1941, pp. 623–9, or Burrows and Atwood, *Propagation of Radio Waves*, p. 426, Fig. 56 (with double abscissa, $2p$).

surface; the Earth radius a is actually about 6400 km, but this figure can sometimes be increased by about four-thirds to take account of atmospheric refraction (see below, § 5.2); thus the supplementary attenuation is given by a series of exponentials (curve 2, Fig. 4.12a) as a function of the abscissa:

$$d_c = \frac{d\,(\text{km})}{\sqrt[3]{\lambda\,(\text{km})}} \quad \text{("reduced" or "critical" distance).} \tag{4.20}$$

It can be seen that it begins to be appreciable when d_c is greater than about 100, and becomes exponential when the stations disappear into the "shadow zone" created by the Earth's curvature, i.e. for d_c from about 1000 upwards, say (see Table, p. 79). It is sometimes approximated by the simple formula:

$$\frac{0{\cdot}62}{\sqrt[3]{\lambda\,(\text{m})}}\ \text{dB/km}. \tag{4.20a}$$

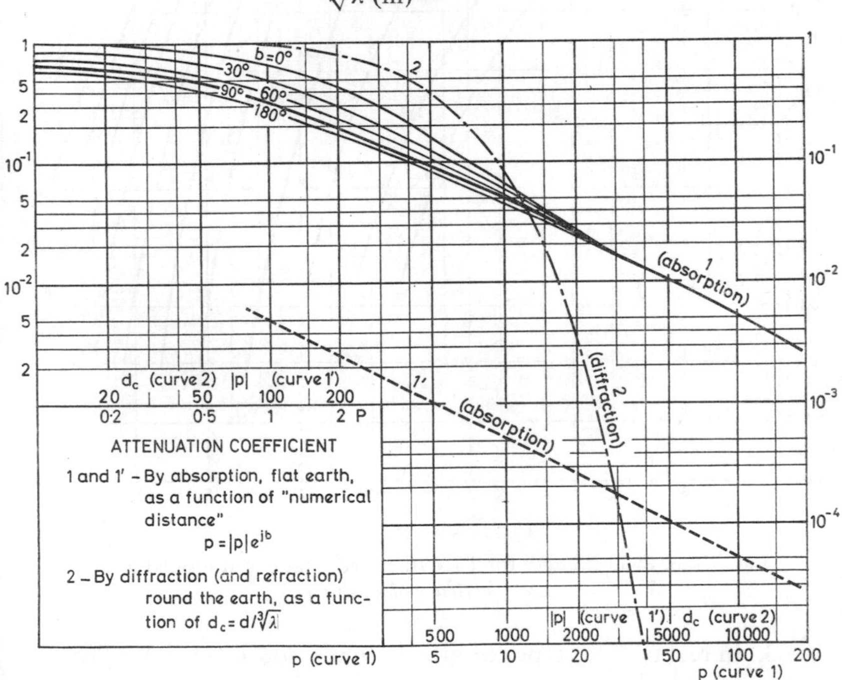

FIG. 4.12a. Theoretical formulae for absorption and diffraction by the Earth

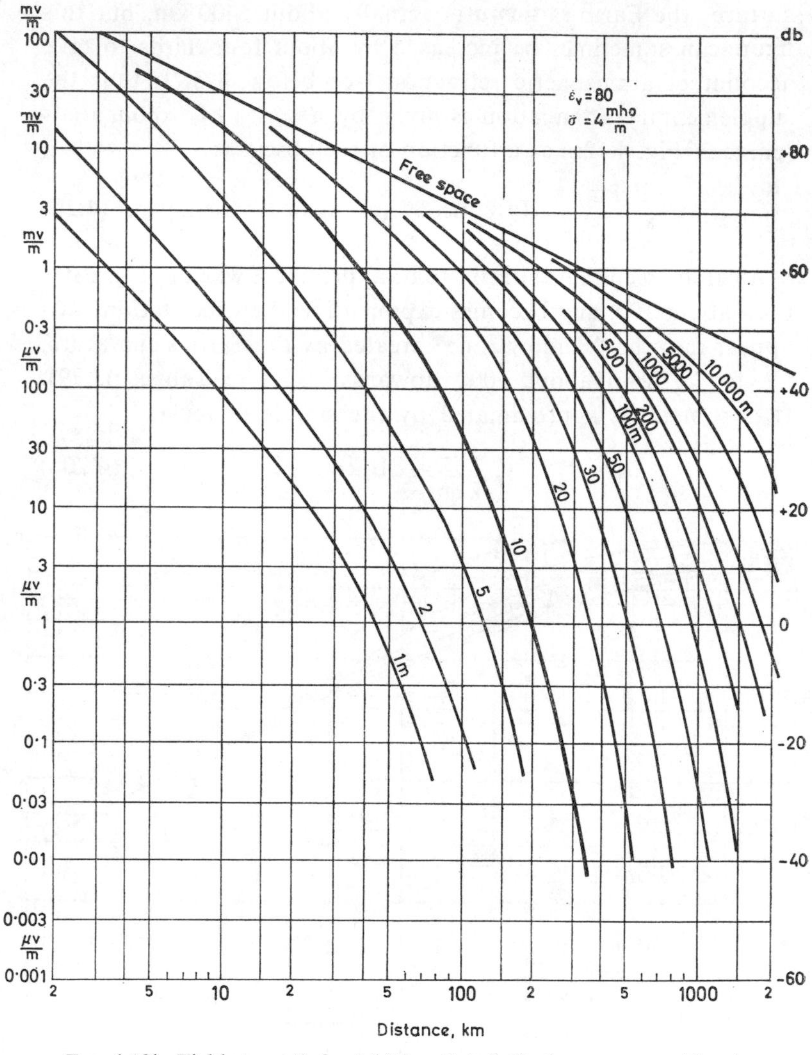

FIG. 4.12b. Field strength for 1 kW radiated. Stations at ground level. SEA—*vertical polarization*

3. In reality, both types of attenuation *(absorption* and *diffraction)* are combined in a complex fashion and the total resultant attenuation is obtained by a series expansion whose terms are tedious to calculate and converge slowly; the numbers against

the curve in Fig. 4.3 are the numbers of terms necessary to obtain an error of less than 1%; some authors think of these terms as distinct "modes" of propagation having a sort of physical existence. In order to facilitate practical applications, we have expressed the results in the form of several graphs prepared in advance for various types of terrain. The wavelength (or frequency) is the parameter for the curves, the abscissa is the distance and the ordinate is the field for a radiated power of 1 kW; if the two scales are logarithmic, the simple law for the decrease of the field in free space according to formula (4.16) is represented by a straight line of slope -1 (depending on the scales) and passing through the point: $x = 10$ km, $y = 30$ mV/m. The curves for the ground wave at various frequencies lie below this line, and the shorter the wavelength and the worse the conductivity of the ground, the lower they lie.

	Values of d for $d_e = 1000$				
λ (km)	10	1	0·1	0·01	0·001
d (km)	2150	1000	470	215	100

Usually, one sticks to *vertical polarization* and to a few typical kinds of terrain: *sea, cultivated land, dry land*; graphs for such cases are shown in Figs. 4.12b and 4.13, in which the curves relating to frequencies less than 30 Mc/s are taken from the most recent official documents;† those relating to higher frequencies are the average of the results of several authors; they are, moreover, of less interest, as we shall see below.

Horizontal polarization is useless at long and medium waves over high-conductivity terrain: the field strength is much too low. It does become usable at metric wavelengths and especially over low-conductivity ground. Figure 4.14 shows the field strength‡ (there is still a marked disadvantage compared with vertical

† C.C.I.R., Los Angeles, 1959.
‡ After Burrows, *loc. cit.*, p. 428.

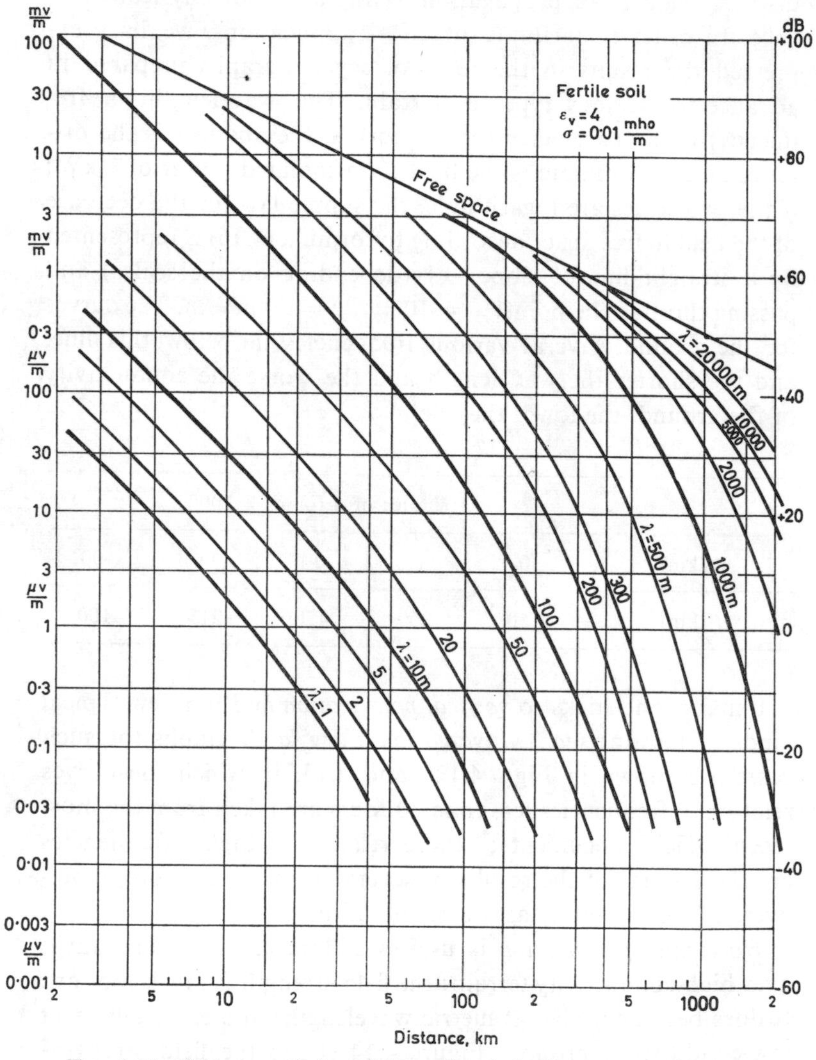

FIG. 4.13. Field strength for 1 kW radiated. Stations at ground level. LAND—*vertical polarization*

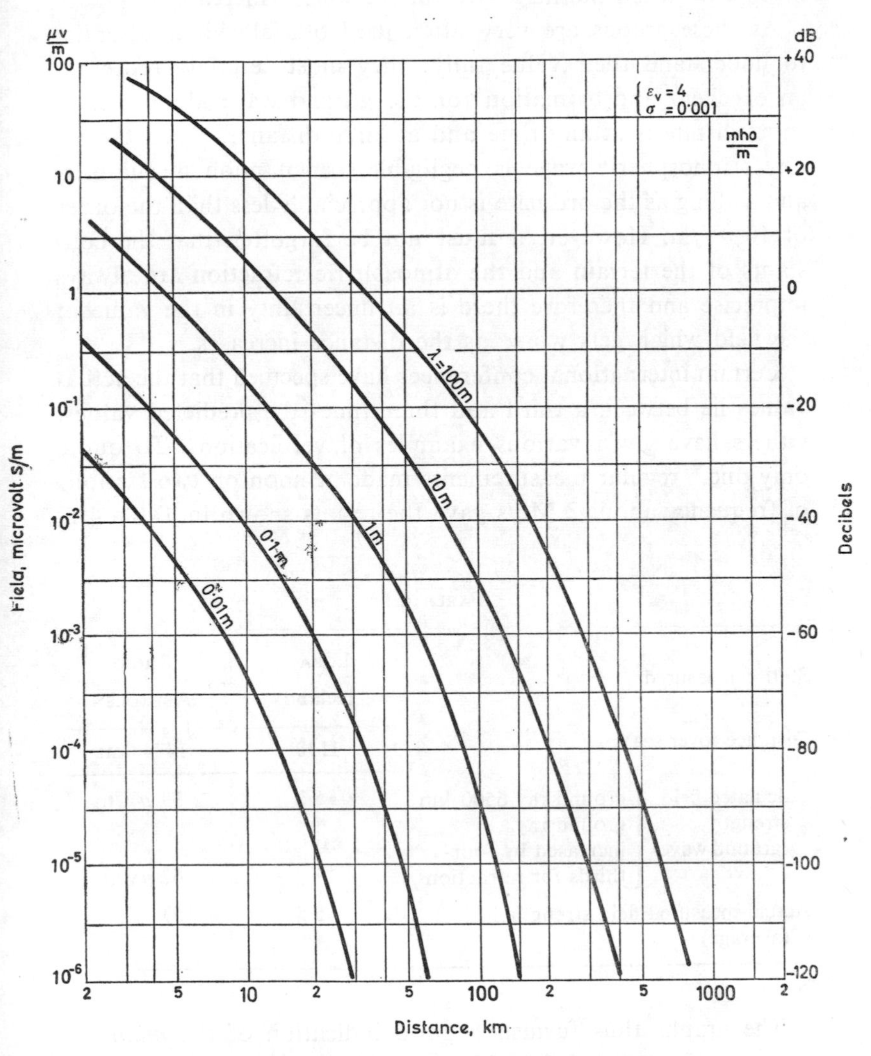

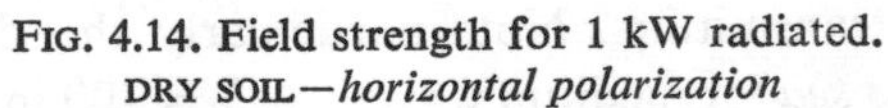
FIG. 4.14. Field strength for 1 kW radiated.
DRY SOIL—*horizontal polarization*

polarization, but, as we shall see later, this disadvantage rapidly disappears when the height of the stations is increased).

As these graphs are very often used officially, it is essential to understand their value fully: they most certainly represent an excellent approximation for the ground wave alone, i.e. at short distances at any time and at large distances when the indirect ionospheric wave is negligible, i.e. at noon in summer, and as long as the ordinate is not appreciably less than the order of 10 μV/m. However, it must not be forgotten that the constants of the terrain and the atmospheric refraction are always imprecise and therefore there is an uncertainty in the value of the field which gets worse as the distance increases.

Certain international conferences have specified that the actual values lie between a third and three times the predicted values. Others have given various examples of verifications. To quote only one,† regular measurements made at noon on two stations of frequency about 2 Mc/s gave the results shown in Table 4.1.

TABLE 4.1

Station measured		LORAN Iceland	Radio ABERDEEN
Distance (over sea)		1130	1474 km
Calculated field strength (ground wave)	ground ray 6500 km	5·2	39 μV/m
	ground ray, increased by four-thirds for refraction	17	62 μV/m
Actual measured field strength (average)		5·5	53

The graphs thus furnish a good indication of the *minimum* range of the transmitters. Note particularly the difference between *land* and *sea* for medium waves: at $\lambda = 200$ m for example, a limiting field of 10 μV/m corresponds to a range of 1000 km

† C.C.I.R., London, 1953, doc. 97.

over sea, and 180 km over cultivated land (it would be 60 km over very dry land).

This difference is less for long and very long waves for which long ranges can be guaranteed over both land and sea, and also for very short waves, which on the other hand suffer a considerable attenuation, even over sea. (But we shall see later that the slightest increase in height greatly modifies this attenuation.) We can now understand the technical motives behind the allocation of frequency bands: wavelengths of 1000–2000 m are much sought after by broadcasting authorities to provide at all times and over all types of terrain a "national" service; whereas, wavelengths of 100–200 m, which are not appropriate to such a service, are still very useful for maritime or coastal services, fishing boats, etc. (see § 8.2).

4.4.3. Propagation over inhomogeneous terrains, especially mixed (land–sea) trajectories

It has been assumed so far that the nature of the ground was the same all the way along the trajectory (ε, σ independent of distance and depth).

It is obvious that this simplification is, in fact, very rare; most of the time, over large distances, the nature of the ground must vary; we have, alternately, ground which is more or less dry, more or less cultivated, more or less hilly, or even with some parts of the trajectory over land and others over sea; this latter case is of particular interest for coastal services or broadcasting in maritime regions interspersed with islands or peninsulas. One can also have stratified ground, i.e. consisting of several layers of different conductivities.

What is the law of propagation of the field in these cases? It is very difficult indeed to calculate.

The only case that can be treated with any precision is the case where a wave, after following a path d_1 over a surface of conductivity σ_1, passes over a surface of conductivity σ_2 and

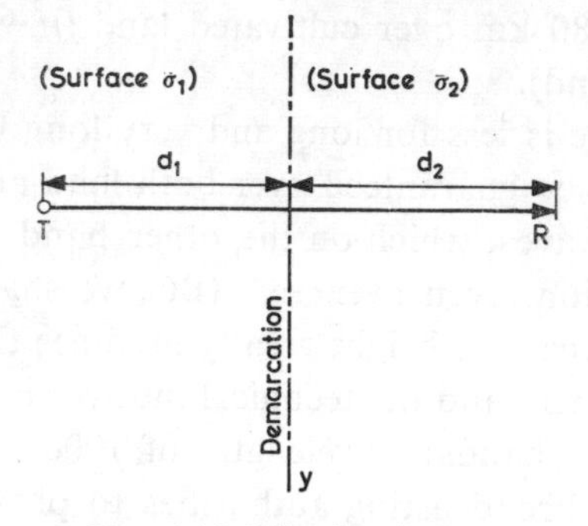

FIG. 4.15. Mixed trajectory

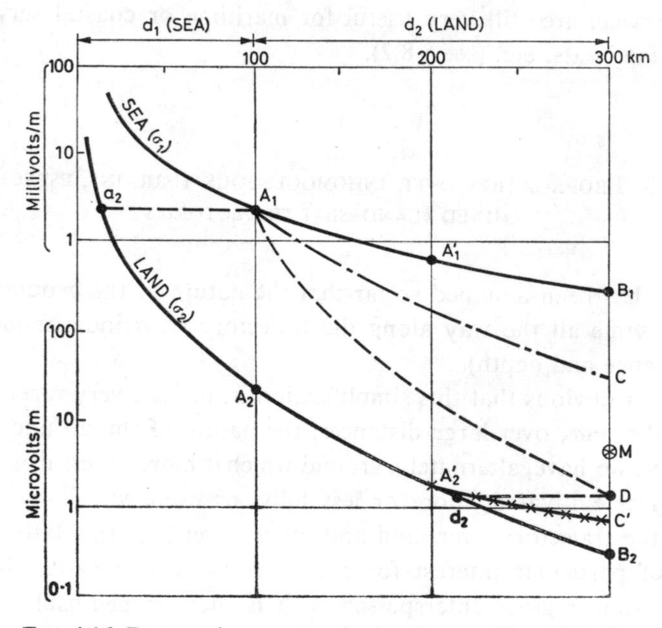

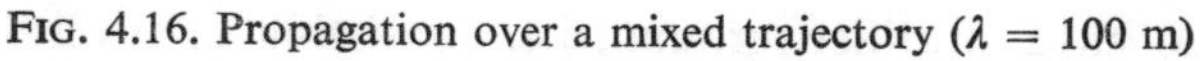

FIG. 4.16. Propagation over a mixed trajectory ($\lambda = 100$ m)

sea $d_1 = 100$ km

land $d_2 = 200$ km

A_1C Eckersley's method (vertical translation of A_2B_2);
A_1D Somerville's method (horizontal translation of a_2d_2);
$A_2'C'$ inverse trajectory (200 km land + 100 km sea), by Eckersley's method (vertical translation of $A_1'B_1$);
M geometric mean of CC', by Millington's method

executes a new trajectory d_2 (the demarcation line being assumed perpendicular to the trajectory and infinitely long (Fig. 4.15).

It is then a question of combining the known laws of attenuation $E(d, \sigma)$ for the two parts of the trajectory.

An attempt was made by Eckersley (1930) who firstly used the curve $E(d_1, \sigma_1)$ and then curve $E(d_2, \sigma_2)$, displacing the second *vertically* to coincide with the first at the point $d = d_1$. From this we obtain Fig. 4.16. Curve A_1B_1 which relates to the terrain σ_1 is applicable up to the point A_1 whose abscissa is d_1; curve A_2B_2, which relates to the terrain σ_2, is applicable beyond that, except that it is raised by the amount $A_2A_1 = B_2C$ to give the point C at abscissa (d_1+d_2). (The method of the equivalence of field strengths or power at the demarcation.)

Another method, due to Somerville, consists in fitting the second curve, not by a vertical displacement but by a horizontal displacement: in Fig. 4.16, translation $a_2A_1 = d_2D$, giving the point D for the abscissa (d_1+d_2) (a result very different from the preceding one if the curves 1 and 2 are very different).† (The method of equivalence of distance at the demarcation.)

As these methods were not entirely satisfactory, a third was proposed by Millington:‡ it consists in applying Eckersley's method twice over, i.e. once in the direction of effective transmission T to R, and once in the reverse direction from R to T, and taking the *geometric mean* of the results obtained. The reciprocity principle is thus satisfied, which is not the case for the other two methods. In Fig. 4.16, the method operates by taking the point A_2' corresponding to d_2 over land, followed by a translation of $A_1'B_1$ to $A_2'C'$, giving the point C' corresponding to the reverse trajectory; we then arrive at the point M, the geometric mid-point of CC'.

This method leads to two conclusions which are at first sight quite remarkable.

The first is that the field after the total transmission of d_1 over

† In order to use Somerville's method the distance is plotted on a linear scale in Fig. 4.16. With a logarithmic scale the result would be different, the horizontal translation changing the ratio of the distances and not their sum.

‡ Millington, *Proc. I.E.E.*, pt. III, Jan. 1949 and July 1950.

sea plus d_2 over land may be greater than it would have been after traversing only d_2 if the transmitter had been situated on land; this is the case in Fig. 4.16 in which the point M is above the point A_2'. According to this, therefore, it may be advantageous to install a transmitter on an off-shore island in order to serve a coastal zone.

The second is that in passing from land to sea, the field may increase with distance up to a point, with a rapid variation of

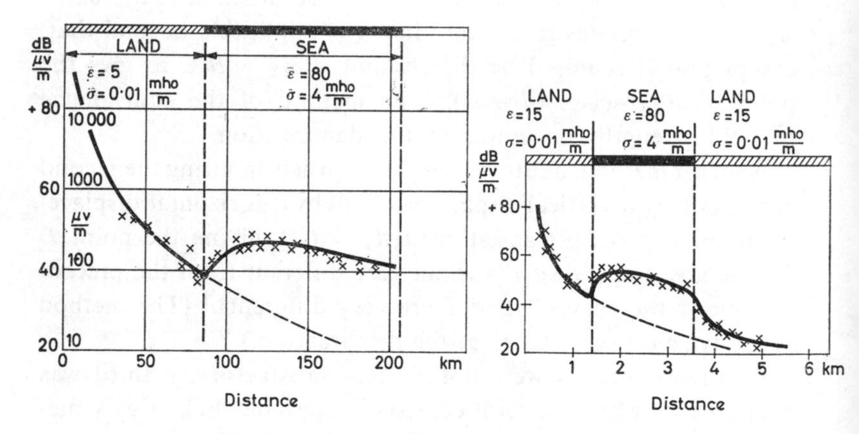

FIG. 4.17. Propagation over mixed land–sea trajectories. The curves are calculated and the points experimental. (Millington, *Proc. I.E.E.*, pt. III, July 1950, pp. 213–15.)

amplitude and phase (this is not visible in Fig. 4.16, but could be easily deduced by an appropriate construction).

These conclusions excited the curiosity of experimenters, who tried several times to verify them and succeeded in doing so. The first was confirmed by some B.B.C. results,† according to which two transmitters produced more or less equivalent fields when one of them was situated on land and the other 50 miles further away on the sea. As regards reinforcement on passing from land to sea, this was carefully measured by Millington him-

† Kirke, *Proc. I.R.E.*, May 1949, pp. 489–96.

self at 3·13 Mc/s and 75 Mc/s in a boat in the Straits of Dover, and by Elson at 1·1 Mc/s in an aircraft† (Fig. 4.17).

Numerous other verifications were reported at the conferences of the International Consultative Committee on Radiocommunication in 1953, 1956 and 1959 and variants of the methods were proposed by many countries.‡ It was finally concluded that Millington's method "is in good agreement with theory and is suitable for the majority of actual cases".

This type of calculation for a change in the nature of the terrain can, naturally, be extended to two or more successive changes; for instance, if the transmission path passes over an island or a strip of low conductivity ground in the middle of some of high conductivity, it is true to say that the adverse effect of the island is less at a great distance than it would have been if the transmitter had been situated on it. If there is a large number of successive strips, alternately good and bad conductors, their effect depends on their width: if the latter is not great compared with the wavelength it is the mean value that matters; but if the width of each strip is large compared with the wavelength (for example, 50 to 100 times), the field regime is established on each one, with a sort of "reflection loss" at each boundary (especially if the change is abrupt). The absorption can then be worse than if the whole path had the worst conductivity.

Lastly, all this is on the assumption that the "strips" of ground extend infinitely on both sides; in reality, it often happens that the zones of poor conductivity are narrow; in this case a sort of "contour" is produced and the field can be formed again further on with its normal value; this happens around the region of Paris for a wavelength of 300 m.||

† *Nature*, 22 Jan. 1949 and 16 July 1949.

‡ One other solution is to calculate Sommerfeld's "numerical distances" separately for the different parts of the trajectory [from eqn. (4.17)] and add them together to obtain the total "numerical distance". (See, for example, Argirovic, *Ann. Télécomm.*, June 1953, pp. 212–24, and Suda, *Wir. Eng.*, Sept. 1954, pp. 249–51.) But the C.C.I.R. came to the conclusion (1959, Report 141, doc. 445) that "outside its limited field of application", this method "gives large errors".

|| P. David, *Revue générale d'Électricité*, 13 May 1933, pp. 623–30, Fig. 5, field strength map of Poste Parisien.

Theoretically, the case of a stratified soil can be compared to the "launching" of a guided wave in a dielectric layer over a metallic surface,† but the application to propagation over the ground is difficult.

4.5. Stations at any height

So far, we have assumed either that the stations were sufficiently elevated for the optical formulae to apply (§ 4.3), or that they were at ground level (§ 4.4); we now come to the case where they are at any height: this is obviously more difficult.

4.5.1. Short distances. Quasi-flat Earth

At short distances, where the Earth's curvature is not important, we might expect a progressive shift from the formulae of § 4.3 (direct wave plus reflected wave) to those of § 4.4.2 (surface wave). This shift is not easy to express. The most constructive method seems to be that of Norton;‡ in the expression for the field, he preserves the three terms:

direct wave	e_1	formula (4.2),
reflected wave	e_2	formula (4.3),
surface wave	$e_3 = e_1 \times A$	(A being the attenuation coefficient, formula (4.18a),

but replaces the parameters p and b (numerical distance and phase) by new corrected parameters P and B which take height into account; if φ is still the angle of arrival of the reflected ray at the ground and $\mathscr{C}$ is the reflection constant of § 3.2.3 (a function of the nature of the ground, the frequency and the

† See, for example: Wait *et al.*, *I.R.E. Trans.* AP1, 1953, pp. 9–12 and AP3, 1954, pp. 144–8; Brick, *Proc. I.R.E.*, June 1955, pp. 721–8; Fernando, *Proc. I.E.E.* part B, May 1956, pp. 307–18; Barlow, *Proc. I.R.E.*, July 1958, pp. 1413–17.

‡ *Proc. I.R.E.* Oct. 1936, pp. 1367–87; Sept. 1937, pp. 1192–236; Dec. 1941, pp. 623–39.

polarization), we have now

$$A' = f(P, B)e^{j\psi},$$

with

$$P \cdot e^{jB} = p\left[1+\frac{\sin\varphi}{\sqrt{\mathscr{C}}}\right]e^{jb}. \tag{4.21}$$

Starting from this transformation and obtaining the values of the function $f(P, B)$ as in § 4.4.2, we can obviously make an accurate determination of the field resulting from the sum of the three waves in the region where they are all appreciable.†

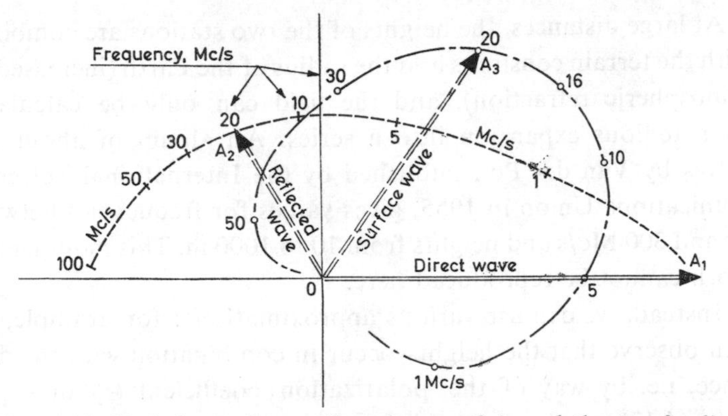

FIG. 4.18. Variations of the reflected wave OA_2 and the "surface" wave OA_3 with respect to the direct wave OA_1.
Vertical polarization over sea:

$d = 5$ km, $h_1 = 10$ m, $h_2 = 50$ m.

Variable frequency indicated in megacycles per second on the curves

The rapid phase variation of the reflected wave and amplitude variation of the surface wave sometimes produce a very odd variation in the resultant field; an example is given in Fig. 4.18.

But this region is actually very narrow. At shorter distances

† Certain approximations can, in any case, be made, for example the following, on which Bullington bases a whole series of graphs (*Proc. I.R.E.*, Oct. 1947, pp. 1121–36):

$$A' = \frac{-1}{1+j(2\pi d/\lambda)(\sin\phi+\sqrt{\mathscr{C}})^2}.$$

or greater heights, the "surface" wave becomes negligible (since $P > p$) and we are left with only the direct and reflected waves, in the interference zone, as in § 4.3. On the other hand, for greater distances or lower heights, the reflection coefficient tends to (-1) and the first two terms cancel each other out, leaving only the surface wave; since φ also tends to zero, P tends towards p and formula (4.21) no longer suffices to take account of the elevation of the sites; this is all the more so in the "diffraction zone".

4.5.2. Large distances. Effect of the Earth's curvature

At large distances, the heights of the two stations are combined with the terrain constants and the radius of the Earth (increased by atmospheric refraction), and the field can only be calculated by a tedious expansion into a series. An album of about 200 plates by Van der Pol, published by the International Telecommunications Union in 1955, gives values for frequencies between 30 and 300 Mc/s and heights from 10 to 1000 m. This monumental work cannot be reproduced here.

Instead, we can use various approximations: for example, we can observe that the heights occur in combination with the distance, i.e. by way of the polarization coefficient $\mathcal{C}_V$ or $\mathcal{C}_H$ of eqn. (4.17);

$$\left.\begin{aligned} &\text{"numerical" heights } q = \frac{2\pi}{\lambda}\sqrt{\mathcal{C}}\times h, \\ &\text{and} \qquad K = \left(\frac{\lambda}{2a}\right)^{1/3}\frac{1}{\sqrt{\mathcal{C}}} \qquad (a = \text{radius of Earth}). \end{aligned}\right\} \tag{4.22}$$

The factor $A_3\,(q_1, q_2, K)$ which multiplies the field of the surface wave then becomes approximately the same for the different terms in the series, symmetrical in q_1, q_2 and independent of d. In other words, the effect of raising the height of each of the ends can be calculated *separately*, which can be interpreted by observing that elevation manifests itself first of all by a change in the aerial constants (radiation resistance, ground losses, polar diagram), and then by a reduction in the effect of the ground in the part of the path lying between the station and the optical horizon where propagation takes place "in space".†

Sometimes the correction factor $A_3(q)$ or "height-gain" relative to each station is still complicated to calculate. Three zones should be distinguished, according to the values of q, i.e. of h.

4.5.3. "IMMEDIATE PROXIMITY" ZONE

In the first zone, which we can call the "*immediate proximity*" zone, elevation manifests itself more by the change in the aerial constants (radiation resistance‡ and polar diagram) than by any actual modification in the propagation; the field multiplication factor is simply:

$$A_3(q) = |(1+jq)|, \tag{4.23}$$

which is usually > 1, but may be less than unity when the imaginary part of q is positive, i.e. for vertical polarization over a good conductor, where η reduces to $-j\sigma\lambda$ and therefore:

$$\mathcal{C}_T = \eta = \frac{-j\sigma\lambda}{-\sigma^2\lambda^2} = \frac{j}{\sigma\lambda};\|$$

this expresses the disappearance of the factor 2 representing the effect of the "image" in the ground (§ 4.4.2).

The limit of this zone is at $q = 1$, i.e. at height:

$$(H_V)_1 = \frac{\lambda}{2\pi}\cdot\frac{1}{\sqrt{\mathcal{C}_v}} \quad \text{and} \quad (H_H)_1 = \frac{\lambda}{2\pi}\cdot\frac{1}{\sqrt{\mathcal{C}_H}}, \tag{4.24}$$

given by Fig. 4.19. It can be seen that this is always less than λ except for long wavelengths over sea; in other words, especially with horizontal polarization, the slightest elevation is sufficient to cause the aerial to be regarded as no longer "at ground level" and the "height" correction to become important.

† Chireix, *Bull. Société francaise des Électriciens*, 1946.

‡ There are in existence relatively simple formulae giving the variation in radiation resistance with height above an infinitely conductive surface; but the hypothesis of infinite conductivity is not acceptable for very short waves where the ratio h/λ is appreciably large.

|| This reduction in field strength for a small elevation of the aerial has been verified experimentally; see, for example, Eckersley and Millington, *Proc. Phys. Soc.*, 1 Sept. 1939, pp. 805–9.

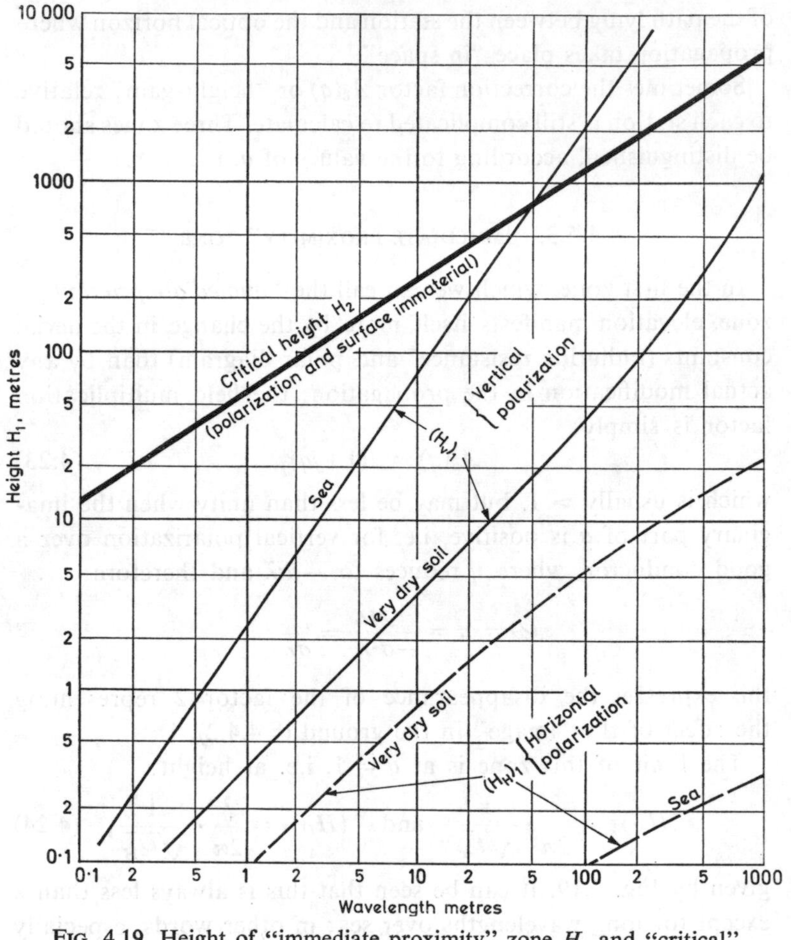

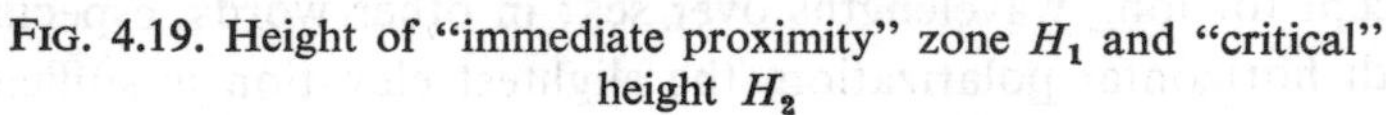
FIG. 4.19. Height of "immediate proximity" zone H_1 and "critical" height H_2

4.5.4. SECOND ZONE. "CRITICAL" OR "NATURAL" HEIGHT

There is a second zone in which increasing the height reduces the ground losses over an increasing distance, bringing about an almost proportional increase in the field strength. The preceding formula reduces to

$$A_3(q) = jq. \tag{4.25}$$

The limit of this zone occurs at a certain value H_2 of h, known as the *critical* or *natural* unit of height, independent of polarization and type of soil:

$$H_2\,(\text{m}) = (30 \text{ to } 50)\,\lambda^{2/3} \text{ (metres).} \qquad (4.26)$$

This, too, is shown on Fig. 4.19. It can be seen that for vertical polarization at long wavelengths over sea, this second zone overlaps the first for heights of the order of 1000 m or more, which is hardly of any practical interest. In all other cases the distinction is quite clear-cut; there is therefore an appreciable region, which is very frequently used in practice, (for example, between $h = 1$ m and $h = 100$ m for metric waves) in which the field strength increases *proportionally to the height of the aerial:* this result is analogous to that found in paragraph 4.3.1 [formula (4.9)] for the flat earth case in the "lower lobe" of the vertical diagram.

4.5.5. Large heights. Height gain curves

In any part of the path where the height is greater than the critical value H_2, the presence and nature of the ground no longer have any effect; the wave propagates as it would in free space. The reinforcement of the field strength is thus practically the same as it would be if we reduced the distance in the formula for the attenuation A, in § 4.4.2, in proportion; the expression for A_3 in terms of h and K [formula (4.22)] is rather laborious to calculate (Hankel functions) but it indicates a practically exponential variation. In the third zone, therefore, the field strength increases *more rapidly* than the height.

Figures 4.20, 4.21 and 4.22 show, for the three zones, the values of the "height correction" factor for various wavelengths and various terrains (dry ground† and sea) for both polarizations.

It can be seen immediately that for vertical polarization at metric (and sometimes decametric) wavelengths this factor is quite large in the majority of practical cases, where the heights of the aerials can easily be several metres or tens of metres,

† An urban zone is practically equivalent to "dry ground".

particularly in aircraft. With horizontal polarization, over dry ground, the effect is greater still; over a high-conductivity surface (sea) it would be even more pronounced, so much so that the difference between horizontal and vertical polarizations, which

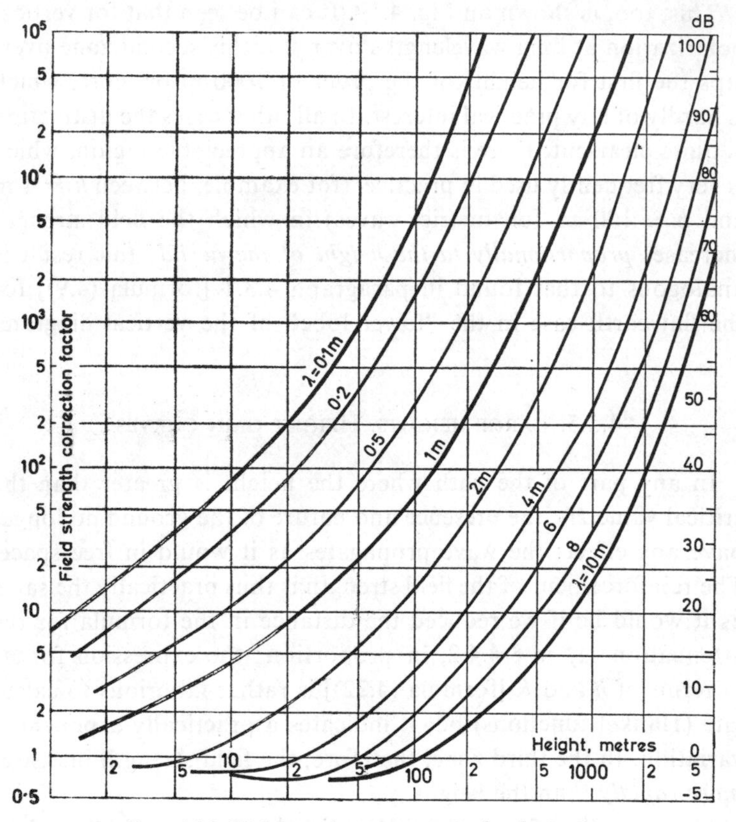

FIG. 4.20. Height gain. SEA—*vertical polarization*

precluded the use of the former at ground level, soon vanishes; and in some cases, either can be used at will.

We shall be discussing later the application of these formulae and graphs, but a numerical example will help to fix ideas.

Suppose that a television service is required on a wavelength $\lambda = 6$ m at a distance of $d = 100$ km over *land*, with vertical polarization. The graph (Fig. 4.13) indicates a field strength of

0·03 μV/m for 1 kW radiated power for ground stations. This is much too weak to ensure comfortable reception: at least 100 μV/m is necessary and it would be inconceivable to try to obtain this by increasing the power. But suppose we raise the height

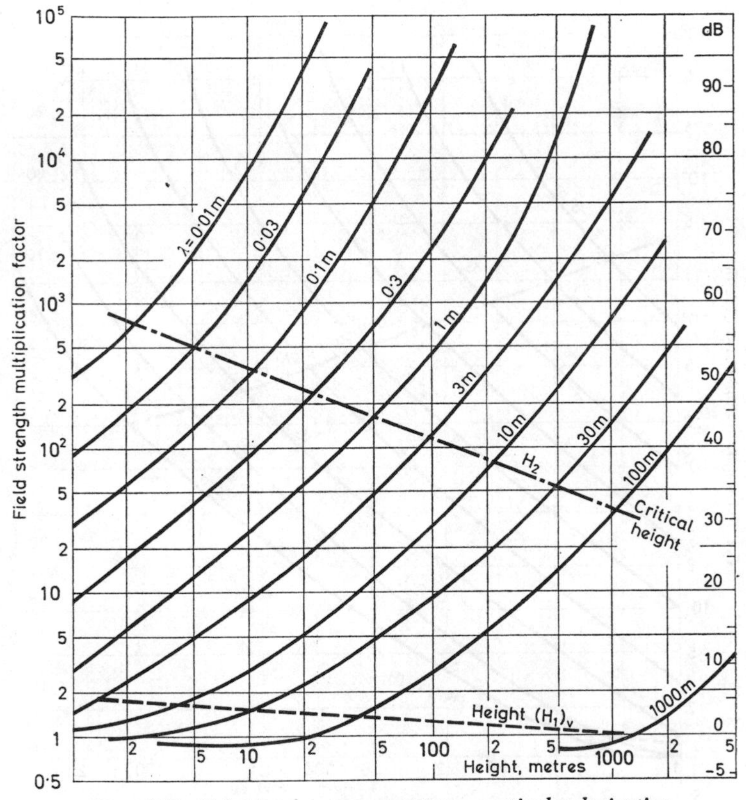

FIG. 4.21. Height gain. DRY GROUND—*vertical polarization*

of the transmitter to $h_1 = 300$ m (Eiffel Tower) and that of the receiving aerial to $h_2 = 10$ m. Applying formula (4.1), we see that for these heights the optical range is

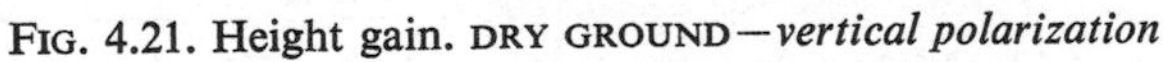

$$4{\cdot}1(\sqrt{300}+\sqrt{10}) = 83 \text{ km}.$$

The required range is greater than the optical range; we are therefore in the "diffraction zone" and the effect of elevation can be calculated from the graph of Fig. 4.13 (vertical polarization): the elevation of 300 m produces a gain of 250 (approx.) and the 10 m elevation a gain of 4 (approx.), giving a total of 1000. The field strength for 1 kW of radiated power is therefore 30 μV/m.

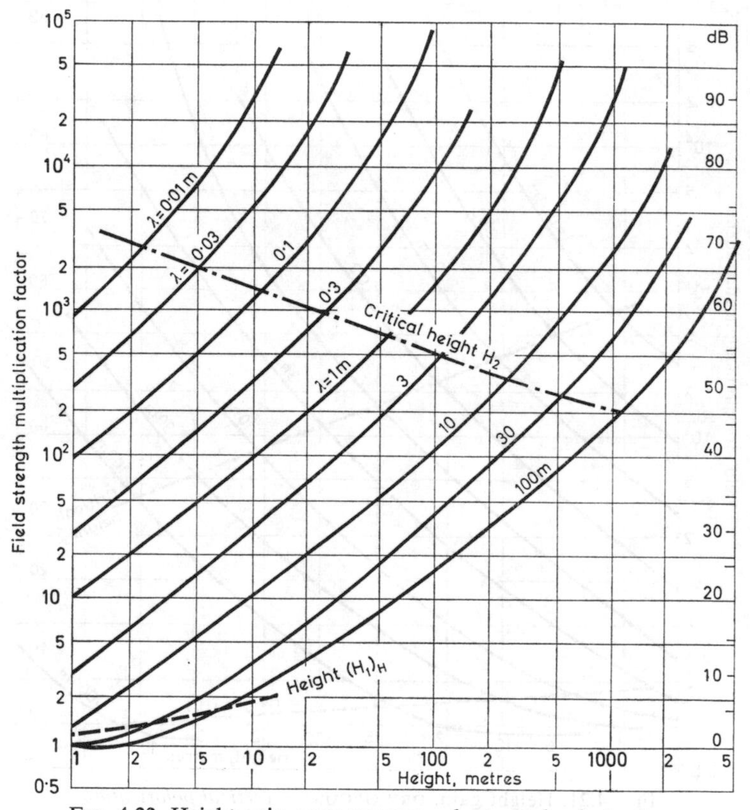

FIG. 4.22. Height gain. DRY GROUND—*horizontal polarization*

Raising the transmitted power to 11 kW would produce the 100 μV/m required for comfortable reception.

With horizontal polarization, the conclusions would be more or less the same: the field strength would be about 10 times smaller (see Fig. 4.14), but the height gains would be of order 1000 and 20, giving, finally, 30 μV/m.

4.5.6. Fishback's method. Stations at the critical height

The preceding method is not the only possible one; for example, one can calculate the field assuming the two stations to be at a given height different from zero—the "critical height" H_2,

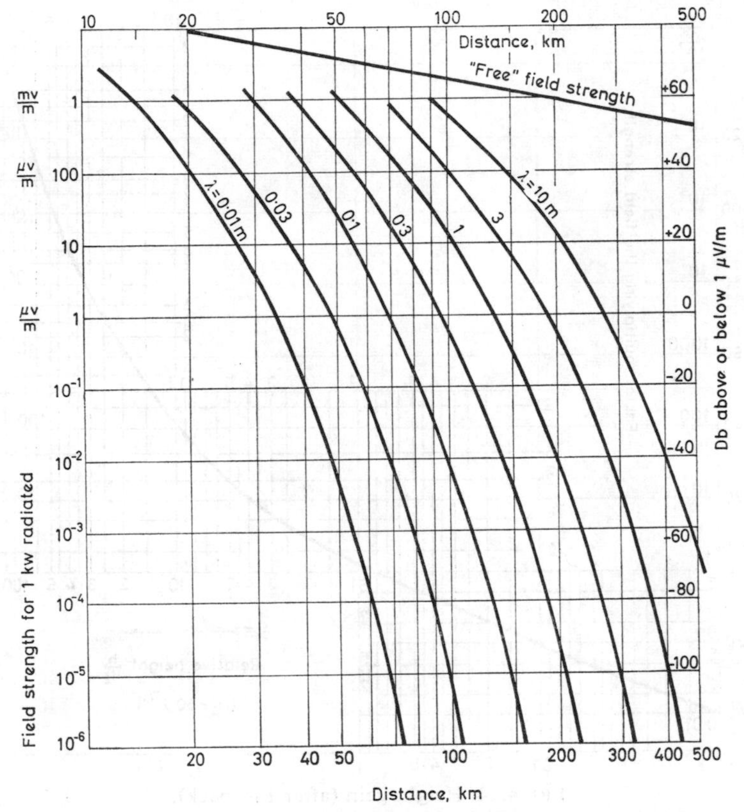

Fig. 4.23. Variation of field strength in the diffraction zone between stations at the "critical height" H_2 (Fishback)

for instance—and then apply a correction for the different heights. The method is obviously advantageous if, in fact, the heights are of the order of H_2: it gets round the need to calculate first a "ground field strength" which is much too weak and then multiply it by an enormous correction factor.

The results found by Kerr and Fishback† are as follows:

First, the field is evaluated in terms of the distance for stations situated at the critical height H_2; Fishback uses a single curve, taking as abscissa a "natural unit of distance" in terms of the wavelength. We have put it back into the usual form (Fig. 4.23).

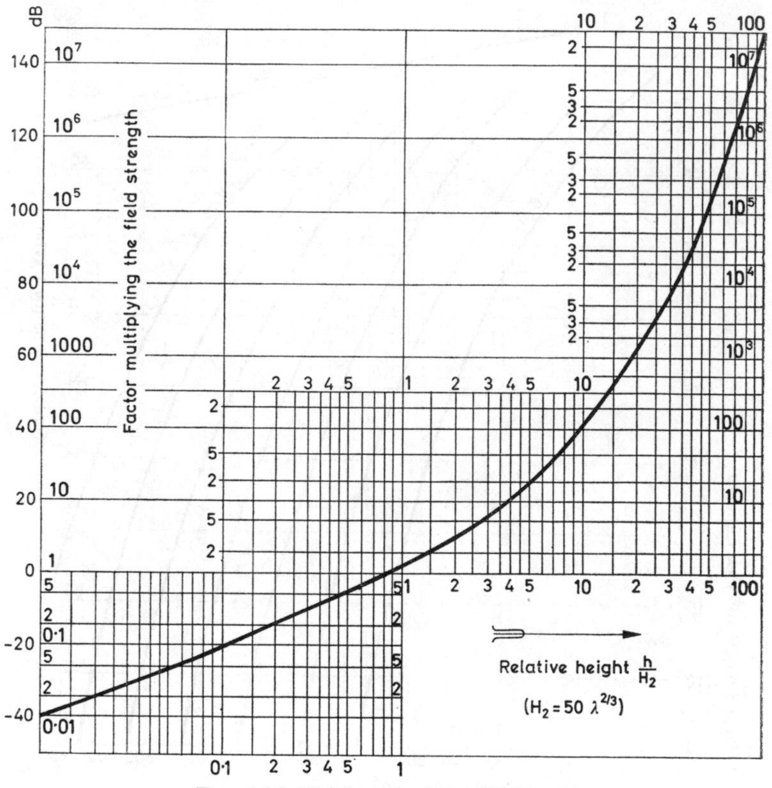

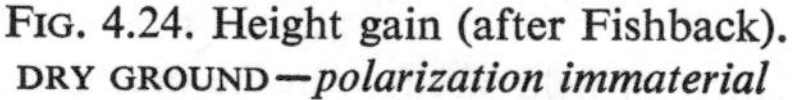

FIG. 4.24. Height gain (after Fishback).
DRY GROUND—*polarization immaterial*

† Kerr, *Propagation of Short Radio Waves*, § 2.14. The "natural unit of distance" is L (km) = $29\sqrt{\lambda}$ (m) and the attenuation given by figure 4.46.1 corresponds to the formula:

$$A = 2\sqrt{\pi\frac{d}{L}}\times e^{-2\cdot 02d/L}.$$

It will be noted that (as indicated in Fig. 4.24) the height factor is not quite equal to 1 for $h = H_2$.

The field is then multiplied by the "height factor" given by Fig. 4.24 as a function of h/H_2.

In principle, these curves are valid for either polarization over dry ground.

By way of example, let us take the preceding problem: $\lambda = 6$ m at 100 km; Fig. 4.23 gives a field of 300 μV/m (approx.) "at the critical height", which is $H_2 = 170$ m (approx.); the transmitter is therefore at $h_1 = 300\text{ m} = 1{\cdot}77\, H_2$, the receiver at $h_2 = 10\text{ m} = 0{\cdot}059\, H_2$; according to Fig. 4.24 the corresponding corrections are +8 and −25 dB, i.e. a total of −17 dB, giving a field strength of 43 μV/m, instead of 30 (above); bearing in mind the theoretical and graphical approximations in both cases, it must be admitted that the orders of magnitude are in good agreement.

4.6. Intermediate zone. Near the horizon region

It must be remembered that throughout the preceding section we assumed that we were dealing with the "diffraction zone", i.e. with stations situated below the optical horizon. The exponential increase in field strength with height ceases when we approach the horizon region, i.e. when the heights h_1, h_2 satisfy formula (4.1)

$$d\ (\text{km}) = 4{\cdot}1[\sqrt{h_1\,(\text{m})} + \sqrt{h_2\,(\text{m})}].$$

In this intermediate zone the field strength must be "fitted" to the value calculated for the "interference zone" (§ 4.3).

The exact calculation of this "fit" is particularly difficult: the series expansions contain a large number of complicated terms.

Several authors have simplified the problem by various hypotheses regarding the relative values of the heights h_1, h_2, sometimes assuming them to be very different,† sometimes, on the other hand, equal: in the latter case Chireix‡ indicates that if the sta-

† Burrows and Atwood, *Radio Wave Propagation*, p. 419, formula (190), assume $h_1 < H_2$ and $h_2 > 40\, H_2$; they need seven abacus or special graphs, impossible to reproduce here.

‡ Chireix, *Bull. Société française des Électriciens*, 1946.

tions are exactly at horizon range, then, depending on the ratio of their height to the critical height, $H_2 = 50\,\lambda^{2/3}$ [formula (4.26)], the additional attenuation with respect to the field strength in free space is shown in Table 4.2.

TABLE 4.2. *Attenuation (with respect to free space) at limit of optical visibility*

For $\frac{h_1}{H_2} = \frac{h_2}{H_2} =$		$\frac{1}{4}$	$\frac{1}{2}$	1	2	4
Additional attenuation	ratio	$\frac{1}{31}$	$\frac{1}{22}$	$\frac{1}{18}$	$\frac{1}{11}$	$\frac{1}{8}$
	dB	−30	−27	−25	−21	−18

However, it appears that the only way to determine the field strength in this zone is to interpolate graphically somehow or other between the calculated values—in the interference zone by applying the method of § 4.3 and in the diffraction zone by starting with the ground-level field and applying the height corrections, as shown above.

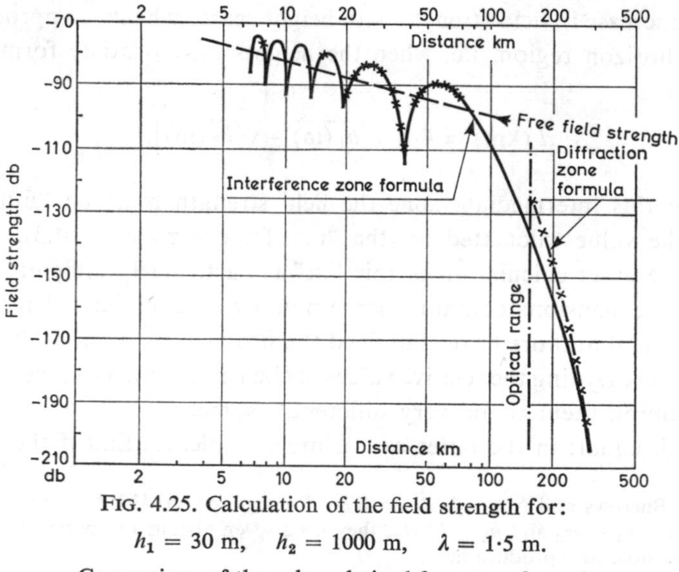

FIG. 4.25. Calculation of the field strength for:
$h_1 = 30$ m, $h_2 = 1000$ m, $\lambda = 1{\cdot}5$ m.
Comparison of the values derived from two formulae

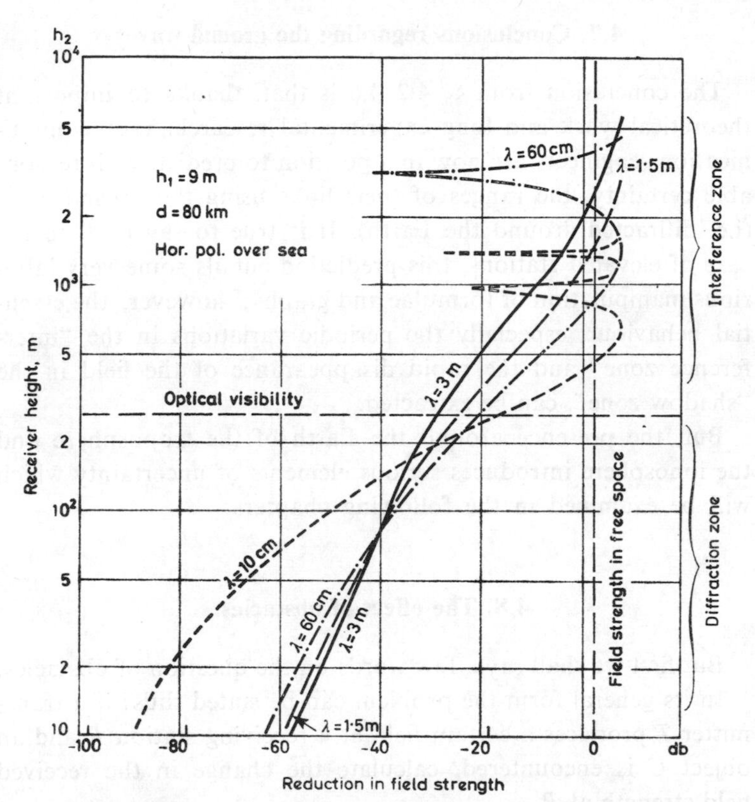

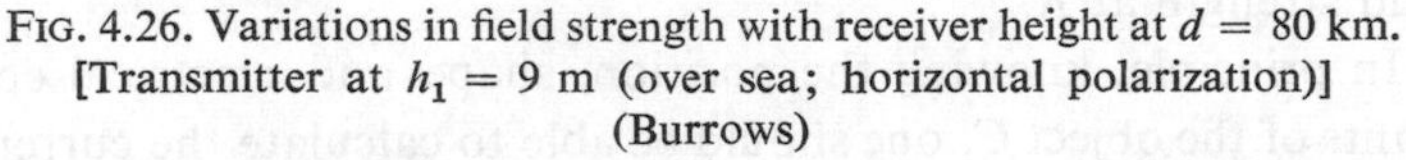

FIG. 4.26. Variations in field strength with receiver height at $d = 80$ km. [Transmitter at $h_1 = 9$ m (over sea; horizontal polarization)] (Burrows)

Figure 4.25, after Burrows, shows that this approach is quite reliable, the zone in question being narrow and the curves to be fitted together being not too far apart.

Figure 4.26 gives in a different form another version of the preceding considerations—it is concerned with the variation of the field (abscissa) as a function of the receiver height (ordinate) for a fixed distance (80 km over sea, horizontal polarization) and a given transmitter height (9 m).

The four curves are valid for $\lambda = 0{\cdot}1$, $0{\cdot}6$, $1{\cdot}5$ and 3 m: only the lower lobes of the diagram are drawn to show the regular decrease in the diffraction zone beyond the optical horizon.

4.7. Conclusions regarding the ground wave

The conclusion from §§ 4.2–4.6 is that, thanks to important theoretical work and long experimental research, radiocommunications engineers are now in a position to predict, with reasonable certainty, the ranges of their links using the *ground* wave (i.e. diffracted around the Earth). It is true to say that, in the case of elevated stations, this prediction entails some very laborious manipulation of formulae and graphs;† however, the essential behaviour, especially the periodic variations in the "interference zone" and the rapid disappearance of the field in the "shadow zone", can be extracted.

But, the presence around the Earth of the troposphere and the ionosphere introduces serious elements of uncertainty which will be examined in the following chapters.

4.8. The effect of obstacles

But first we shall say a few words on the question of obstacles.

In its general form the problem can be stated thus: if a transmitter T produces a certain field at a receiving station R and an object C is encountered, calculate the change in the received field strength at R.

In principle, knowing the position, shape and electrical constants of the object C, one should be able to calculate the currents which are set up in it and consequently the amount re-radiated to R. But, in fact, the calculation is intractable in this general form: certain restrictive hypotheses have to be made about the geometry of triangle TCR and the shape and dimensions of the object C.

T and R might be effectively far apart and the perturbating obstacle somewhere "between them"; this is how we have already considered the case where the object is a perfectly smooth plane

† Various short-cuts in these manipulations can be found in: Burrows and Atwood, *Radio Wave Propagation*, pp. 55–63, 99–110, 433–53; Kerr, *Propagation of Short Radio Waves*, § 2.16; Bullington, *Proc. I.R.E.*, Oct. 1947, pp. 1121–36.

surface of separation, producing a quasi-optical "reflection" (§§ 3, 4.3.1); we then alluded to the case where this surface was irregular (§ 4.3.2) and where the quasi-optical reflection merged into a "diffusion" in all directions, and then to the case where the surface was spherical (§ 4.3.3) and introduced a divergence coefficient; finally, we also studied diffraction round an imperfectly conductive Earth, when T and R lay in the vicinity of the surface (§§ 4.4, 4.5).

But, besides these classical "obstacles", there may perchance exist between the transmitter and receiver various features of the surface (hills, cliffs, ...), man-made structures (buildings, walls, metallic framework, ...), vegetation (trees, bushes, ...), atmospheric opacity (clouds, rainstorms, ...) and we want to know to what extent the received field will be attenuated and in particular, the size and extent of the "shadow zones" created by diffraction round an obstacle with dimensions comparable with the wavelength.†

Another aspect of the problem arises if the obstacle, although small, is situated in the immediate vicinity of the transmitter or receiver; this leads to appreciable induced currents and directional effects associated with the "aerial + obstacle" combination; this is the case for stations situated near a pylon or on a ship or aircraft where it would be quite misleading to study the aerial independently of the way it was sited.

"Electromagnetic detection" (radar) has introduced a third important case, in which the receiver is near the transmitter and one is interested only in the fraction of energy returned by the obstacle in the direction common to both. In order to calculate the range of a radar, one has first to define the "equivalent surface" of the obstacle σ as that of an ideal reflector which would capture all the energy which reaches it and re-emit it into space in a perfectly isotropic manner.

Even with the restrictions imposed by these hypotheses, the

†Furutsu has shown that this problem is analogous to that of a non-homogeneous terrain (§ 4.4.3) and that a *common* solution to them can be found (*J. Rad. Res. Lab.*, **3**, Tokyo, 1956, p. 331; **4**, 1957, p. 135; **6**, 1959, p. 71).

problem of calculating the effect of an obstacle is, in general, only possible if we assume the obstacle to have a simple geometric shape such as a thin wire, a sphere, a cylinder, a semi-plane, etc.

For the more or less elaborate shapes of the majority of real obstacles, we therefore have to be content with approximate or purely experimental evaluations.

We shall now review the principal results.

4.8.1. Filiform obstacles

The obstacle may be a one-dimensional metallic conductor (i.e. in the other dimensions it is very small compared with a wavelength)—for example, a metal telegraph pole or pylon, a ship's rigging, a telephone or power line, "window",† or an assembly of similar structures forming a framework, a grille, etc.

The current induced in such a conductor by the primary field from the transmitter is calculated using the formulae of the preceding sections or those relating to "receiving aerials". The secondary field re-radiated is then deduced.

At any point in the vicinity, this secondary field will be added vectorially to the primary field, i.e. the resultant field will depend on their relative phase, which in turn will depend on the difference in the distances of the given point from the transmitter and obstacle, producing a series of hyperbolic zones of alternate attenuation and reinforcement (with phase perturbation) (see Plate 1).

Naturally, the perturbating effect of the obstacle will be a maximum at these resonances, i.e. when the length of the obstacle is in the region of a half-wavelength (if it is isolated in space) or a quarter-wavelength (if it is earthed). Obstacles of this size will rarely be encountered at long waves; however, at medium or short waves, it will often be the case for masts or pylons—this

†Thin strips of metal falling slowly through the air—a device used during the 1939–45 war to jam enemy radars by creating false echoes. A thousand or so strips would simulate a large bomber, but, their individual weights being only a fraction of an ounce, each aircraft could carry enough to "sow" hundreds of "false aircraft" in its wake.

is what makes goniometry so tricky on board ships. At metric waves, telegraph poles, street lamps, trees, even the operator's body, may happen to be resonant and give appreciable re-radiation at surprising distances.

It is this fact that is used so successfully in radar. If we calculate the "equivalent reflecting cross-section of a dipole, according to the definition of the preceding section, we find that it is slightly greater than the "capture cross-section" defined in § 1.1.2, formula (1.23), because the definition is different (in this case it is a question not of useful power but of power re-radiated by the short circuited dipole), giving:

$$\sigma = 0{\cdot}78\lambda^2 \dagger.$$

If there are several conducting obstacles in the field, we take the sum of the re-radiated fields; in this way we can deal with grilles, "window", etc.‡

4.8.2. Spherical obstacles

The case of a sphere can also be treated quite rigorously—it has, in fact, been done already, on a different scale, for diffraction round the Earth.

Suppose the stations are situated on opposite sides of a spherical hill of radius *a;* we can apply the formulae for the "shadow zone" due to the Earth's curvature, especially formula (4.19); we then obtain a supplementary attenuation of

$$A = \frac{260}{a\,(\text{km})^{2/3} \cdot \lambda\,(\text{m})^{1/3}}\ \text{dB/km}.$$

For example, if $a = 8$ km, then for a wave $\lambda = 1000$ m the loss is 6·6 dB/km and for $\lambda = 1$ m the loss is 65 dB/km, an

†This value is in agreement with Burrows and Atwood, p. 45; but Kerr (p. 465) gives the slightly different value of $0{\cdot}88\lambda^2$.

‡The case of thin strips falling at random has been calculated: see Van Vleck, *J. App. Phys.* **18,** March 1947, p. 274. For the case of grilles, see, for example, Esau *et al.*, *Hfr. Techn. El. Ak.*, **53**, 4, April 1939, 113–15; Wessel, *ibid.*, **54**, 2, Aug. 1939, 62–9; Moullin, *J.I.E.E.*, March 1944, pp. 14–22.

enormous and rapidly prohibitive attenuation, due to the fact that the ray grazes the ground over the whole of its path (we shall see later that a shield of the same height—in this case, 250 m—placed in the middle of the trajectory causes far less trouble.

This formula has been verified by McPetrie *et. al.* for hills of approximately cylindrical shape and wavelengths of 3 cm to 11 m.†

If the sphere acts as a "radar target", its "equivalent cross-section" σ, defined in § 4.8, can be evaluated (σ is the cross-section of an object which would give the same reflected field strength if re-radiating uniformly in all directions); it is convenient to compare this cross-section with the apparent cross-section of the sphere (πa^2), taking as a variable the ratio a/λ of the radius to the wavelength.‡

This gives curve 1 of Fig. 4.27, on which it can be seen that the re-radiation is very weak as long as the radius of the sphere is small compared with the wavelength; it increases rapidly as a approaches $0{\cdot}1\ \lambda$; the equivalent cross-section passes through a maximum (equal to 3·8 times the apparent cross-section) at $a = 0{\cdot}17\ \lambda$, which is the resonance of the sphere. Then, after various oscillations, the equivalent cross-section tends to become equal to the apparent cross-section when the radius of the sphere is greater than the wavelength.||

One approximate formula giving this variation for $a \ll \lambda$ is Rayleigh's formula:

$$\frac{\sigma}{\pi a^2} = 1{\cdot}4\times10^4\times\left(\frac{a}{\lambda}\right)^4;$$

it is shown dashed in Fig. 4.27, curve 2.

(This decrease in the echo returned when the obstacle is small

†*Proc. I.E.E.*, III A, **93**, no. 3, 1946, pp. 527–30.

‡See, for example, Mie, *Ann. Phys.*, 1908, p. 377; Stratton, *Electromagnetic Theory*, 1941, p. 563.

||One might be tempted to apply this formula to the calculation of "radar echoes from the Moon"; but, for the wavelengths used, the Moon is not an infinite conductor and its surface presents appreciable irregularities. Hence the uncertainty in the value of the result.

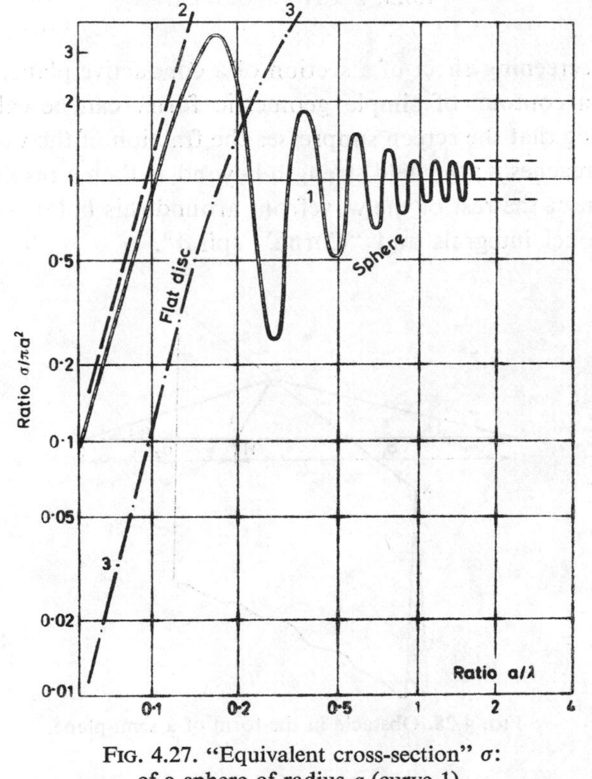

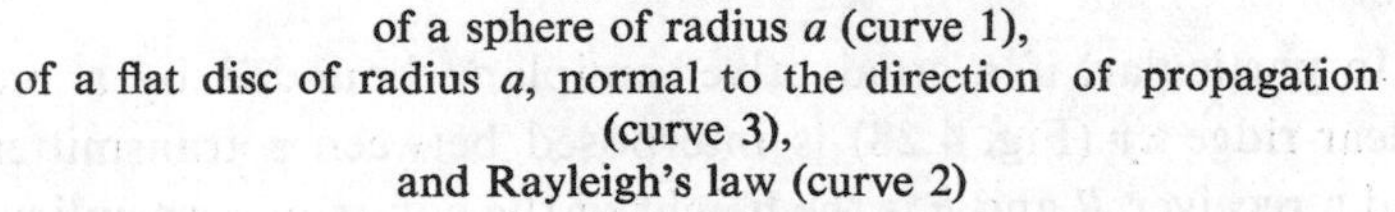
FIG. 4.27. "Equivalent cross-section" σ:
of a sphere of radius a (curve 1),
of a flat disc of radius a, normal to the direction of propagation (curve 3),
and Rayleigh's law (curve 2)

compared with the wavelength is not peculiar to a sphere: it is found for any shape.)

The problem has also been treated for a sphere with finite conductivity or simply a dielectric, which is of interest in "diffusion" or absorption by the atmosphere.†

†See, for example, Hart and Montroll, *J. App. Phys.*, April 1951, pp. 376–86.

4.8.3. Sections of planes

The screening effect of a section of a conductive plane, bounded by a contour of simple geometric form, can be calculated assuming that the screen suppresses the fraction of the wavefront which reaches it; the field strength beyond is then a result of diffraction of the rest of the wavefront around this hole; it is given by Fresnel integrals and "Cornu's spiral".

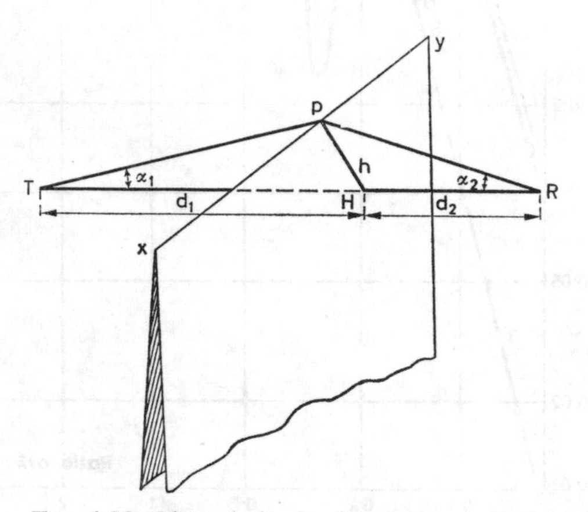

FIG. 4.28. Obstacle in the form of a semi-plane

In particular, if a conductive semiplane bounded by a rectilinear ridge *xy* (Fig. 4.28) is interposed between a transmitter *T* and a receiver *R* and *h* is the height of the common perpendicular *PH*, d_1 and d_2 the distances of *T* and *R* from the foot of the perpendicular *H*, α_1 and α_2 the angles *PTH* and *PRH*, then the attenuation produced in the field by the presence of the screen is given† in Fig. 4.29 as a function of the auxiliary variable:

$$v = \pm h\sqrt{\frac{2}{\lambda}\left(\frac{1}{d_1}+\frac{1}{d_2}\right)} = \pm\sqrt{\frac{2h}{\lambda}(\alpha_1+\alpha_2)}, \qquad (4.27)$$

counted positive if the point *H* is outside the screen (i.e. if the

†Burrows and Atwood, *Radio Wave Propagation*, p. 464, Fig. 4; *Variant.*, pp. 6, 68–9.

screen does not mask the line of sight TR), negative if the point H is in the screen (i.e. if the screen hides T from R).

It can be seen that in the former case the presence of the screen may produce a reinforcement of up to 1·18 (by suppressing the

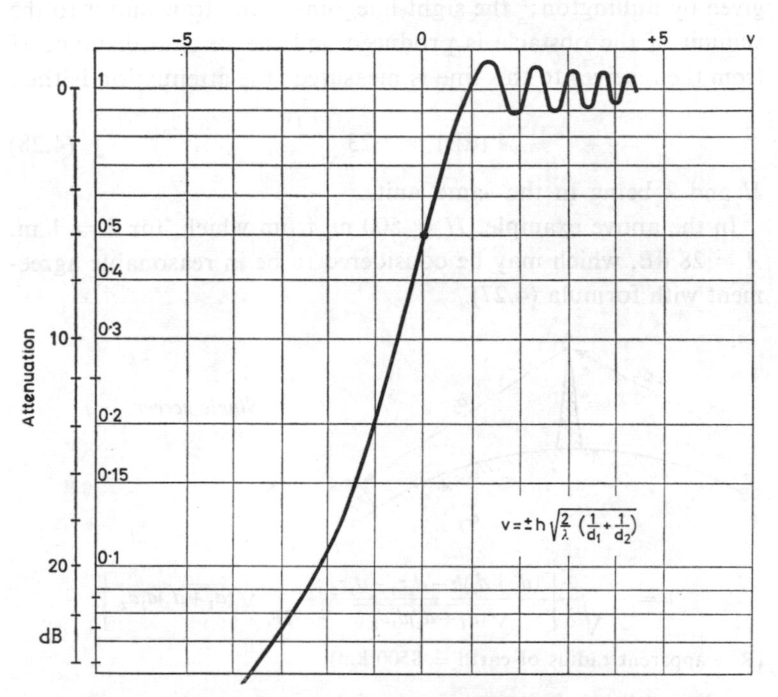

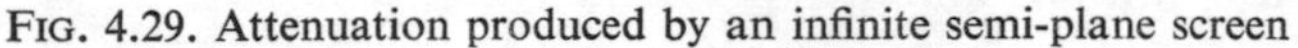

FIG. 4.29. Attenuation produced by an infinite semi-plane screen

second Fresnel zone); in the second case, there is always an attenuation which tends towards:

$$A = \frac{0 \cdot 225}{v};$$

in the intermediate case, $h = 0$, the attenuation factor is 1·5, which is natural since the screen then masks exactly half of space.

By way of example, let us take again (see § 4.8.2) a transmitter and a receiver 4 km apart, separated by a hill 250 m high, but assuming that this obstacle (situated in the middle) is relatively thin and representable by a "wall" rather than a sphere.

The present formula will then give, for $\lambda = 1000$ m, $v = -0{\cdot}36$, whence attenuation = 9 dB, and for $\lambda = 1$ m an attenuation of 34 dB. (These attenuations are much lower than those found in § 4.8.2.) For short waves, we can also use an empirical formula given by Bullington;† the sight-line joining the transmitter to the summit of the obstacle is produced and the vertical distance H from the receiver to this line is measured; the attenuation is then

$$A\ (\mathrm{dB}) = 1{\cdot}23\sqrt{\frac{H}{\lambda}}, \tag{4.28}$$

H and λ being in the same units.

In the above example, $H = 500$ m, from which, for $\lambda = 1$ m, $A = 28$ dB, which may be considered to be in reasonable agreement with formula (4.27).

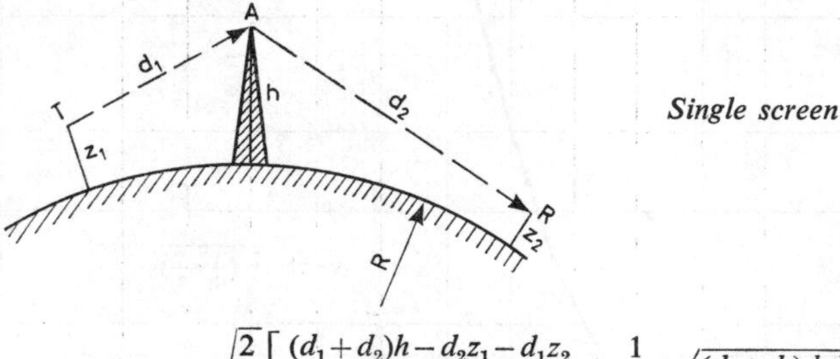

$$v = -\sqrt{\frac{2}{\lambda}}\left[\frac{(d_1+d_2)h - d_2z_1 - d_1z_2}{\sqrt{(d_1+d_2)d_1d_2}} + \frac{1}{2\mathrm{R}}\sqrt{(d_1+d_2)d_1d_2}\right]$$

(R = apparent radius of earth = 8500 km).

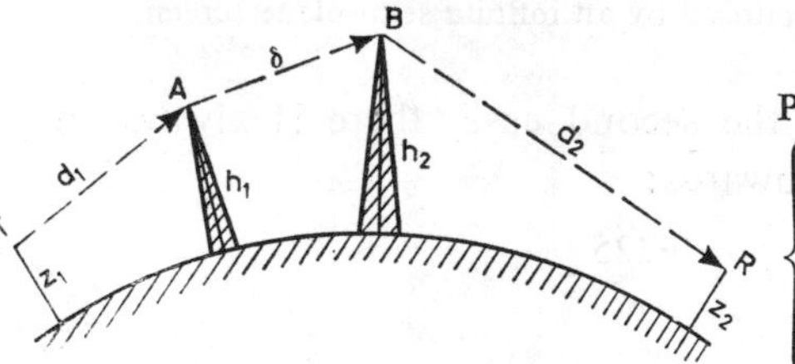

Two screens

Putting:

$$\begin{cases} \mathrm{D} = d_1+\delta+d_2, \\ k_1 = \dfrac{\mathrm{D}(d_2+\delta)d_1}{2\mathrm{R}} \\ k_2 = \dfrac{\mathrm{D}(d_1+\delta)d_2}{2\mathrm{R}} \end{cases}$$

k_1, k_2: negligible at short distances

FIG. 4.30. Sacco's formulae for the effect of screens (proposed to the C.C.I.R., Stockholm, 1948, p. 287)

$$v = -\sqrt{\frac{2}{\lambda}}\sqrt{\frac{[\mathrm{D}h_1-(d_2+\delta)z_1-d_1z_2+k_1]\,[\mathrm{D}h_2-d_2z_1-(d_1+\delta)z_2+k_2]}{\mathrm{D}d_1d_2}}.$$

†Bullington, *Proc. I.R.E.*, **38**, Jan. 1950.

We shall study this point again later (§ 8.5.3).

Sacco has generalized formula (4.27) to include the case where the stations are at a given height above the Earth, taking account of its curvature, and also the case where two screens in succession, of heights h_1 and h_2, interrupt the path of the wave. Figure 4.30 gives his formula for the parameter v of Fig. 4.29; it can easily be verified that they reduce to formula (4.27) in the case of a single screen between two stations on flat ground.

Other cases which have been studied include that of two parallel half-planes,† a flat strip bounded by two parallel ridges,‡ and a tri-rectangular trihedron ("corner reflector"). This last case is of interest because it is the preferred shape for reflectors for trials or radar beacons; if L is the common length of the edges of the trihedron, θ the angle which its axis of symmetry makes with the arrival direction of the waves, the equivalent cross-section is||

$$\sigma = \frac{4\pi L^4}{3\lambda^2}(1 - 0{\cdot}0076\theta^2).$$

The case of a flat plane bounded by a circle of radius a is interesting from two points of view: it can be considered as either a circular disc or a circular aperture in a plate. If it is normal to the direction of propagation and centred on the line TR, the problem can be treated simply by considering the Fresnel zones, whose radius at distance d from the transmitter is:

$$r_n = \sqrt{n\lambda d}.$$

The equivalent cross-section of the reflecting disc is then given* by curve 3, Fig. 4.27: if the disc is large compared with the wavelength, the ratio $\sigma/\pi a^2$ is proportional to the square of a/λ and tends towards:

$$\frac{\sigma}{\pi a^2} \rightarrow 4\pi^2 \left(\frac{a}{\lambda}\right)^2$$

†Cheney and Watson, *J. App. Phys.*, May 1951, pp. 675–9.

‡ Megaw, *J.I.E.E.*, III. A, **93,** 1, 1946, p. 79; Burrows and Atwood, *Radio Wave Propagation*, p. 88.

||Burrows and Atwood, *Radio Wave Propagation*, p. 45.

*Norton and Omberg, *Proc. I.R.E.*, **35,** Jan. 1947, p. 4.

(it is thus virtually a case of specular reflection); on the other hand, if the disc is small compared with the wavelength, $\sigma/\pi a^2$ decreases as the fourth power of a/λ and we have Rayleigh's law.

If the disc is not normal to the direction of propagation, but oblique, it acts also as a frame (magnetic dipole) and the re-radiation may be slightly increased.

The treatment of the case of a circular hole in a plate is similar;† it can be shown that energy passes easily through the aperture as long as the wavelength is less than πa, but longer waves are very rapidly blocked. This can be extended to slits of any shape and to arrays of holes or slits which constitute the well-known micro-wave type of aerial.‡

4.8.4. Cylinders, paraboloids, ellipsoids, etc.

Some obstacles can be represented approximately by cylinders with circular or elliptic sections—for example, an aircraft fuselage, or the hull of a ship. It is therefore of interest to see how their presence modifies the field in their immediate vicinity (i.e. for a receiving aerial protruding from their surface). The calculation, done as long ago as 1921 by Mesny,|| explains perfectly the "quadrantal" deviations observed in goniometers so placed.

This case is also a generalization of the "filiform conductor" envisaged in § 4.8.1—i.e. a *thick* mast or pylon. For example, when a radar aerial is shielded in certain directions by a mast, funnel or gun turret on a ship, what reduction is there in its efficacy and what order of magnitude of deviations can be expected?

The problem has been exhaustively treated by Megaw;* the

†Rocard, *Onde Électrique*, July 1946, pp. 288–98; Vasseur, *Onde Électrique*, Jan.–March., 1952, pp. 1–10, 55–71, 97–111; Levine *et al.*, *Phys. Rev.*, Oct. 1948, pp. 958–74.

‡For example, Simon, *Annales Radio-El.*, July 1951, pp. 205–43.

||Mesny, *Radio-Review*, 1921; Sinclair, *Proc. I.R.E.*, June 1951, pp. 660–5.

*Megaw, *J. I.E.E.* III A, **95**, March 1948, pp. 97–105.

formulae are too complicated for us to reproduce here, but his conclusion, which is well confirmed by experiment, is that the perturbating effect of the mast decreases very quickly when the wavelength and "masking angle" decrease, so that it is possible in general to tilt the aerials sufficiently for the perturbation at microwave frequencies to be negligible.

If the cylinder is considered as a "radar target", it has been possible to calculate its equivalent cross-section for the case where the diameter D and length L are large compared with the wavelength and the axis is parallel to the electric field at normal incidence: it is then found† that

$$\sigma = \pi \frac{DL^2}{\lambda}.$$

According to this formula, a cylindrical rocket of length $L = 3$ m and diameter $D = 0{\cdot}5$ m, when viewed broadside in the most favourable orientation by a 10 cm radar, would have an equivalent cross-section of about 135 m²; but in general it would be detected obliquely under far less advantageous conditions.

The equivalent reflection cross-section of a paraboloid of revolution has also been calculated‡ and it is found to be better than that of the sphere with the same radius of curvature as the apex; that of an ellipsoid has also been calculated: for a surface with radii of curvature r_1, r_2, the cross-section is practically the same as for a sphere of radius $a = \sqrt{r_1 r_2}$. A long, thin ellipsoid ultimately behaves like a dipole, a fact which has been known since the dawn of radio.

4.8.5. Miscellaneous obstacles. Buildings, etc.

The obstacles usually encountered are not as geometrical as these but have very odd shapes; their conductivity, too, is finite, sometimes quite small. Their "transparency" or their "equivalent reflecting cross-section" can, therefore, only be determined by experiment.

†Burrows and Atwood, *Radio Wave Propagation*, p. 45.
‡Horton and Karal, *J. App. Phys.*, May 1951, pp. 575–81.

The following are a few indications as regards the most interesting practical cases.

Walls, houses and buildings have little effect on long waves unless they contain a metal framework, such as reinforced concrete; in this case, the field strength in the interior of a house may be reduced to about a tenth of its value outside (−20 dB).

As the wavelength decreases, absorption by the walls increases, but the effect of apertures becomes more important; for medium and short waves the field strength is always lower inside houses, especially on the lower floors (an effect which is aggravated by the increase in artificial interference); in a large town the attenuation is a much more rapid function of distance (this has already been mentioned in Table 2.1 when attributing a very low "conductivity" to urban areas); large screening effects are observed behind tall monuments and skyscrapers.

When we come to the metric wavelengths of television, various assessments have shown† that the field strength may be reduced by several decibels by trees or isolated houses, by 10–20 dB in the streets of an average town, 25–35 dB in the streets of New York. It is even smaller inside houses, especially on the lower floors and may be influenced by parasitic reflections from nearby metallic objects (window blinds, furniture, etc.).

At decimetre and centimetre waves the only objects which can be regarded as "transparent" are thin screens of trees without foliage, wooden-framed windows, thin wooden partitions; but ordinary masonry walls and screens of leafy trees are practically opaque.‡ (We shall be discussing later the transparency of the atmosphere itself.)

Hills or mountains produce reflections or "shadows" whose magnitude may sometimes be evaluated roughly by the formulae of § 4.8.2, etc.

†See, for example, Dufour, measurements in Switzerland, *Techn. Mitt.*, Dec. 1948, pp. 241–8; Bullington, measurements in U.S.A., *Proc. I.R.E.*, **38**, Jan. 1950; Clifford, *Radio-Electronics*, May 1950, pp. 30–1.

‡See, for example, McPetrie and Ford, *Proc. I.E.E.*, IIIA, **93**, 3, pp. 531–8; Head, *Proc. I.R.E.*, June 1960, pp. 1016–20; Rice, *Bell Syst. Tech. J.*, 1959, pp. 197–210.

4.8.6. Usefulness of certain obstacles. Passive reflectors. Radar targets

The preceding examples give the impression that obstacles are always a nuisance as they introduce undesirable additional attenuation into transmissions.

However, this is certainly not true. There exist cases, sometimes fortuitous, sometimes intentional, where the energy reflected by an obstacle in the direction of the receiver can reinforce, or even replace, the energy received in its absence.

Thus it can happen that the presence of mountain ranges reinforces the field in valleys oriented in the direction of propagation, which act like "guides".

The interposition of an elevated crest causes "knife-edge" diffraction to occur which can, at metric and shorter wavelengths, augment the direct field-strength attenuated by the Earth's curvature (see § 8.5.2).

Similarly, a *passive reflector* (plane, spherical or parabolic mirror), situated on top of a hill, in sight of both stations, can reflect to the receiver an energy greater than that which it receives directly behind a mask (see § 8.5.3).

Meteor trains in the upper atmosphere can also fulfil this function to an extent which can be used at decametre wavelengths (see § 6.3.7).

But the most interesting example is undoubtedly that of *radar targets*; it is well known that the detection and location of aircraft, ships and other moving objects is possible using the energy reflected by these obstacles and returned by them towards an appropriate receiver.

In order to evaluate conveniently the range of a radar (see § 8.5.4), the Americans introduced the simple, arbitrary concept of characterizing the target by its *equivalent cross-section* σ, assuming that the target captures the power, in the primary field, which passes through this cross-section and re-radiates completely and isotropically.

The radar cross-section can be calculated for obstacles of simple geometric form: such as a half-wave dipole (§ 4.8.1); for a conducting sphere of radius r greater than the wavelength λ, it is simply (§ 4.8.2):

$$\sigma = \pi r^2.$$

For the tri-rectangular trihedron of § 4.8.3, if the three edges are of length $L > \lambda$, the radar cross-section in the direction of the axis of ternary symmetry is

$$\sigma = 0{\cdot}289\, L^2.$$

These shapes are sometimes used to calibrate radar sets. But the majority of actual targets (ships, aircraft, rockets, etc.) have much more complicated geometrical shapes which are not amenable to calculation. We can only get to know their radar cross-section by experiment (and compare it with that of unwanted fixed obstacles, such as mountains, cliffs, sea waves, etc.).

Numerous measurements made during the 1939–45 war have given the following orders of magnitude:†

"Radar cross-section" for small aircraft: $\sigma = 4\text{–}10\ \text{m}^2$.‡

"Radar cross-section" for large bombers: $\sigma = 40\text{–}80\ \text{m}^2$.

These are mean values consistent with a serious probability of detection: but, naturally, the cross-section fluctuates greatly, depending largely on the angle at which the aircraft is viewed and on the way it is moving: sometimes "mirror-type" reflections may be obtained from a fairly flat portion of the fuselage or wings, producing a much more intense radar return (like the reflection from an object which "catches" the sun).

If there is a group of N aircraft, the radar cross-section increases slowly with N.

These figures are valid for all wavelengths shorter than the resonance due to some particular feature of the aircraft; when this resonance is reached, i.e. about $\lambda = 10\text{–}50$ m for example, the vibrating part can act as a dipole and the formula of § 4.8.1

†Burrows and Atwood, ***Radio Wave Propagation***, p. 47; Kerr, ***Propagation of Short Radio Waves***, §§ 6.4 and 6.5.

‡Probably less for modern supersonic jet aircraft, which are more streamlined and have no propeller.

shows that the radar cross-section can be greater still. The first radars did operate in the 10–15 m wavelength region. But shorter waves are preferred for reasons of propagation and aiming precision.

As regards ships, the experience of the American Naval Research Laboratory is summed up in the formula:

$$\sigma = (0{\cdot}01\text{–}0{\cdot}1)\,\frac{(\text{width})^2\,(\text{height})}{\text{wavelength}}\,.$$

The "width" means the apparent width in the *present* position of the ship; the empirical coefficient ranges from 0·01 for a submarine to 0·1 for a cruiser or liner. The radar cross-section thus varies from 40 m^2 for a submarine (on the surface) to several thousand square metres for a large ship; the "mirror" effect mentioned above for aircraft can momentarily increase the radar cross-section (according to Burrows) to over 10 million square metres for a ship which is detected broadside on. (It must not be forgotten, in interpreting these figures, that they represent the equivalent cross-section of an *isotropically* re-radiating reflector; now in this particular case, the reflection from the target is almost entirely in the direction of the radar set: the "equivalent" cross-section can, therefore, far exceed the actual cross-section.)

In certain regions, the radar range calculated from the above values is, in fact, appreciably reduced by the presence of parasitic echoes which are either effectively *fixed*†—such as hills, cliffs, rocks, surrounding buildings—or clearly *variable* in a random manner—sea waves.

It is difficult to evaluate the intensity of these echoes; however, because they are such a nuisance, efforts have been made to analyse them; by way of indication, we might mention that, according to certain German documents, the radar cross-section of 2½ acres of cultivated land would be about 5 ft^2 at 10 cm wavelength; for 2½ acres of pine forest it would be about 6 ft^2; this seems small but it must be borne in mind that there can be many

†Although, in fact, fluctuating slightly, due to variations in speed over the transmission and reception paths, or to small movements of parts of the obstacle—such as leaves of trees, etc.

acres of ground in the field of a radar set. As for sea waves, their regular pattern engendered the hope that their diffusing effect could be calculated by likening them to sinusoidal ripples, screens made up of drops of water, etc.;† some recent results have been noted in §§ 3.2.6 and 3.3. For very oblique incidence (as in the case of navigation radar), the order of magnitude of the reflection coefficient is from -30 to -70 dB for a wavelength of 10 cm, i.e. for every 10,000 m^2 of surface "illuminated" by the beam, the equivalent radar cross-section would be $\sigma = 10\text{–}0{\cdot}001\ m^2$; here again, the vast area of sea "illuminated" by the beam is sufficient to explain the fact that wave echoes ("clutter") mask targets of small dimensions.

Rain, storms,‡ meteorites, birds can all constitute obstacles detected by radar sets; echoes of mysterious origin (known as "angels") are even observed sometimes: these can be attributed either to birds or insects or to discontinuities in the atmosphere.

†See especially Kerr, *Propagation of Short Radio Waves*, §§ 6.6–6.12 and 6.21; also Blake, *Proc. I.R.E.*, March 1950, pp. 301–4.

‡Doc. 94 of the Conference of the C.C.I.R., London, 1953, on the detection of rain and storms; a good summary is given by Voge, *Onde Électrique*, March 1953, pp. 145–8.

CHAPTER 5

THE ROLE OF THE TROPOSPHERE

IN THE foregoing discussion, we have assumed that the atmosphere is perfectly transparent, like a perfect, homogeneous dielectric; under these conditions it takes no part either in direct visibility links or in those which follow the Earth's curvature.

But in fact, as long ago as the dawn of wireless, the success of transatlantic radiocommunication proved that this simplifying hypothesis was false and that it was necessary to take account of the ionization of the upper atmosphere—rarefied layers at heights of 100–600 km, known for this reason as the *ionosphere*. Its effect will be studied later.

But the lower layers—the unionized troposphere and stratosphere—were still thought to be of no interest. People were content to curse the electric charges which they bore (in clouds or air masses at various temperatures) and which recombined in flashes from time to time, producing "atmospherics" which are a nuisance in radiocommunication. We shall return to this point later.

It was not until 1930 that Jouaust and others invoked the troposphere to explain abnormal ranges obtained at metric wavelengths; in effect, due to the gases and water vapour in it and the variations in pressure and temperature with altitude, this layer possesses a dielectric constant and therefore a refractive index (§ 5.1) slightly greater than unity near the ground and gradually decreasing and tending to unity as the altitude increases. An "electromagnetic ray" launched obliquely therefore undergoes a progressive refraction and curves downwards; thus a certain amount of the radiated energy is directed back towards the Earth and follows its curvature more easily. This effect is observed in optics, as exemplified by the well-known increase in geographical range of lighthouses and the exceptional phenomenon of the "mirage". Moreover, it can easily be verified that the method used by students of optics to allow for it—increasing the Earth's radius by a factor of about $\frac{4}{3}$—is also correct for electromagnetic waves and this technique was included in the formulae for the ground wave, as indicated in §§ 4.3, 4.4.2, etc. Admittedly, ranges calculated thus could be increased—perhaps by 20 or 30%—by more intense refractions, especially in summer; this used to happen quite often for air-to-ground links.

The matter rested there until about 1941–2. At that time, operators who used metric radars regularly noticed that, in some cases, the "abnormal" ranges exceeded the bounds of probability—sometimes for such long periods that they themselves could be classed as "normal". In this way radar sets on the English coast used to detect the French coast below the horizon; radars on the Tunisian coast followed landing craft as far as Sicily; and over warm seas (the battleship *Richelieu* at Dakar in 1941, the Persian Gulf, the Indian Ocean) exceptional ranges were quite regularly observed. The record is held by the Bombay radar installation on $\lambda = 1{\cdot}5$ m which sometimes detected ships at 700 nautical miles (further than from Land's End to John o'Groats) and the coast of Arabia at 1500 nautical miles.

This anomalous propagation was attributed to *superrefraction* caused by very rapid variation of the temperature and humidity of the air in the immediate vicinity of the surface of the sea (100 ft or so); it was assumed that the corresponding variation in the refractive index created a "duct" in which the waves were "trapped" as in a waveguide, an explanation which leaves one mystery unsolved, because, even in a guide, such a low attenuation as a function of distance is quite unheard of.

The development of powerful transmitters at metric wavelengths for f.m. and television revealed another surprising phenomenon. It was noticed that beyond the horizon the rate of decrease of the field strength, after a certain distance, no longer followed the exponential law indicated by theory (§ 4.4.2, point 2) for which the shorter the wavelength, the more rapid is the decrease.† Instead, the field strength followed a $1/d^4$ law, approximately. Never wanting for imagination, the majority of the propagation theorists immediately attributed this unforeseen phenomenon to "diffusion" by the troposphere due to its "turbulence" or lack of homogeneity. Other explanations were also forthcoming.

The level thus obtained is too weak to be usable in radar, and perhaps even for picture transmission; but it may be sufficient to produce unexpected interference at distances of the order of 300–600 miles.

These two types of unpredicted propagation are naturally of great interest to radio engineers; in an attempt to understand and predict them, they want to study the fine structure of the atmosphere—sometimes called, with slight exaggeration, the "microstructure"—which is not always known with sufficient precision by the meteorologists.

As it happens, radar technologists, having discovered in the meantime that clouds and showers of rain or snow are detectable by radar, find themselves in a position to furnish the meteorologists with a new means of analysing and further studying the troposphere.

Thus has a new branch of science been born—radiometeorology—in which the propagation expert and the atmospheric specialist help each other, the one seeking to deduce new theories from what is known about the atmosphere, whilst the other fills in the gaps in his knowledge using the phenomena of propagation. The field of study thus opened up should keep them busy for a long time.

In what follows, we shall summarize briefly the various aspects of the role of the troposphere.

† The attenuation in the terrestrial diffraction zone is theoretically $0{\cdot}62\lambda^{-1/3}$ (m) dB/km.

5.1. Review of meteorological ideas

The structure of the atmosphere influences wave propagation by way of the refractive index n equal to the square root of the relative dielectric constant of the air with respect to a vacuum. n is given by the formula:

$$(n-1)\times 10^6 = \frac{77\cdot 6}{T}\left(p+\frac{4810e}{T}\right), \tag{5.1}$$

where p is the pressure in millibars, e is the partial pressure of water vapour in millibars, T is the absolute temperature in °K.

$(n-1)$ is often expressed in units of one-millionth (10^{-6}) which are called M-units.

The "troposphere" is the lower layer of the atmosphere, where clouds can exist. Its upper limit ("tropopause") is at a height of about 6 km at the poles and about 18 km at the equator.

On average, in this layer the pressure p, temperature T and humidity e (partial pressure of water vapour—a few per cent of p)† decrease regularly with increasing altitude:

pressure p by about 1 mm (1·3 mb) per 11 m,
temperature by about 1 degree C per 200 m,
humidity by about 1 mb per 300 m.

(See Fig. 5.1, curves 1, 2—about noon, fine weather.)

This linear variation characterizes the so-called *standard* or *normal* atmosphere.

The latter, which corresponds to the average atmosphere for various seasonal and geographical conditions in temperate latitudes is defined by a linear decrease in n with altitude, with a gradient of

$$\frac{dn}{dh} = -0\cdot 039\times 10^{-6} \text{ per metre.} \tag{5.2}$$

Recent work has shown that, even on average, the linear law corresponding to the "standard" atmosphere cannot be taken as

† It can also be defined by the "dew point", i.e. the temperature T_s at which saturation would occur if cooling took place at constant pressure. The more humid the air, the nearer T_s is to T.

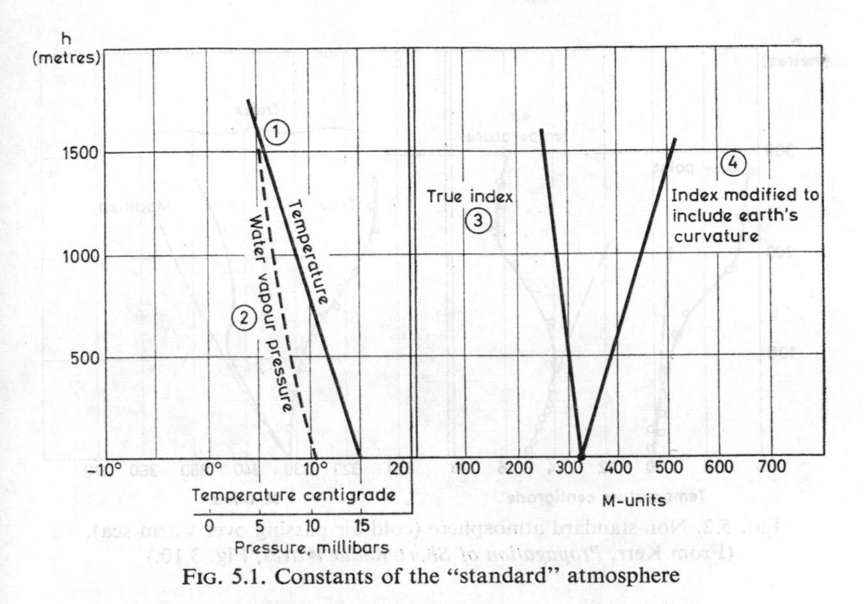

FIG. 5.1. Constants of the "standard" atmosphere

applicable to the whole of the troposphere. Its sphere of validity does not extend beyond the first few kilometres of the atmosphere. In 1959 the C.C.I.R. proposed the adoption, for work which involves a more extensive region of propagation, of a "fundamental reference atmosphere", defined—on the basis of the most recent experimental data—by the relationship:

$$n(h) = 1+289\times10^{-6}\,e^{-0\cdot136\,h}, \tag{5.3}$$

h denoting the height in kilometres above sea level. In the first kilometre, this reference atmosphere is almost exactly equivalent to a "standard" atmosphere.

But, it must be noted that the regular variations which characterize the "standard" atmosphere or the "fundamental reference atmosphere" are rather exceptional, especially at low altitudes, below 5000 ft, for example.

In fact, there exist in the atmosphere, convection currents due mainly to variations in the temperature and humidity of the ground (or sea) which bring about exchanges between air and ground and ultimately determine the air temperature. These exchanges are facilitated by *turbulence* of the air and they engender

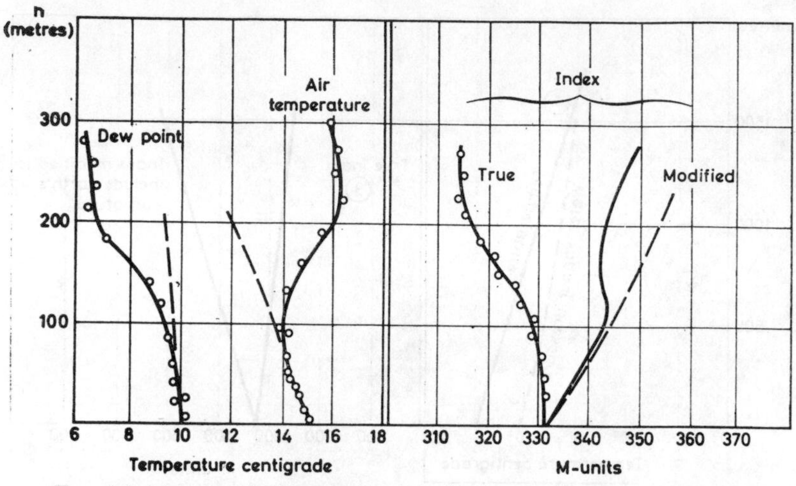

FIG. 5.2. Non-standard atmosphere (cold air passing over warm sea). (From Kerr, *Propagation of Short Radio Waves*, Fig. 3.10.)

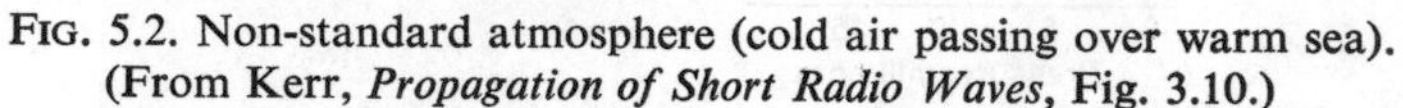

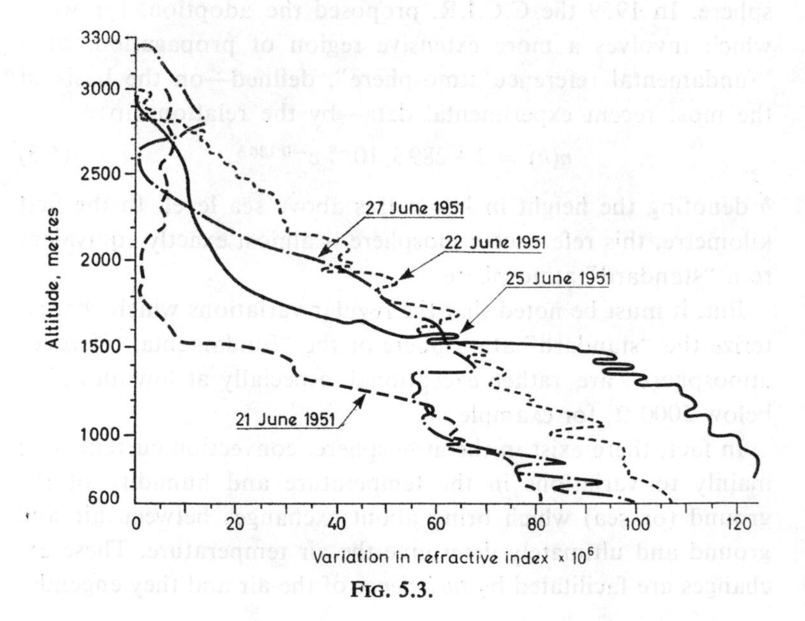

FIG. 5.3.

vertical movements accompanied by more or less adiabatic compressions and expansions.

Horizontal displacements of air masses above the surface ("advection") also have a considerable influence on the temperature and humidity distributions, particularly near the coast.

The presence of two air masses, one above the other, with different physical properties can give rise to the formation of layers with a very marked discontinuity in altitude. This is the case, for instance, when a mass of hot ("tropical") air passes below a pocket of cold ("polar") air. The discontinuity surface, whose inclination to the horizontal is usually of the order of a fraction of a degree, is called a "front". Discontinuities in altitude can also arise from the falling of an air mass to a lower level ("subsidence"): the most characteristic situation engendered by subsidence is the "sea of clouds", i.e. a turbulent, cloudy region surmounted by a calm, limpid atmosphere.

In consequence of these various phenomena, the temperature and humidity distributions in the lower layers of the atmosphere—and therefore the distribution of the refractive index—often deviate markedly from the "standard" (or "reference") decay laws.

Figures 5.2 and 5.3 show examples of variations which can be observed in practice at certain times. The second figure corresponds to measurements carried out in an aircraft with a special low inertia measuring instrument: a number of small irregularities are thus shown up.

5.2. Atmospheric refraction. Equivalent models for propagation calculations

Consider an atmosphere in which the refractive index n varies with the altitude h. "Electromagnetic rays" which would be rectilinear in a vacuum or a homogeneous atmosphere become curved. If the Earth's curvature can be neglected, to a first approximation the equation to these curves is none other than Snell's law.

If there are two media (1) and (2) with refractive indices n_1 and n_2, separated by a plane surface (Fig. 5.4), it is well known that the angles which the incident ray in medium (1) and the refracted ray in medium (2) make with the normal to the surface of discontinuity satisfy the relation:

$$n_1 \sin i_1 = n_2 \sin i_2 . \tag{5.4}$$

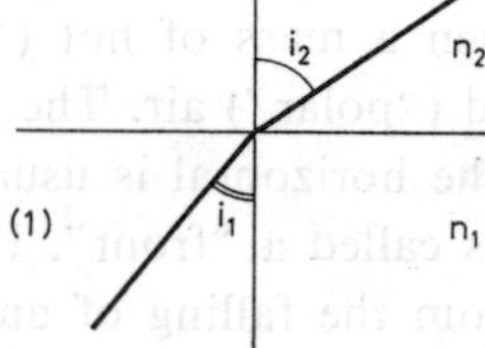

FIG. 5.4. Refraction at a discontinuity surface

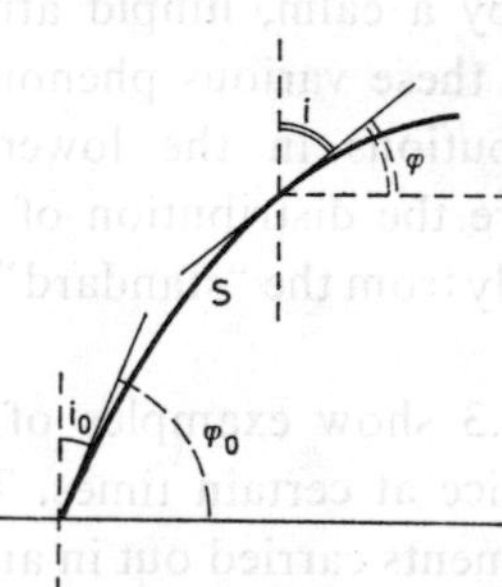

FIG. 5.5. Refraction in an inhomogeneous medium

If, now, the index varies continuously (Fig. 5.5) in a direction parallel to a vertical axis, *Oh*, the equation of a ray can be derived directly from the preceding relationship and can be written:

$$n(h) \sin i(h) = \text{constant}. \tag{5.5}$$

If we consider the inclination φ to the horizontal of the tangent to the ray, we have

$$n(h) \cos \varphi(h) = n(O) \cos (\varphi)_O . \tag{5.6}$$

This is Snell's law for a medium with a continuously varying refractive index.

The curvature of a ray (the reciprocal of the radius of curvature) at a given point has the value

$$\sigma_R = \frac{d\varphi}{ds}, \tag{5.7}$$

the limit of the ratio of the angle $d\varphi$ between the tangents at two points to the distance ds between them, as ds tends to zero, evaluated along the radius (s is a curvilinear abscissa).

We find, differentiating (5.6) with respect to s,

$$\frac{dn}{ds}\cos\varphi(h) - n\sin\varphi(h)\frac{d\varphi(h)}{ds} = 0, \tag{5.8}$$

whence

$$\sigma_R = \frac{dn}{ds\sin\varphi}\,\frac{\cos\varphi}{n} = \frac{dn}{dh}\,\frac{\cos\varphi}{n}, \tag{5.9}$$

since

$$dh = ds\sin\varphi. \tag{5.10}$$

In the Earth's atmosphere, n only differs from unity by a few parts in 10,000. However, we are nearly always dealing with links between points at altitudes which are much smaller than the length of the trajectory. The electromagnetic rays following this trajectory are only slightly inclined to the horizon: φ is small and $\cos\varphi$ is of the order of 1. Thus

$$\sigma_R \simeq \frac{dn}{dh}. \tag{5.11}$$

This curvature is reckoned algebraically. It is positive if it is directed towards the ascending vertical. The Earth's curvature (radius of curvature a) is thus negative and equal to

$$\sigma_E = -\frac{1}{a}. \tag{5.12}$$

The curvature of a ray relative to the Earth is therefore

$$\sigma_{RE} = \sigma_R - \sigma_E = \frac{dn}{dh} + \frac{1}{a}. \tag{5.13}$$

It can be demonstrated that, if the Earth is replaced by an earth of radius a', different from a, surmounted by an atmosphere in which the index n' is different from n, the relative curvature being preserved:

$$\frac{dn'}{dh}+\frac{1}{a'}=\frac{dn}{dh}+\frac{1}{a}, \tag{5.14}$$

then the calculation of the field strength leads to the same result in both cases.

(1) Consider the case of the *normal atmosphere.*

$$\frac{dn}{dh}\simeq -0{\cdot}25/a,$$

a, the radius of the Earth, being about 6500 km.

It can be seen that in a "standard" atmosphere, the electromagnetic rays are curved downwards by refraction: they consist of arcs of circles of radius $4a$.

But the standard atmosphere can be replaced by a vacuum ($n' = 1$, $dn'/dh = 0$), on condition that the Earth's radius a is changed to a radius a' such that

$$\frac{1}{a'}=-\frac{1}{4a}+\frac{1}{a}=\frac{3}{4a},$$

$$a'=\frac{4a}{3}. \tag{5.15}$$

More generally, if the temperature and humidity in the lower atmosphere vary regularly with altitude, such that n is a linear function of h:

$$\frac{dn}{dh}=\text{constant}=-\frac{1}{\theta a} \quad (\theta \text{ positive or negative}), \tag{5.16}$$

then we can make the same transformation with a terrestrial radius a' defined by

$$\frac{1}{a'}=-\frac{1}{\theta a}+\frac{1}{a},$$

$$a'=\frac{\theta}{\theta-1}a=ka. \tag{5.17}$$

Since a' must be positive, θ must actually be > 1. The corresponding atmospheres are said to be "of standard type" (the "standard" atmosphere has $\theta = 4$, $k = \frac{4}{3}$).

(2) In the case of an atmosphere in which the index varies with height h according to an arbitrary law, the calculation of the field strength above a spherical earth could be reduced to the calculation of the field strength above a plane earth (a infinite) by replacing the refractive index in the lower atmosphere by an index N, such that

$$\frac{\mathrm{d}N}{\mathrm{d}h} = \frac{\mathrm{d}n}{\mathrm{d}h} + \frac{1}{a}, \tag{5.18}$$

$$N = n + \frac{h}{a} = n + 0{\cdot}157h\ (\mathrm{m})\ 10^{-6}. \tag{5.19}$$

N is called the "modified index".

Instead of N, the equivalent parameter

$$M = (N-1) \times 10^6$$

is often used. M is called the "modulus of the modified refractive index". It is expressed in M-units, which we have already defined above. A change of one M-unit corresponds to a change of one part in a million in the modified index N.

In a "standard" atmosphere, M will vary linearly with n but with a positive vertical gradient (Fig. 5.1):

$$\frac{\mathrm{d}M}{\mathrm{d}h\ (\mathrm{km})} = \left(-\frac{1}{4a} + \frac{1}{a}\right) 10^6 = \frac{3}{4a} \times 10^6 = 118. \tag{5.20}$$

Note. It can be established rigorously that in the general case of an arbitrary atmosphere where the refractive index n would be a function of three coordinates and not of altitude h alone, the curvature of electromagnetic rays at any point is given by the relation

$$\sigma_R = |\mathrm{grad}\ n| \cos \varphi, \tag{5.21}$$

where φ denotes the complement of the angle i between the ray and the vector grad n. Formula (5.11) relating to an atmosphere stratified in horizontal layers is therefore also valid for an atmosphere with spherical stratification, taking account of the Earth's curvature.

5.3. The influence of the fine structure of the troposphere on propagation

We have just seen how we can reduce the study of propagation in a "standard" atmosphere to propagation in a vacuum by making a correction to the Earth's radius. But the standard atmosphere itself is only an "average" atmosphere from which the real atmosphere often deviates a long way. In the following paragraphs we shall be studying the influence of the fine structure of the real troposphere—which is in a perpetual state of flux. We shall see that the effect is noticeable at ultrashort wavelengths—especially decimetre, centimetre and millimetre waves. On the other hand, for decametric and longer waves it is generally sufficient to employ the "standard" correction to the Earth's radius.

5.3.1. Fading over links which are in direct line of sight

It might be thought that transmission over links in direct line of sight with nothing in the way would be absolutely stable, the received field strength being equal to the theoretical field strength for propagation in free space. This is not the case and we shall attempt to give the reasons.

Suppose then, that we have a link in direct line of sight, not masked by the roundness of the earth, between a transmitter T and a receiver R.

The field at R is the result of interference between the direct field and the field reflected from the ground. Depending on the relative phase of these two fields, the interference gives rise to regions of reinforced field strength—when the direct and reflected fields are in phase—and regions of reduced field strength when they are in opposition. If E_0 denotes the amplitude of the direct field and ϱ is the reflection coefficient of the ground, the corresponding resultant field strengths are, respectively:

$$E_0(1+\varrho) \quad \text{and} \quad E_0(1-\varrho).$$

An attempt could obviously be made to locate the receiver in a region of maximum field strength; unfortunately, the temperature, humidity and pressure of the air, factors which affect the refractive index of the atmosphere, vary from point to point and vary with time at any given point: the real atmosphere is not homogeneous, as we have seen, and its structure is continually changing. The amplitude and phase of the direct and reflected fields at a point will not therefore be the same from one moment to the next and regions of strong and weak field strength will be interchanged.

We can, alternatively, consider the "electromagnetic rays" between the transmitter and receiver. In a homogeneous space they would consist of straight lines; in an inhomogeneous space they will be curved† and the shape of these curves (and therefore their length) will vary with time in synchronism with the evolution of the meteorological conditions in the path of the wave. As the phase of the electromagnetic field along the direct or reflected path increases by 2π per wavelength, a small change in the relative lengths of these paths (a few centimetres or decimetres for the wavelengths under consideration) may suffice to transfer the receiver from a favourable region into a region of low field strength, i.e. to produce considerable variations of level in the receiver.

In general, every time a number of waves following different paths from the transmitter interfere at the receiver, there is a risk of fluctuations in the received field strength. This can occur even without reflections from the ground: inhomogeneities in the atmosphere sometimes split the direct trajectory between the transmitter and receiver into several distinct transmission paths. To quote a particular example of this—it so happens that there exist between two elevated atmospheric layers discontinuities in refractive index large enough to produce sizeable reflections, at least at nearly grazing incidence.

The most frequent cause of fading (since it is present at all times), and also the one which creates the largest falling off in

† This curvature is, however, quite small. It has been established experimentally that the angle of incidence of the waves can vary by about $\pm 0{\cdot}5$ degree in the vertical plane and $\pm 0{\cdot}1$ degree in the horizontal plane.

reception level, is interference between the direct ray and rays reflected (or diffracted) by the ground. It can be eliminated, at least to a great extent, if all obstacles are removed, not only from the direct line joining the transmitter T and the receiver R, but also from the first Fresnel zone on both sides of this line (Fig. 5.6). This zone is defined, it will be recalled, as the locus of points M in space such that the optical path TMR differs from the direct path TR by one half-wavelength at most. It is an ellipsoid of revolution with foci T and R ($MT+MR = TR+\lambda/2$). The radius r' of a right section of the ellipsoid at distances d_1 from

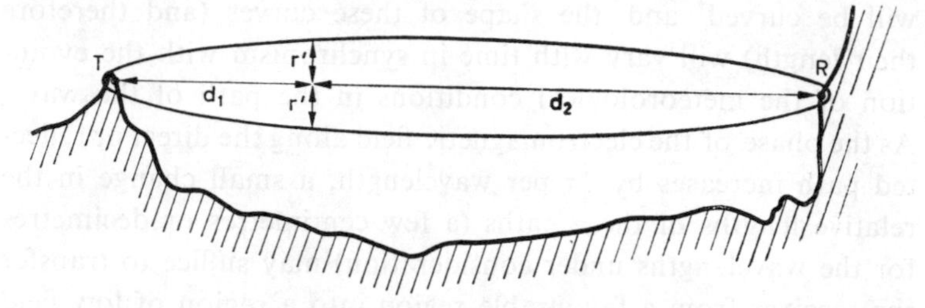

FIG. 5.6. Clearance of the first Fresnel zone

the transmitter and d_2 from the receiver can easily be shown to have the value:

$$r' = \sqrt{\frac{d_1 d_2 \lambda}{d_1+d_2}}. \tag{5.22}$$

Its maximum value, at the mid-point of the link, is equal to $\sqrt{\lambda d}/2$, where d is the length of the link.

We mentioned in § 5.1 certain phenomena giving rise to these discontinuities in altitude (the passage of "fronts", "subsidence").

Experience has shown that the variation Δl in the relative difference in length between different propagation paths, over an optical transmission path, does not, in general, exceed a few decimetres, or at most a few metres. The corresponding variation in relative phase, $(2\pi/\lambda)\Delta l$, is therefore small and practically negligible at decametre and longer wavelengths. On the other hand, it becomes appreciable at shorter wavelengths, and the shorter the wavelength the more important it is. This explains why the degree of fading and its relative duration increase with frequency—and also, in general, with distance over paths with comparable clearance, profile and climate (since Δl tends to increase.)

The distribution of fading usually exhibits a strong seasonal variation. At centimetric and decimetric waves the months of July and August are the most troublesome in the majority of cases, at any rate, over terrestrial trajectories, where a large amount of fading occurs during the hours of darkness (10 p.m. to 5 a.m.).†

Over analogous transmission paths operating at similar frequencies, the spread in field strength at the receiver depends to a large extent on the siting of the stations, the topography and the climate. Long links are sometimes more stable than short links.

Figure 5.7 shows some typical statistical distribution curves for received field strength over ultrashort wave links, corresponding to periods of the year when fading is most prevalent.

Notice the good relative stability of the Chasseral–Mt. Africa link, in spite of the distance of 100 miles.

In addition to the clearance of the first Fresnel zone, there are other factors which are favourable to the stability of the received signal. We mention the following, in particular: a poor reflection coefficient from the ground (due to the nature of the terrain or its relief, maritime trajectories being least favourable), the presence of natural masks attenuating the reflected ray and, lastly, transmission over mountainous regions where the turbulence of the air causes mixing of the atmosphere (cf. the Chasseral–Mt. Africa link, Fig. 5.7).

The reader will find in an article by one of the authors‡ a detailed study of the techniques that can be adopted to improve the stability of links, particularly that of wide band radio beams at

† In some regions, however, such as California, fading is quite common in winter and during the day.

‡ J. Voge, *Onde Électrique*, June 1954, pp. 491–8.

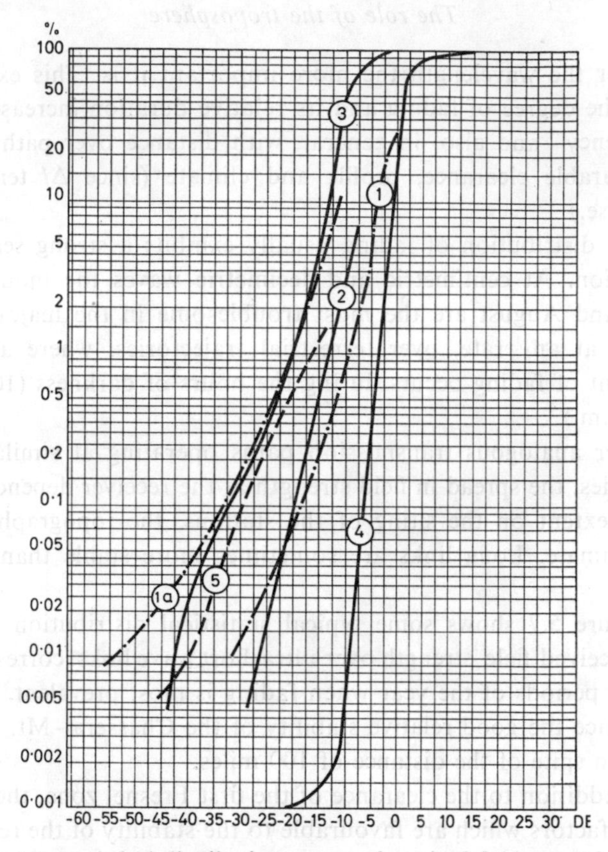

FIG. 5.7. Statistical distribution curves of received field strength for the times of year when fading is most prevalent.

Abscissa: level of field strength in decibels with respect to the theoretical level for free space propagation.

Ordinate: percentage of time during which the field strength remains below the level shown in the abscissa.

1 and 1a: limiting dispersion curves for distributions observed over the New York–San Francisco radio cable for the most unfavourable month (frequency: 4000 Mc/s; 107 sections of mean length $32\frac{1}{2}$ miles (52 km), the longest section being 58 miles (93 km); total distance 2940 miles (4700 km).

2: Mt. Chasseral–Mt. Africa link, for the most unfavourable four weeks of 1952 (frequency: 3000 Mc/s; 160 km).

3: Noyers St Martin–La Herlière link, summer 1951 (frequency: 3150 Mc/s; 76 km).

4: Mont-Diablo–San Francisco (frequency: 1800 Mc/s; 45 km); altitudes of stations: 1140 m and 50 m.

5: Stevns–Copenhagen, April 1952 (frequency: 10,000 Mc/s; 45 km), only case of a maritime trajectory.

ultrashort waves consisting of a chain of relays, each directly visible from the next. In unfavourable cases, a receiver using space or frequency diversity may prove necessary to obtain sufficient stability.

5.3.2. Abnormal propagation over long trajectories not in direct line of sight. Superrefraction

According to the theories of diffraction (§ 4.4.2), the field strength of a transmitter working at metric, decimetric, or centimetric waves should decrease very quickly on passing beyond the horizon. The attenuation in the terrestrial diffraction zone would be approximately

$$\frac{0{\cdot}62}{\lambda^{1/3}\,(\mathrm{m})} \quad \mathrm{dB/km}, \tag{5.23}$$

which corresponds to a reduction in field strength by one-half in 4·5 km at $\lambda = 10$ cm.

However, ranges much greater than the optical range are observed occasionally, in fact, quite frequently in certain regions and at certain seasons. This phenomenon is known as "abnormal propagation". It is interpreted as a kind of channelling of the electromagnetic energy round the Earth. We gave several typical examples at the beginning of this chapter. We shall now present a theoretical explanation.

As we have seen, the use of the modified indices N or M permits us to reduce propagation studies to plane problems. The electromagnetic rays then obey Snell's law in a plane-stratified medium:

$$N_1 \cos \varphi_1 = N_0 \cos \varphi_0 . \tag{5.24}$$

The modified indices N, like the real indices n, are in the neighbourhood of 1. Likewise, for electromagnetic rays slightly inclined to the horizon, $\cos \varphi$ is approximately 1 (and can be replaced by $1-\varphi^2/2$ to a first approximation). If the angles φ are expressed in milliradians, Snell's law then becomes

$$\frac{1}{2}(\varphi_1^2-\varphi_0^2) = M_1 - M_0 . \tag{5.25}$$

Figure 5.8 represents a number of electromagnetic rays emitted from a point at altitude h_0 at various inclinations in a standard atmosphere.

The rays bend upwards. In fact, the curvature of the rays is positive and equal to $\mathrm{d}N/\mathrm{d}h = +3/4a$. This expresses the fact that the true curvature of the rays in a "standard" atmosphere

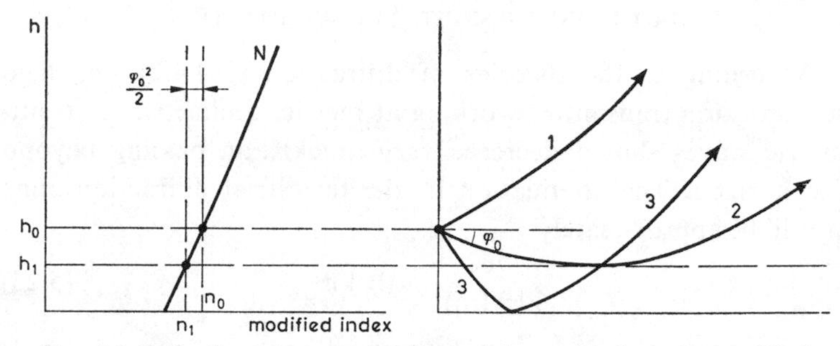

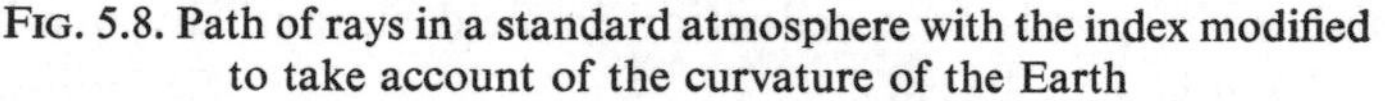

FIG. 5.8. Path of rays in a standard atmosphere with the index modified to take account of the curvature of the Earth

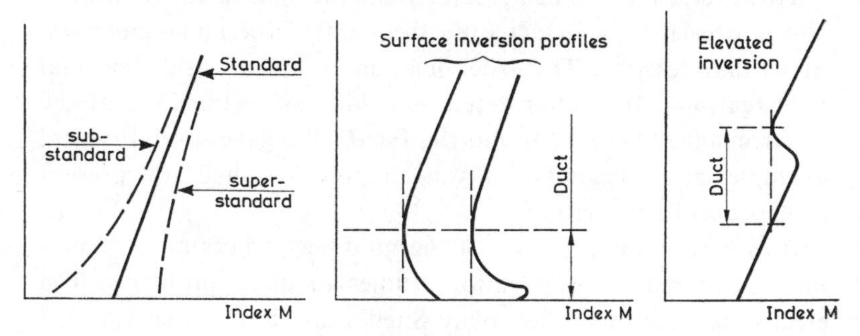

FIG. 5.9. M-profiles in a non-standard atmosphere

is less than the Earth's curvature and, if we return to the flat Earth case, the rays therefore deviate away from the surface. But, as we saw above, the atmosphere is very seldom "standard".

Figure 5.9 shows various types of vertical distribution of M which can be observed (corresponding to various types of distribution of the real index n). It can be seen that these distributions may be substandard or superstandard and that the superstandard distributions can lead in the limit to inversions—that is to layers which are of limited thickness in practice, in which M decreases with altitude. Inversion layers can occur either at ground level or above (elevated inversion).

Refraction is less marked in a substandard atmosphere than in a standard atmosphere as is also the relative curvature of the earth with respect to the rays, and the field strength and range decrease.

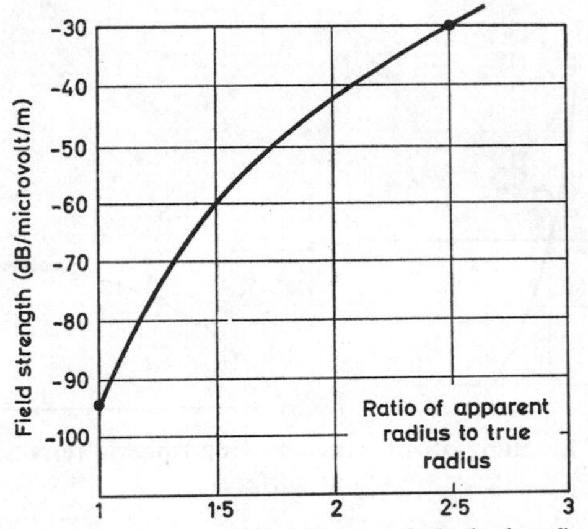

FIG. 5.10. Variation of received field strength (calculated) with apparent radius of the Earth

In a superstandard atmosphere the opposite applies.

Thus, any experimental variation in the range of a radar, for example, can be explained by assuming (if M is not known) that refraction is more or less intense, or, if desired, that the value of the Earth's equivalent radius of curvature ka must be decreased or increased. By way of example, Fig. 5.10 shows the relation between k and the field strength (in decibels with respect to 1 μV/m), calculated for a transmitter at 91·4 Mc/s, equivalent power 22 dB above 1 kW (including aerial gain), altitude 380 ft above ground, received at a distance of 200 miles on an aerial 31 ft above surrounding obstacles;† it can be seen that a variation

† From Note prélim. Lab. Nat. Radio., no. 165, by Sadoun: calculated according to the method of Domb and Price for the Wrotham–Bagneux link.

of 50–60 dB can be easily explained by stretching the apparent radius of the Earth in the ratio of 1 to 3. In other cases, people have gone as far as deducing from the observations an apparent radius 30 times greater than the true radius (180,000 km).†

But the most important phenomena are those produced by inversion layers: electromagnetic rays are then, as it were, propagated in a duct where they are trapped and can escape neither upwards nor downwards.

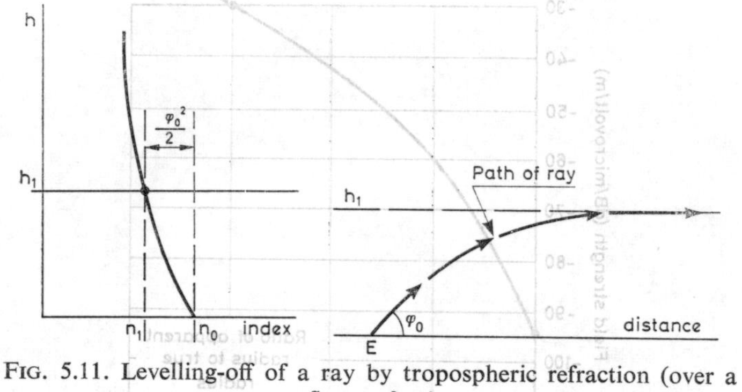

FIG. 5.11. Levelling-off of a ray by tropospheric refraction (over a flat surface)

Let us suppose that the modified index decreases with increasing height (Fig. 5.11). φ will also decrease with increasing height and may even become zero at a height h_1 defined by $M_0 - M_1 = \varphi_0^2/2$. The curvature of a ray is therefore directed towards the ground and it becomes horizontal at height h_1. Beyond this point, the ray continues downwards towards the ground, following a path symmetrical with the rising portion (Fig. 5.12).

Reflection will take place at the ground and the electromagnetic ray will propagate a long way by successive arcs, in the layer between the ground and the height h_1.

Figure 5.12 corresponds to the case of a surface inversion layer of thickness h_M. Analogous reasoning, in the case of an elevated inversion layer (the last case envisaged in Fig. 5.9) would show that the duct in which the rays are trapped then extends over

† See Doc. 195 (Italy) of the Assembly of the C.C.I.R., Geneva 1953.

the zone indicated on the graph. The phenomenon of inversion layers channelling electromagnetic waves—known as "abnormal propagation" or "superrefraction"—explains the exceptional ranges observed at ultrashort wavelengths and mentioned at the beginning of this chapter.

The analogy with waveguide propagation is obvious and it greatly impressed the microwave specialists; however, we must not forget one essential difference. In the metallic guide, reflection

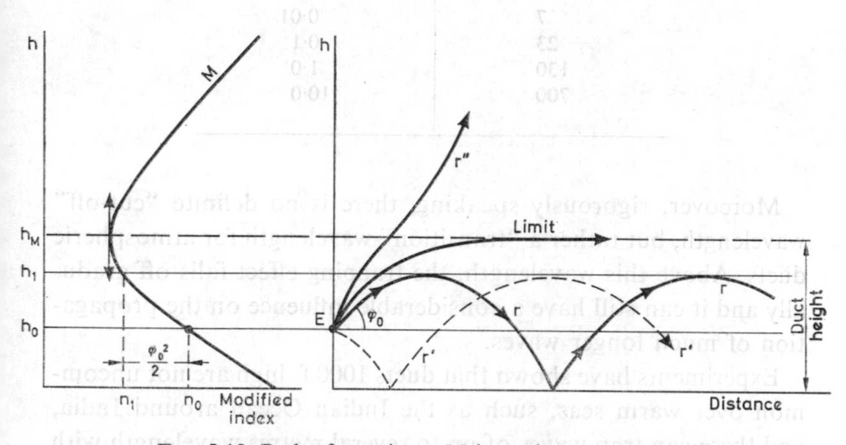

FIG. 5.12. Path of rays in an M-inversion zone ("duct")

occurs at the walls whatever the inclination of the rays; in an inversion layer, only rays leaving the transmitter with small enough inclination φ_0 (such as r and r' in Fig. 5.12) can be "guided". The trajectory of the other rays, such as r'', does not curve down as far as the horizontal tangent and they escape from the duct; similarly, certain rays coming from higher up can enter the duct. In practice, only rays inclined at less than 1° or 1·5° above the horizon benefit from abnormal propagation. As in metallic guides, a rigorous study of the phenomenon reveals the existence of a maximum wavelength above which guiding cannot occur; but, whereas in a metallic guide this "cut-off" wavelength is of the same order as the dimensions of the guide, in this case it is much less than the thickness of the atmospheric duct, as shown in Table 5.1 (which must be considered as giving orders of magnitude only). In fact, theory, shows that the cut-off wavelength is of the order of the product of the thickness of the guide and the sine of the maximum angle of inclination of the trapped rays.

TABLE 5.1

Height of "duct" (h_M) (m)	Maximum trapped wavelength (m)
7	0·01
23	0·1
130	1·0
700	10·0

Moreover, rigorously speaking, there is no definite "cut-off" wavelength, but rather a "transition" wavelength for atmospheric ducts. Above this wavelength, the trapping effect falls off gradually and it can still have a considerable influence on the propagation of much longer waves.

Experiments have shown that ducts 1000 ft high are not uncommon over warm seas, such as the Indian Ocean around India, and these can trap waves of up to several metres wavelength with incredibly low attenuations so that fantastic radar ranges have been achieved. 100 ft ducts are much more common. In the Mediterranean, during summer, abnormal propagation is observed for more than 70% of the time.

The meteorological conditions which favour the formation of inversion layers are those which lead to a very rapid decrease in n with height (with a gradient which is greater than $1/a$ in absolute value). Referring to formula (5.1), we see that this corresponds to a rapid reduction in pressure or humidity or an increase in temperature ("temperature inversion") with height. In practice, only the humidity and temperature can have sufficient effect. The most usual causes of inversion layers are, over land, nocturnal cooling of the ground, which creates a temperature inversion, and, over sea, the passage of warm dry air over colder water. The phenomenon of subsidence can give rise to an elevated inversion layer (dry air above a more humid sea of clouds, for example).

"Superrefraction" does not always have a favourable effect: it may permit "leakages" from military communications or mutual interference between quite distant sources, such as television transmitters or airfield control towers, operating on the same frequency. It can also, as the counterpart of the trapping mechanism, lead to a failure to detect aircraft flying at high altitude above the duct or inversion layer. Very little work has been done on this question, important though it is;[†] according to Voge, the "direct" ray is more easily stopped than the ray "reflected by the ground", so that, if it is the only ray, the field may turn out to be reduced by 20–25 dB or more, even to the point of total disappearance for intervals of a few minutes to several hours;[‡] on the other hand, if there is a sizeable reflected ray, as is normally the case over sea, the attenuation of the direct ray may reduce its amplitude to approximately the same as that of the reflected ray, giving the possibility of very rapid variations ("fading") due to beating between them.

Systematic observations of the variations may throw light on their possible origin: it is a fashionable research topic at the moment.

Another such is "radio meteorology"—the meteorological study of inversion layers and the possibility of *forecasting* them, which would also permit us, incidentally, to forecast abnormal propagation. The importance of this aspect is obvious in the military applications of radar. Thus, towards the end of the Second World War, instruction manuals to this end were distributed to observers and meteorological officers. Figure 5.13 gives a few examples of the "coverage diagrams" thus obtained;[*] the distribution of the field strength is presented as a function of

† Price, *Proc. Phys. Soc.*, **61**, 1 July 1948, pp. 59–78; Voge, Note prélim. Lab. Nat. Radio., no. 167, 1953.

‡ Barsis *et al.*, N.B.S. Report 2494, 1 May 1953.

* Joint Communication Board, Report J.A.N.P. 102, Washington, 1944; *Tropospheric Propagation of Radio Meteorology*, Central Rad. Prop. Lab. Nat. Bureau of Standards, Washington, 1946.

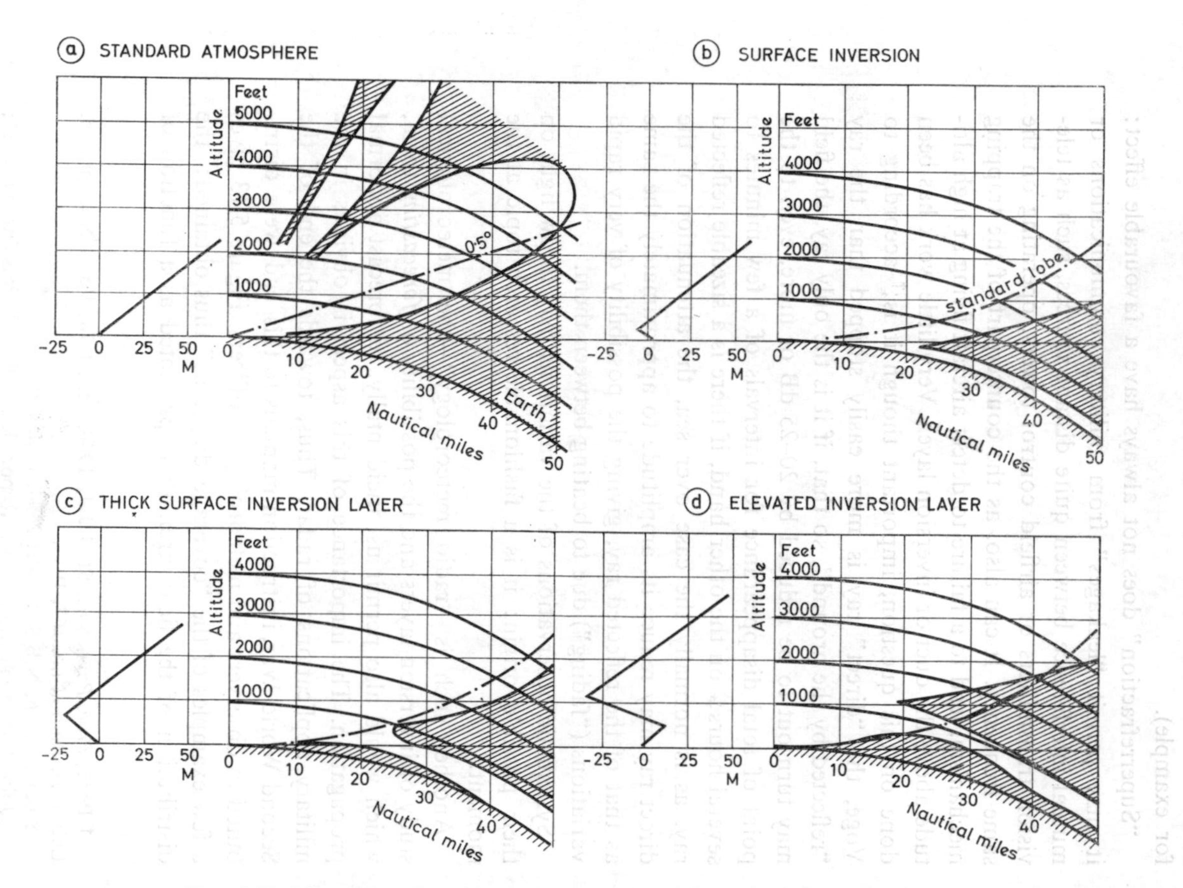

FIG. 5.13. The effect of meteorological conditions on radar coverage diagrams

distance, taking account of the Earth's curvature, the scale of heights being exaggerated by a factor of 40/1.

Diagram (a) relates to a standard atmosphere (*M* increasing linearly with height) for a 1·5 m radar situated at a height of 100 ft above sea level: the lobes, calculated as in § 4.3, indicate a low probability of detection in the shaded regions, particularly the lowest one.

Diagram (b) represents an M-inversion in the immediate vicinity of the ground and the consequent formation of a duct extending for more than 50 nautical miles: the upper limit of the blind region is also displaced away from its old position (broken line) in a direction such as to reduce the size of the region.

Diagram (c) represents the situation for a thicker inversion layer: since the rays trapped by the duct only return after a certain distance, there is a "region of silence".

The transition is obvious with diagram (d) which represents the case of an elevated inversion layer: the "region of silence" at ground level extends so far that only an elevated duct is possible.

Research directed towards the forecasting of these abnormal propagations has continued since 1944 but the practical conclusions have only been published in reports with restricted circulation.

5.3.3. Normal propagation over long trajectories not in direct line of sight

The long range propagation associated with superrefraction in ducts may be common in some areas but it is still, on the whole, exceptional.

Prolonged observations, starting in 1946, of powerful metric and decimetric stations (radars, television, f.m. radio) have provided the experimenters with another surprise. In fact, it turns out that the decrease in field strength beyond the horizon, after starting off very steeply, in conformity with diffraction theory (§§ 4.3, 4.5) suddenly changes its form and, beyond a certain point, has only a relatively small slope: however, the field strength is still a long way below the free space value (by about 40–100 dB).

Figure 5.14 gives some idea of the variation.

The received field strength well beyond the horizon is nearly always subject to rapid fluctuations (the mean period, nearly proportional to wavelength, is of the order of a few tens of seconds at 3 m and a fraction of a second at 10 cm). Therein lies a marked difference from the abnormal propagation field strength. Figure 5.15 shows recordings of field strength taken over the same trajectory in periods of normal propagation (Figure 1) and abnormal propagation. The "abnormal" field strength is far more

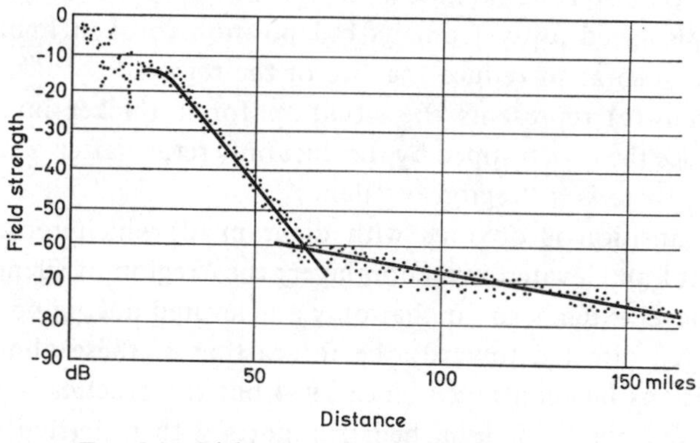

FIG. 5.14. Field strength as a function of distance

$\lambda = 9$ cm. Transmitter $h_1 = 230$ m. Receiver $h_2 = 10$ m.

(U.R.S.I., Zurich, 1950)

stable and slightly higher, (it may be quite close to the field strength which would exist at the same distance in free space).

The rapid fluctuations of the normal field often follow a distribution law well known in statistics, namely, Rayleigh's law (§ 1.2.1). Longer term variations in field strength are also observed: these are characterized, for example, by the distribution of the horary median values, are relatively moderate and have, moreover, a tendency to decrease with distance. Figure 5.16 indicates, as a function of distance, the horary median field strength values exceeded for the given percentage of time. They are valid for transmitting and receiving aerials situated at heights of 300 m and 10 m respectively, (typical of a television trans-

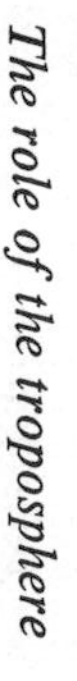

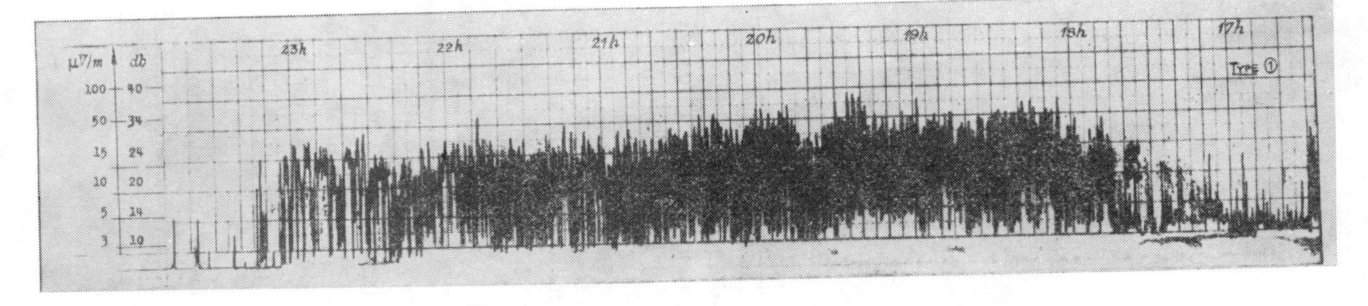

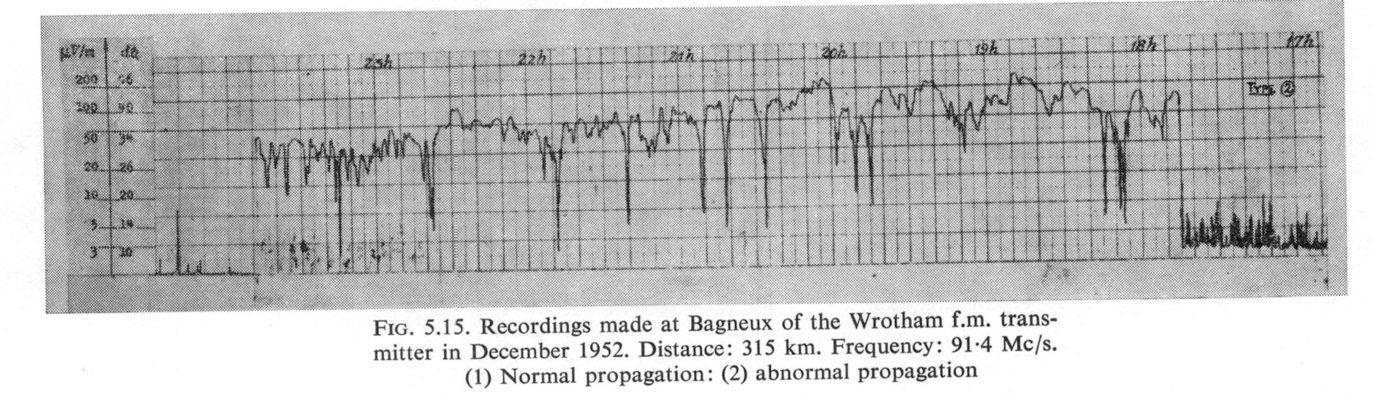

FIG. 5.15. Recordings made at Bagneux of the Wrotham f.m. transmitter in December 1952. Distance: 315 km. Frequency: 91·4 Mc/s. (1) Normal propagation: (2) abnormal propagation

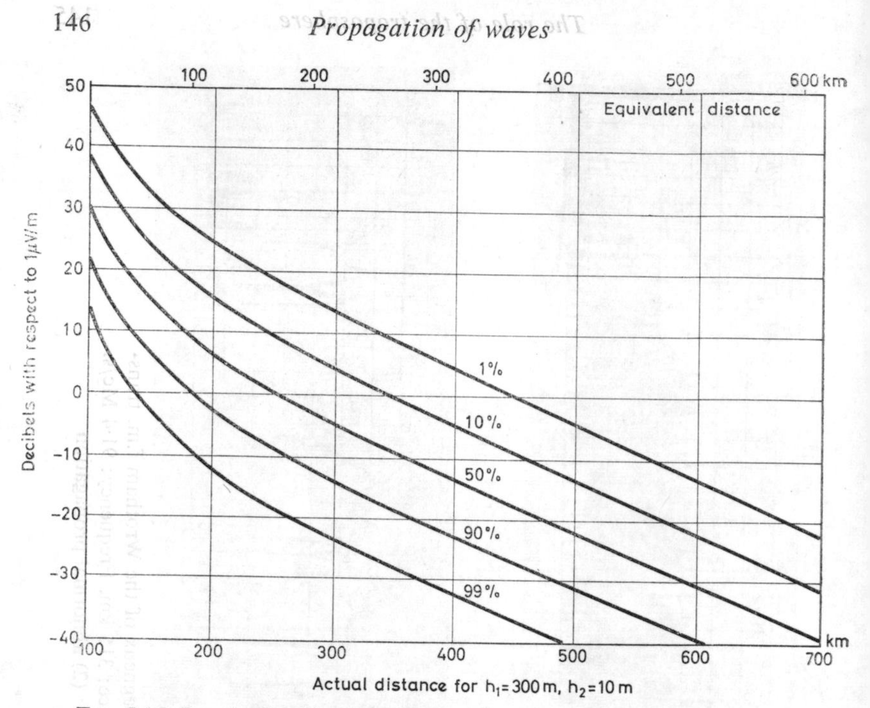

FIG. 5.16. Received field strength in decibels with respect to 1 μV/m, for a power of 1 kW radiated by a vertical or horizontal half-wave dipole. Distances (in km) are to be read from the bottom scale for transmitter and receiver heights of 300 m and 10 m respectively. The top scale gives the equivalent distance beyond the optical limit, which corresponds to the actual distance for aerials on the surface of a spherical Earth

mission) on the one hand, or both at ground level on the other.

The curves corresponding to 50%, 10% and 1% of the time and to heights of 300 m and 10 m were adopted by the C.C.I.R.† (Los Angeles, 1959): they are intended mainly for the study of mutual interference between stations working on the same frequency, which is a serious inconvenience if it occurs for 1–10% of the time.

In order to establish a link, a far higher probability would be necessary, say, 90–99% of the time. We have therefore, added the corresponding curves based on the hypothesis of symmetrical

† C.C.I.R. Recommendation No. 312.

distribution of values (in decibels): 10 and 90% on the one hand, 1 and 99% on the other, with respect to the median value. This hypothesis seems acceptable to a first approximation.

If the transmitting and receiving aerials are at any other height above the ground, the curves can still be used by referring to the scale corresponding to having both aerials at ground level, provided the true distance d is replaced by an equivalent distance d_e equal to the length of the "blind" part of the trajectory, i.e. the distance between the horizons of the transmission and reception aerials. If the heights of the aerials above ground level are h_1 and h_2, this would give

$$d_e\,(\text{km}) = d\,(\text{km}) - 4{\cdot}1\,(\sqrt{h_1\,(\text{m})} + \sqrt{h_2\,(\text{m})}). \qquad (5.26)$$

A few additional remarks on the curves of Fig. 5.16 are appropriate here.

(a) The curves are valid for non-mountainous terrestrial trajectories. The observed field strengths for maritime trajectories lie above the given curves, the difference amounting to, perhaps tens of decibels.

(b) The indicated fields correspond to median values in terms of the position of the receiving station. The field distribution around this median value for a group of receiving stations in a given region is nearly "normal logarithmic" (see § 1.2.1). The difference between the 50 and 90% curves for a group of stations may vary by 6–14 dB (average, 10 dB) depending on the degree of irregularity of the terrain.

(c) The curves correspond to data collected in the United States and western Europe, i.e. in areas of temperate climate.

(d) To a first approximation the curves are valid for frequencies in the range 40–600 Mc/s. Precise measurements seem to reveal a slight decrease in field strength as the wavelength decreases (approximately as $\sqrt{\lambda}$, on average, up to 4000 Mc/s).

(e) Recent experiments have extended the distance scale well beyond 700 km out to about 1500 km at metric and decimetric waves, over land and sea. There does not seem to be any severe drop in received field strength at a particular distance, but merely a steady decrease.

Inspection of the curves, Fig. 5.16, shows that the decrease in field strength is much less than that predicted by diffraction theories. It follows roughly a $1/d^4$ law, corresponding to a slope of the order of 0·1 dB/km, instead of a steep exponential decay of $0{\cdot}62\lambda^{-1/3}$ dB/km. The consequences of this are important. The field strength for normal propagation beyond the horizon at ultrashort wavelengths is great enough to establish permanent links of several hundred miles provided high-power transmitters (of the order of 1–10 kW) and highly directional aerials are used (parabolic mirrors several tens of feet in diameter are employed). The aerial gains are, however, less than those which would be observed on optical trajectories, for reason which we shall discuss later. The loss of gain becomes appreciable when the theoretical gain exceeds 30 or 40 dB.

The rapid fluctuations in field strength are annoying but they can be greatly reduced, if not completely eliminated. If the receiver has two aerials placed far enough apart, either vertically or perpendicular to the direction of propagation (minimum spacing: 25 wavelengths), there is practically no correlation between the instantaneous variations in the received signals. A remarkable reduction in the amount of "rapid fading" is possible using diversity reception, which combines the signals received by two, or sometimes three or four, aerials disposed in this way. Other possibilities are frequency diversity (using two frequencies separated by a few tens of megacycles per second) and angular diversity (using two receiving aerials whose beam axes are displaced symmetrically in azimuth, by a fraction of a degree, in practice, with respect to the axis of transmission).

One important point concerning the practical utilization of this mode of propagation to set up point to point links is the bandwidth which can be used for transmission of signals of different types without prohibitive distortion. Attempts to receive short duration pulses, multiplex telephony or television appear to show that the effects of distortion are practically negligible in frequency bands of several megacycles per second at distances of 200 miles. Beyond this, the usable band would decrease very rapidly with distance; but it could be increased to some extent by employing very high gain aerials. Transmission of up to a hundred telephone links has already been achieved over hundreds of miles. Links operating between Florida and Cuba and between France and Algeria, with a relay point in the Balearic Islands, carry television programmes.

The advent of "trans-horizon" links† makes possible the solution of hitherto insoluble communications problems (wide-band transmission across an arm of the sea or in sparsely populated areas where it would not be feasible to set up a large number of

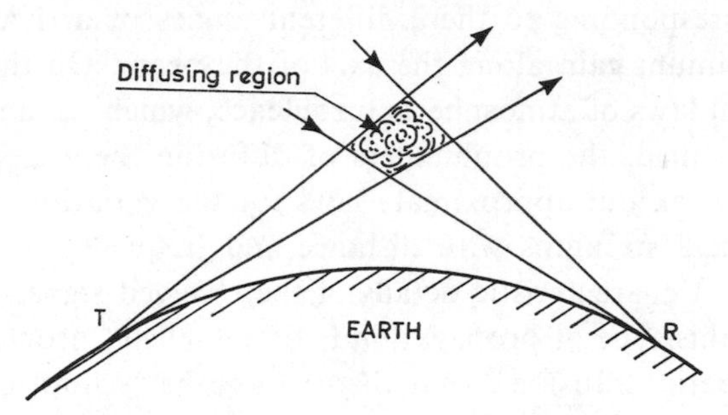

FIG. 5.17. Diffusion by tropospheric turbulence

optical-range relay posts only a few tens of miles apart: such is the case in the French African territories). Frequencies in use at present range from 150 to 2000 Mc/s or even 4000 Mc/s.

Since 1950, a large number of theories have been published in an attempt to explain the relatively high field strengths (compared with the diffraction field strength) which are observed in "trans-horizon" transmissions. The phenomenon is often attributed to a slight diffusion of the waves in that part of the troposphere which is simultaneously in direct optical view by the transmitter and receiver (Fig. 5.17). The diffusion is assumed to be due to atmospheric turbulence which creates random eddies and thus small, essentially fluctuating, local inhomogeneities.‡

† See Voge and Du Castel, *Écho des Recherches*, no. 35, 1959, pp. 22–3, for a method of setting up a "trans horizon" link project, starting from propagation data.

‡ X., *Proc. I.R.E.*, Jan. 1960, pp. 30–44 (section 2).

The received signal thus arises from the summation of elementary contributions due to diffusion from an extended region of the atmosphere. This would explain why the rapid fluctuations are distributed according to Rayleigh's law, which is applicable whenever the field results from the superposition of a very large number of components having phases distributed at random with respect to each other. It would also explain the loss of aerial gain: the received field being derived from the summation of elementary fields arriving by diffusion at different site angles and azimuths, the apparent gain of the aerial is a sort of average gain corresponding to these different angles of arrival and not the maximum gain along the axis of the beam. On the basis of statistical laws of atmospheric turbulence, which seem to be relatively justified, the propounders of diffusion theories have been able to work out approximate laws for the variation of the observed field strengths with distance and frequency as well as a number of characteristic details of the received signals. For this reason, this type of propagation is often called "propagation by tropospheric diffusion". But there is another school of thought which says that partial reflections from a large number of elements with slightly discontinuous quasi-horizontal surfaces such as are frequently encountered in the troposphere (cf. Fig. 5.3) play a dominant part: partial reflection replaces diffusion by turbulent elements.† Actually, the supporters of the two theories are not perhaps as far from each other as they appear at first sight and one can imagine the vortex elements of the partisans of diffusion, initially spherical, assuming a horizontally elongated shape which makes them very similar to the quasi-horizontal elements of the supporters of partial reflection.

As examples of other advanced explanations, we must mention diffusion by irregularities in the Earth's surface,‡ and also the effect of the single inhomogeneity due to the mean decrease in the refractive index with altitude.

† Du Castel, Misme, Spizzichino and Voge, *Réflexions partielles dans l'atmosphère et propagation à grande distance*, Éditions de la Revue d'Optique, Paris 1960.

‡ Voge, Note Prélim. Lab. Nat. Radio. no. 187, May 1955.

Finally, we should point out that radiometeorological correlations have been observed between the horary median values of the received fields and parameters which are a function of the atmospheric refractive index: the vertical gradient of the index between the ground and a height of 1000 m ($\Delta_n = n$ (1000 m) $-n(0)$), or the surface index ($n_s = n(0)-1$). The C.C.I.R. recommends the application of the following corrections if the curves of Fig. 5.16 are to be used in areas other than temperate zones, characterized by values of Δ_n or n_s which are assumed to be known (from meteorological statistics): the correction to the 50% curve is

$$-0{\cdot}5(\Delta_n \times 10^6 + 40) \text{ dB}$$

or, if Δ_n is unknown,

$$0{\cdot}2(n_s \times 10^6 - 310) \text{ dB}.$$

The application of the correction to the 1, 10, 90 and 99% curves is more dubious. The C.C.I.R. has already prepared maps† giving $\Delta_n \times 10^6$ for various times and various seasons, in numerous parts of the world. However, the corrections indicated above must be applied with caution until their validity in a particular area has been verified by direct propagation experiments.

5.4. Absorption by the atmosphere

Apart from its refracting and diffusing properties, the atmosphere can obviously have *absorptive* properties, due, particularly, to the presence of water or water vapour in the form of clouds, fog (droplet diameter 0·02 cm), rain (drops from 0·01 to 0·6 cm), hail (hailstones of 0·4 cm and sometimes more).

It is well known that for visible light, and also for infrared, the absorption due to the water vapour can be quite large; when electromagnetic waves of shorter and shorter wavelength came into use, the question therefore arose as to the point at which this absorption would become important.

† Report no. 147 of the C.C.I.R. (Los Angeles, 1959).

A considerable number of theoretical and practical papers have appeared dealing with this question.[†]

In the early days of metric waves, aviators had already observed ill-defined regions of opacity. Subsequently, the influence of rain, fog, etc., was methodically and patiently explored; the effect of the constituent gases of the atmosphere was also calculated; there are molecular resonances due to changes in the energy levels of the electrons under the influence of the incident wave; the molecules behave like small electric (or, in the case of oxygen, magnetic) dipoles.

The essential results, now well confirmed, are shown in Fig. 5.18: frequency and wavelength are plotted in abscissa, attenuation in decibels per kilometre of path length in ordinate.

Water vapour absorption bands appear at 0·15 and 1·35 cm and oxygen bands at 0·5 and 0·25 cm. Attenuation due to rain (in dB/km) is nearly proportional to the rate of precipitation (in mm/hr).

Although there is bound to be some uncertainty about the density and frequency of rain and fog, it is clear that perturbation effects are seldom noticeable at wavelengths above 0·10 m: on the other hand, they do become important around 0·03 m and would appear to be prohibitive at millimetre waves, except in certain narrow bands lying neatly between the absorption bands of the gases.

This possibility has not yet been exploited very much and the wavelengths of communication equipment or radars only very rarely fall below 3 cm. However, it may be desirable to operate at millimetre waves in order to improve the accuracy of a radar; the bands of minimum absorption around 8 mm, or when techniques are available, around 3 mm, will then be chosen. Sometimes, on the other hand, it is desirable to have a relatively large

† See, for example, Franz, *Hfr. Techn. El. Ak.*, May 1940, pp. 141–3; Quarles, *El. Eng.*, April 1946, pp. 209–15; Ryde, *J.I.E.E.* **3**, A, 1946, pp. 101–3 (Radiolocation Convention); Lamont, *Proc. Phys. Soc.* 1 Dec. 1948, pp. 562–9; Burrows and Atwood, *Radio Wave Propagation.* p. 50, etc.: Kerr, *Propagation of Short Radio Waves*, chaps. 1 and 8; Saxton *et al.*, *Proc. I.E.E.*, Jan. 1951, pp. 26–36 (pt. 3).

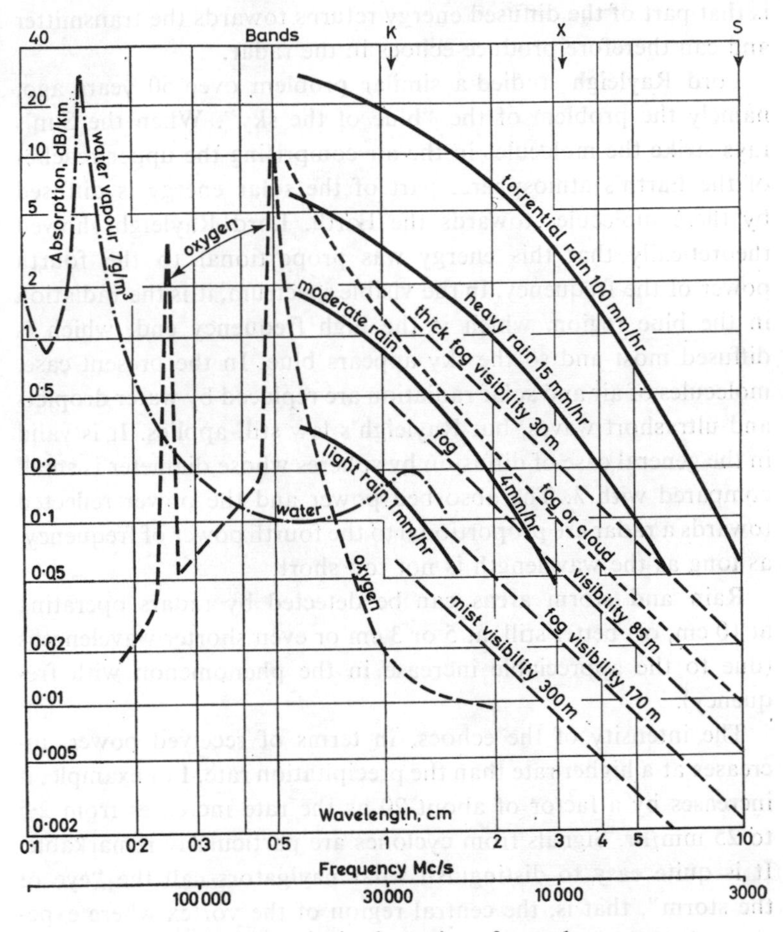

FIG. 5.18. Atmospheric absorption of very short waves

attenuation as a function of distance, for example, in the case of short range military communications equipment where it would be embarrassing if signals were intercepted at greater distances: a wavelength on the edge of an intense absorption band, such as the oxygen band at 5 mm, is used.

Absorption by raindrops is due more to a diffusion phenomenon (the incident energy being reflected in all directions) than to the phenomenon of absorption proper by the liquid. The result is that part of the diffused energy returns towards the transmitter and can therefore produce echoes in the radar.

Lord Rayleigh studied a similar problem over 50 years ago, namely the problem of the "blue of the sky". When the Sun's rays strike the molecules in the air comprising the upper reaches of the Earth's atmosphere, part of the solar energy is diffused by these molecules towards the Earth. Lord Rayleigh showed theoretically that this energy was proportional to the fourth power of the frequency. In the visible spectrum, it is the radiation in the blue region, which is the high frequency end, which is diffused most and so the sky appears blue. In the present case, molecules of air and solar radiation are replaced by water droplets and ultrashort waves, but Rayleigh's law still applies. It is valid in the general case of diffusion by spheres whose diameter is small compared with λ. The absorbed power and the power reflected towards a radar are proportional to the fourth power of frequency, as long as the wavelength is not too short.

Rain and storm areas can be detected by radars operating at 10 cm, or, better still, at 5 or 3 cm or even shorter wavelengths (due to the appreciable increase in the phenomenon with frequency).

The intensity of the echoes, in terms of received power, increases at a higher rate than the precipitation rate. For example, it increases by a factor of about 20 as the rate increases from 2·5 to 25 mm/hr. Signals from cyclones are particularly remarkable. It is quite easy to distinguish what navigators call the "eye of the storm", that is, the central region of the vortex where experience has shown that the sky is blue and cloudless.

Clouds and mist which contain very fine water droplets clearly produce less attenuation than rain and in general they are not visible by radar. However, it must be remembered that a cloud can contain large drops which do not reach the ground, either due to evaporation or because they encounter a rapid rising current. For this reason, clouds which are invisible to the naked eye sometimes give large radar echoes, whereas large cumulus clouds are not detected.

The clouds which can be most easily detected by radar are,

fortunately, the cumulo-nimbus clouds. These are the clouds which give rise to the most intense precipitations, particularly in tropical climates; the violent lightning discharges which build up inside them, or between them and the ground, and the enormous turbulence which they create, make aerial navigation, especially at night, very dangerous. Furthermore, these clouds, which can develop vertically as much as 6–8 miles in tropical regions very quickly lose their shape; their tops can rise at a rate of several hundred yards per minute and a cloud can collapse in less than half an hour. In some cases, on the other hand, especially over sea, they may persist for hours. In the detection of cumulo-nimbus clouds at a distance, as well as rain and thunderstorms, radar can render important service to aerial navigation, especially at night. It has also allowed considerable progress to be made in the meteorological study of the mechanisms of formation of atmospheric precipitations—natural or man-made.

One last point deserves a mention: in many cases, the use of circular polarization seems to reduce the intensity of rain echoes appreciably. This device is used to eliminate the masking effect produced by a rainy zone in the detection of ships and aircraft.

CHAPTER 6

THE ROLE OF THE IONOSPHERE

6.1. General

We now come to the most delicate part of the problem: the role of the upper atmosphere.

By 1880, and therefore before the birth of radiocommunications, geophysicists had come to the conclusion that the upper atmosphere *could* and *should* be an *electrical conductor*.

It *could* be, because the properties of a rarefied gas, at pressures as low as those obtaining at a height of, say, 100 km, are very different from its properties at ground level; the number of molecules is reduced to about 10^{13} per cm^3 (instead of 3×10^{19}). The gas is subjected to the luminous radiation from the Sun, before it is attenuated by passing through the atmosphere, and is therefore very rich in ultraviolet. It must therefore be strongly ionized; in spite of recombination due to collisions, a free electron density of the order of 10^6 per cm^3 is quite plausible. Allowing for the "mean free path length" (about 1 cm), the calculation described in § 2.1 indicates, at radio frequencies, an electron current which is greater than the displacement current and, hence, refraction and reflection and the possibility that the "ionosphere" might act more or less as a *mirror*.

In any case, the existence of effective currents in this region† had already been accepted as the only possible explanation of the small variations in terrestrial magnetism which are frequency observed, either in isolation or in conjunction with the *polar auroras* (which, actually, occur most often at heights of the order of 100 km).

† The term "layer" was used for a long time. The present tendency is to replace it by "region", as "layer" is thought to be too precise.

As soon as the first transatlantic communications by Marconi (1901) revealed the existence of much better propagation conditions than the theory of diffraction could predict, several physicists (Kennelly in the U.S.A. and Heaviside in England) at once invoked the possibility of reflection in the ionized upper atmosphere—hence the name "Kennelly–Heaviside (K–H) layers" which has been assigned ever since to these regions. The hypothesis was also supported in France by Blondel (1903) and Poincaré (1904). The phenomena of propagation provided a good means of verifying it: with a little imagination, all sorts of experiments were devised to use waves to explore the interesting region and bring us back, in a fraction of a second, information on its height, density, thickness, constitution, etc.† As a result, reflection from the upper atmosphere, regarded at first with reserve (if not scepticism) but gaining ground daily as more and more irrefutable evidence was amassed, gradually became established and is nowadays accorded a degree of "probability" comparable with that of the "hypothesis" that the Earth goes round the Sun.

In the following we shall review briefly:

(a) methods of studying the ionosphere;

(b) the conclusions reached regarding its constitution and "normal" variations;

(c) "accidental" variations and perturbations.

We shall return later to the application to specific propagation problems.

6.2. Methods of studying the ionosphere

How can we study the ionosphere? The ingenuity of the physicist has suggested all sorts of methods, some direct, others indirect.

We have already mentioned the variation of the Earth's mag-

† See, for example, R. Bureau, *L'Ionosphère, carrefour de recherches*, Note 37 of the French National Radioelectricity Laboratory, 1943, and the following: Jouaust, L'Ionosphère, *Revue d'Optique*, 1946; Rawer, *Die Ionosphäre*, Noordhoff, Holland, 1953; Vassy, *Physique de l'Atmosphère*, 1959, and numerous reports of the International Radioscientific Union (U.R.S.I.)

netism which was the first proof of ionization and the existence of appreciable currents in the upper atmosphere.

But there are many other phenomena which can furnish useful information; optics, especially, was used a long time ago and techniques have improved with the perfecting of spectral analysis: thin lines reveal the presence of atoms, bands indicate molecules; their width and displacement give information on the temperature and velocity of the particles. The difference between daylight and the luminosity of the night sky, the crepuscular phenomena when the sun sinks below the horizon, the shadow of the Earth on the Moon during lunar eclipses, all provide new information; the distortion of the paths of meteors proves the existence of winds up to heights of the order of 90 km; finally, polar auroras, which occur at heights of 70–1100 km, give some idea of the variation of the parameters with height: up to now, some 32,000 photographs of 12,000 auroras have been taken.

In acoustics, certain abnormal types of sound propagation over large distances can be used, when explosions occur.

But, of course, radio engineers now possess their own means of investigation.

The most obvious is the observation of natural phenomena: thus, atmospherics, by their form and angle of arrival, indicate propagation "trapped" between the Earth and some reflecting envelope; certain audio frequency "whistlers" slipping from one hemisphere to the other can only be explained by storm interference signals being propagated along the lines of force of the Earth's magnetic field up to heights of 20,000 miles. By measuring the relative variation of "cosmic" noise, coming from the stars, we can deduce the absorption in our atmosphere; and subtracting what can reasonably be attributed to the lower layers (known by other means) from the total, we can calculate the transparency, and, therefore, the composition and density, of the upper layers which are out of reach. Apparatus now exists for carrying out this type of measurement (known as a *riometer*, from "Relative ionospheric absorption").†

† See, for example, *Proc. I.R.E.*, Feb. 1959, pp. 315–20, The riometer.

The quite recent possibility of exploring the upper reaches of the atmosphere by means of rockets and satellites has naturally opened up a whole series of new avenues: sometimes these devices measure certain parameters directly, sometimes other parameters are deduced by observing the intensity, polarization, variation, etc. of the signals which they transmit. However, these methods are still exceptional and the results are fragmentary and we shall not deal with them here.†

The principal method of exploring the ionosphere is still, therefore, the method first used in 1926, of "probing" by receiving and analysing the "echoes" returned by the upper regions of the atmosphere of short pulses transmitted from the ground. The first intermittent experiments using this technique were carried out by isolated workers, but later the work was delegated to special stations, more or less automatic, operating permanently. The interest in the forecasts which they have made possible (see later, § 8.4) has forced the various national services to multiply these stations all over the world; a new effort was authorized on the occasion of the International Geophysical Year (1958), and finally 165 stations in all co-operated to collect and pool an immense mass of information thanks to which the mysteries of the ionosphere began, if not to vanish, at least to be clarified.

We summarize below the essential features of this method and its results.

6.2.1. Vertical probe at normal incidence and fixed frequency

A transmitter T and a receiver R are placed side by side (Fig. 6.1) on the ground (pointing more or less towards the zenith) and a series of very short pulses (order of 10^{-5} sec) is transmitted.‡

† See, for example, the special issue of *Proc. I.R.E.*, April 1960, on Space electronics.

‡ We are leaving out the ingenious method (now discarded) of "frequency sweeping". Obviously, one can also *goniometer* the echoes; we shall be returning to this later.

The output of the receiver is applied to the Y-plates of an oscilloscope; it is analysed using a fast horizontal sweep of known velocity.

The "direct" signal TR produces one pip (Fig. 6.2) and the echo via TIR produces a second pip: their separation T can be measured.

At short waves the experiment is almost invariably successful: at least one very clear echo is obtained.

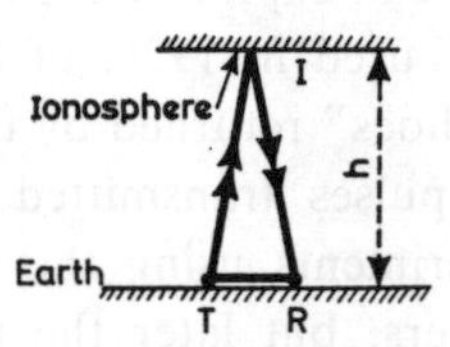

FIG. 6.1. Principle of ionospheric probing

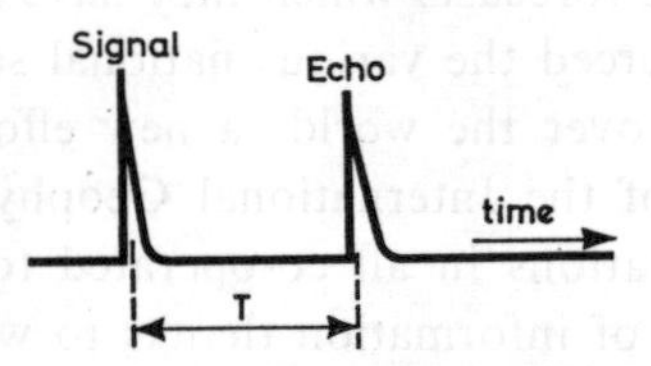

FIG. 6.2. Result of an ionospheric probe

Its *position* is quite stable; the interval T, measured periodically by several stations, remains constant for several minutes, or even hours, above any one country or continent. However, it sometimes changes abruptly at certain times (Fig. 6.3).

We shall return later to this point.

The *amplitude* of the echo is less steady; it often exhibits large, rapid fluctuations, sometimes with a vague periodicity of the order of a second.

The amplitude maxima correspond roughly to the field strength of a ray reflected with the coefficient $R = 1$ and with no appreciable absorption *en route*. From the delay of the echo, T, we can deduce the *virtual height* of the region:

$$h = c \cdot \frac{T}{2},$$

which would correspond to the delay T for the *normal* velocity of propagation, $c = 300{,}000$ km/sec. It is known, moreover, that it must be greater than the height actually reached by the wave because, as the surface of the reflecting layer is not a sharp discontinuity, the wave penetrates it with a *progressive* decrease in its "group velocity", stops and then accelerates on its way back, so that some of the time is "wasted".

The virtual height h thus calculated generally lies between 100 and 700 km.

Thus, there exist in the atmosphere, at these heights, ionized regions, labelled D, E and F, whose lower surface is fairly well

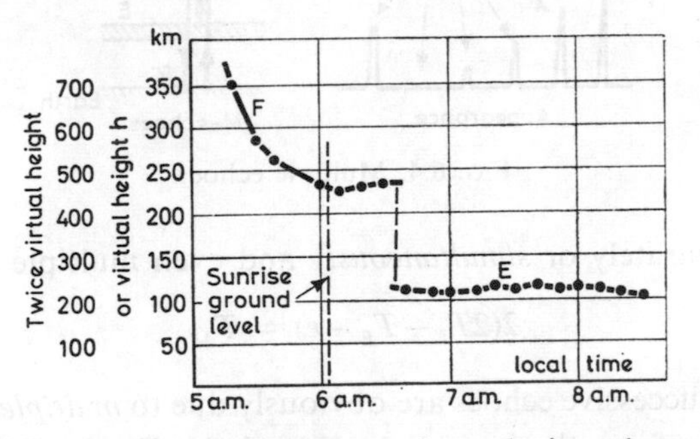

FIG. 6.3. Fixed frequency ionospheric probe

defined and in which the electron density is sufficient to reduce the dielectric constant to zero and render the medium impenetrable (§§ 2.4 and 3.1).

The extent and homogeneity of these regions is striking (without being absolute, as we shall see later).

There are numerous explanations for the amplitude fluctuations: the interior surface of the layers necessarily exhibits irregularities and turbulence which can bring about convergence or deviation of the reflected ray or interference between two neighbouring rays (the ordinary and the extraordinary which we shall deal with later, for example) or rotation of the polarizations, etc. There may also be variable absorption in the atmosphere below the reflecting region.

Finally, it quite often happens that the traces of the echoes on the oscilloscope are *more complicated than this.*

In particular, not just one, but two, three, . . ., up to nine *successive echoes* may be observed (Fig. 6.4) with delays which are exact multiples: T, $2T$, $3T$, . . ., $9T$.

One may also observe delays corresponding to two regions, E and F:

$$T_E, \quad 2T_E, \quad 3T_E, \ldots$$

$$T_F, \quad 2T_F, \quad 3T_F, \ldots$$

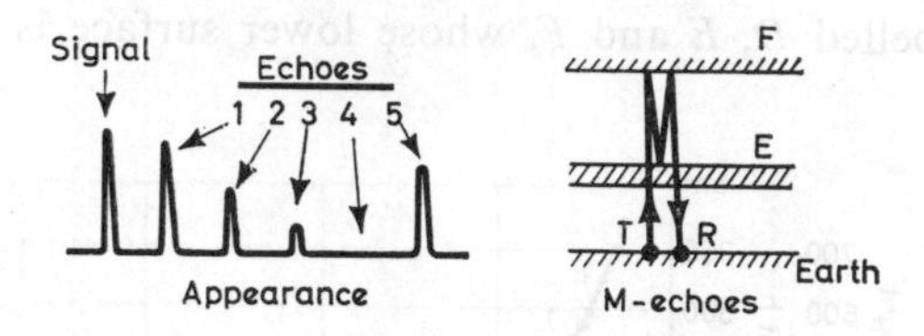

FIG. 6.4. Multiple echoes

either separately or *simultaneously* and even multiple delays:

$$2(2T_F - T_E - \varepsilon) = T_M.$$

These successive echoes are obviously due to *multiple reflections* alternately from the ionized layer and the Earth. (The acoustic analogy has been known for a long time.)

Delays with the value T_M undoubtedly correspond to multiple reflection: once from the F region, next from the *upper surface of the E region,* once again from the F region and back to Earth (Fig. 6.4).

The small difference ε then gives some idea of the *thickness of the E region* (usually quite large, but sometimes only a few miles).

6.2.2. Variable Frequency Probes

Attempts have been made to ascertain the density of the ionized region by repeated probing at a variable frequency.

The frequency of the transmitter and receiver are varied progressively over a wide band (1–12 Mc/s, for example), either *manually*, by adjusting the tuning and noting h as a function of f, or *automatically*, by a periodic servo-controlled rotation of the tuning elements and photographic recording of the results.

Usually, the apparatus consists† of an oscilloscope whose beam is *modulated* by the output current of the receiver and *deflected*, in the Y-direction, by a fast time base scan, as above, and, in the X-direction, by a slow scan proportional to frequency f (or its logarithm).

The beam thus scans the screen with parallel lines (Fig. 6.5) but it is intensified only when a signal reaches the receiver: we

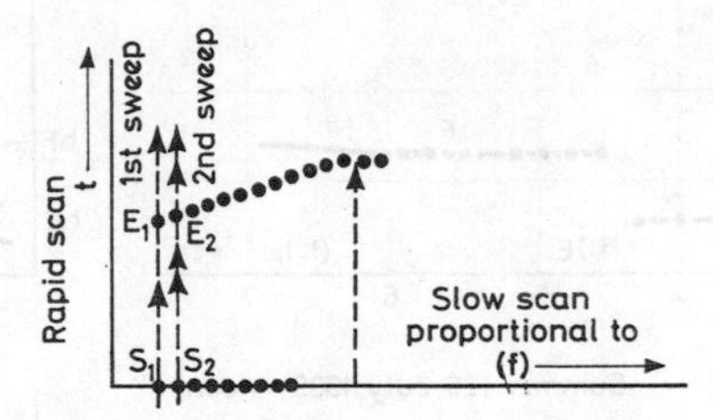

FIG. 6.5. Variable frequency probe

can therefore obtain a trace on an axis of abscissae S_1S_2, corresponding to the "direct pips TR", and a line of dots E_1E_2 corresponding to the "reflected pips TIR", the separation being a measure of the interval T.

Using a suitable control and synchronization system, we thus have directly, by a kind of rudimentary "television", the curve of h as a function of f.

The result is usually of the form given in Fig. 6.6.

The echoes appear with increasing amplitude for $f = 1$ to 3 Mc/s (depending on the transmitter power).

The height is then stable at about 120 km, the E region, with a very slight increase.

Around 4 or 5 Mc/s, a second echo appears at about 250–300 km (F or F_1) and the height increases whilst the E echo stops abruptly ["critical" frequency $(f_c)_E$].

† For more detail, see Plates 2 and 3 and, for axample, Notes prélim. Lab. Nat. Radio., nos. 113, 120 (1948) and 135 (1950).

In winter, this section extends up to about 7 or 8 Mc/s where the echo disappears suddenly to give another *critical frequency* $(f_c)_F$, whereas in summer, the F region may "split" during daytime into two branches, F_1 and F_2.

As these variations are very interesting to follow up, a large number of "probe stations" have been built in order to observe them continuously. They are plotted on a map, Chapter 8.

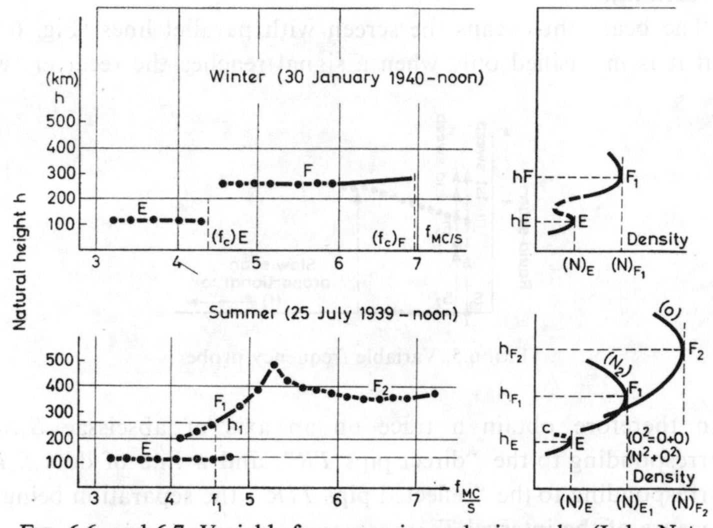

FIG. 6.6. and 6.7. Variable frequency ionospheric probe (Lab. Nat. Radioél. Note no. 6)

The above results can all be easily explained:

1. If a whole frequency band, up to f_1, is uniformly reflected at the same height h, it means that ionization appears abruptly at that height with a density N corresponding to the apparent index becoming zero [formula (2.9)]: in other words, f_1 is less than or equal to the critical value (2.9):

$$f_1 \leqslant f_c = \sqrt{\frac{Nq^2}{4\pi^2 m\varepsilon}}.$$

If the virtual height increases with f, it means that the ionization is progressive and the wave has to penetrate more and more deeply into the region to find its "mirror density".

Finally, if the echo fades and then disappears at a frequency f_1, it means that the ionization maximum of the region has been reached and, the corresponding critical frequency (f_c max) being now less than f_1, total reflection gives way to a quickly attenuated partial reflection.

Thus, to the curves of $h(f)$ (Fig. 6.6), there correspond curves of $N(h)$ *giving the ionization density*, except for certain arcs which are masked by the maxima below them but whose shape is easy to interpolate (Fig. 6.7).†

It can thus be stated that there exist in general *two* zones of maximum ionization:

the E region, with

$$N = (1\text{–}2)\times 10^5 \text{ (daytime)};‡$$

the F region, sometimes subdivided into F_1 and F_2,

$$N = (4\text{–}10)\times 10^5 \text{ (daytime)}.$$

It is easy to understand the existence of an ionization *maximum* in *one* gas, where the rarefaction increases with altitude whilst it is being subjected to increasing ionizing action due to the Sun (ultraviolet light, particles of matter), meteorites, perhaps other cosmic radiation.

The existence of several successive *maxima* in air is explained by the fact that air is a mixture of several gases of unequal densities, the composition of which varies with height.

Finally, the simultaneous existence of echoes from two regions, E and F (Fig. 6.6, bottom, between 4 and 5 Mc/s) means that the density of the lower region is not perfectly uniform in the horizontal plane—it consists of "clouds of strong ionization" (which

† According to some authors, the law of variation of the virtual height h as a function of frequency could be fitted remarkably well by the hypothesis that in each region the ionization increases away from the lower limit h according to a parabolic law:

$$N = N_0(h-h_0)^2,$$

from which it would be deduced that the true reflection height is the virtual apparent height for the particular frequency:

$$f = 0{\cdot}834\, f_c.$$

‡ And number of collisions $\nu = 7\times 10^3$ per second.

reflect waves) alternating with "gaps of weak ionization" (which let them pass through).

2. No echoes are found below about 1–3 Mc/s, for quite a different reason: *absorption* in the trajectory underneath the region.

In fact, there must be here a region of weaker ionization but greater density, and therefore with more molecules and collisions ($\nu = 10^6$): the *D layer*.

As we saw in § 2.2, collisions produce a conductivity

$$\sigma' = \frac{Nq^2}{m} \frac{\nu}{\nu^2+\omega^2},$$

which in turn, as indicated in § 2.4, produces an exponential absorption:

$$e^{-\alpha d} \quad \text{where} \quad \alpha = \frac{\sigma'}{2}\sqrt{\frac{\mu}{\varepsilon}}.$$

For *short waves*, ν^2 being much smaller than ω^2, the exponent α is roughly proportional to λ^2; the absorption therefore increases markedly with wavelength.†

For *intermediate and medium* waves, ν^2 becomes more important than ω^2 and the absorption stops increasing (otherwise, the indirect ray would disappear altogether).

The presence of the ionization density N in the formula for σ shows that *the absorption depends on the amount of sunlight* and must be much greater by day than by night.

6.2.3. Probing at Oblique Incidence

In the preceding methods, the transmitter and receiver were very close together, which facilitated the measurement of the echo delay time T, the transit time for the direct signal *TR* being negligible compared with that of the reflected signal *TIR*.

† A comparison is often made with the absorption of sound by a woollen pad, but this is incorrect, since the absorption of sound by the wool is more marked as the frequency increases, which is the opposite of what happens here.

However, it is quite evident that the kind of transmission we are interested in occurs over large, or even enormous, distances, or in other words, that reflection from the ionosphere occurs not at normal incidence but at oblique incidence or even, sometimes, almost at grazing incidence; forecasts relating to this type of transmission therefore have to be derived from the preceding probings by making a number of hypotheses and calculations which we now summarize.

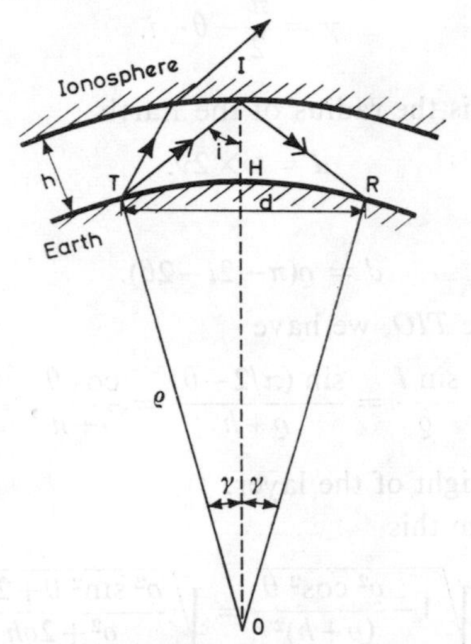

FIG. 6.8. Incidence on the ionosphere

We saw from eqn. (3.18) that if f_c is the critical frequency for normal incidence, the limiting frequency for total reflection, for an angle of incidence i (or its complement φ) is

$$f'_d = \frac{f_c}{\sin \varphi} = \frac{f_c}{\cos i} = f_c \sec i \qquad (6.1)$$

(which is called the "secant law" by the Americans).

Now, in a radio link over a distance d (Fig. 6.8), the wave reaches the ionosphere *obliquely*, with an angle of incidence i which increases with d. The relationship is easy to find.

Let T be the transmitter, R the receiver, I the point of reflection on the ionosphere, O the centre of the Earth, 2γ the geocentric angle TOR, $\theta = ITH$ the angle of the reflected ray above the horizon.

In triangle TIO, the sum of the angles is equal to π and therefore

$$\left(\frac{\pi}{2}+\theta\right)+i+\gamma = \pi,$$

or

$$\gamma = \frac{\pi}{2}-\theta-i.$$

However, if ϱ is the radius of the Earth

$$d = \varrho\times 2\gamma.$$

Therefore

$$d = \varrho(\pi-2i-2\theta). \tag{6.2}$$

But, in triangle TIO, we have

$$\frac{\sin i}{\varrho} = \frac{\sin(\pi/2+\theta)}{\varrho+h} = \frac{\cos\theta}{\varrho+h},$$

where h is the height of the layer.

We obtain from this

$$\cos i = \sqrt{1-\frac{\varrho^2\cos^2\theta}{(\varrho+h)^2}} = \sqrt{\frac{\varrho^2\sin^2\theta+2\varrho h+h^2}{\varrho^2+2\varrho h+h^2}}$$

and as $h \ll \varrho$ we can neglect the terms in h^2, leaving

$$\cos i = \sqrt{\frac{\varrho\sin^2\theta+2h}{\varrho+2h}}. \tag{6.3}$$

We see therefore that for a given θ we can obtain i from (6.3) which gives us d from (6.2) on the one hand and f_d'/f_c from (6.1) on the other.

This shows that if the frequency of the transmission is $f > f_c$, the result will depend on the transmitting angle and the angle of incidence on the layer.

Let us calculate, by means of formulae (6.1), (6.2), (6.3), the transmitting angle θ_0 corresponding to a given ratio of transmitter frequency f to critical frequency f_c, and the corresponding distance d_0.

Any radiation transmitted *at an angle greater than* θ_0 will penetrate the layer and not return (Fig. 6.8).

Only rays transmitted at angle $\theta < \theta_0$ will be reflected; they will come down again at distances d greater than the distance d_0 corresponding to θ_0.

The distance d_0 (the "skip distance") is thus *the lower limit of the range of the indirect ray.*

Below this distance, down as far as the short distances at which the direct ray becomes appreciable (§ 4.4), there is therefore nothing at all—the *zone of silence.*

According to the (purely geometrical) formulae (6.2) and (6.3), the range increases as the transmission angle θ decreases. The greatest ranges should thus be obtained for $\theta = 0$, which gives

$$(f'_d)_{\max} = f_c \sqrt{\frac{\varrho+2h}{2h}},$$

$$d_{\max} = 2\varrho \left[\frac{\pi}{2} - \text{arc cos} \sqrt{\frac{2h}{\varrho+2h}} \right] \simeq \sqrt{8\varrho h};$$

this would correspond to "hops" of 2500 to 5000 km, depending on the height h of the layer.

However, the ray which is transmitted tangentially to the ground follows the surface for a large part of its trajectory, in the course of which it undergoes considerable counter-reflection and absorption (§§ 4.4–4.5). It may thus be severely attenuated, whereas a ray transmitted at a small angle (θ in the neighbourhood of 10°) will be more acceptable at a large distance, for example:

about 1200 km for h = 120 km (E region);
about 3000 km for h = 400 km (F region).

In fact, the actual ranges are a little greater due to the horizontal trajectory in the region (MM', Fig. 6.9) which can be as much as several hundred kilometres.† As far as very large ranges are concerned, they can be interpreted by assuming that the wave executes *several consecutive hops,* for instance, three to five in

† It seems difficult to realize that the path MM' inside the region—even if we assume the thickness to be large—can extend for thousands of miles. This, however, can happen (C.C.I.R. 1959, report 164).

the case of a transatlantic link; abrupt changes in the angle of arrival denote a change in the number of hops.

Clearly, of course, all of this calculation is only approximate. It does not take into account any of the following:

(a) the intersection of the incident and reflected rays at a point J (Fig. 6.9) above the layer (due to the fact that the surface of separation is not a discontinuity and the ray penetrates a certain depth into the layer);

(b) deceleration by reduction of the group velocity in the curved trajectory *ABMM′CD*;‡

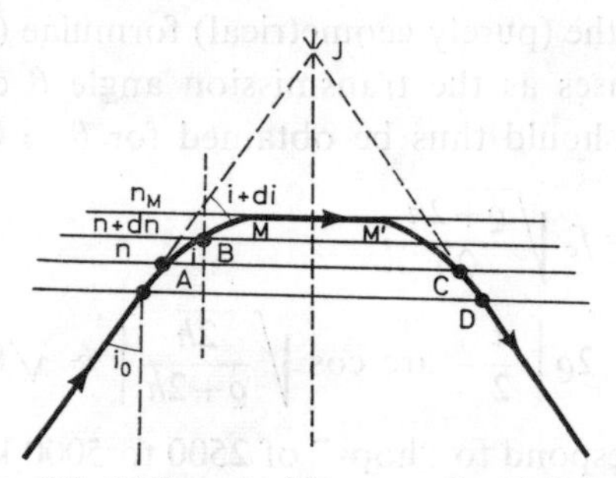

FIG. 6.9. Oblique incidence on the K–H layer

(c) the curvature of the surface of the layer;

(d) the Earth's magnetic field;

(e) experimental observations capable of giving directly the relation between the critical frequency f_c and the maximum usable frequency f'_d.

Finally, the calculation does not take account of *absorption* in regions below the reflecting region; this absorption can be considerable and becomes worse as the trajectory in the layer gets longer, i.e. more oblique.

In order to mitigate these uncertainties and understand better what is going on, "oblique" probes are sometimes set up between a transmitter and a distant receiver; this practical method is

‡ It has been domonstrated that, in most cases, the time taken for the curved trajectory *ABMM′D* can be replaced by the time which would be necessary for the triangular trajectory *AJD* at the normal velocity. This is a fortunate simplification (the theorem of Martin or Breit and Tuve).

equivalent to studying an actual link under its normal conditions, but using certain features (short pulses, etc.) and analysing apparatus (oscilloscopes, orientable directional aerials, etc.) to locate the point of reflection more accurately and evaluate the attenuation in the trajectory.

Obviously, it is necessary to resolve the difficulty of synchronizing the time scales of the two stations so as to be able to evaluate the duration of the trajectory; this can be achieved in various ways, for example by transmitting alternately in both directions, i.e. by "reflecting" the echo from one station back to the other.†

One can thus verify the possible values for the usable frequency, the extent of the zones of silence, the best angles of "arrival" and "departure" in the vertical plane (usually 6–22°).

Finally, signals are sometimes received over quite unexpected distances and from completely unforeseen directions, for examples, at points 1000 or 2000 km from the actual transmitter, or outside the plane of the great circle through T and R, or even from the opposite direction.

These can all be explained by invoking:

(a) unequal attenuations over certain possible trajectories, the shortest being heavily attenuated and only the signals with a long trajectory being perceptible;

(b) diffusion, which we shall discuss later (§ 6.3.5).

6.2.4. PROBING BY "BACK-SCATTER"

Experience has shown that, in the case of oblique probing using retransmission of the signal from the receiver to the transmitter, one can dispense entirely with the second station, the "back-scatter" from the ground being sufficient to return an appreciable amount of energy to the transmitter. Thus, to find out whether a link is possible (at a given frequency) between a

† See, for example, Cox and Davies, *Wir. Eng.*, Feb. 1955, pp. 35–41; Dieminger *et al.*, *Nachr. Techn. Ztg.*, Nov. 1955, pp. 578–86; Delobeau *et al.*, *Ann. Télécomm.*, March 1955, pp. 55–64.

transmitter *T* and a reception zone *R* (Fig. 6.10), there is effectively no need to have a station at *R*: it is sufficient to transmit a beam directed so that on reflection from *I* it reaches the zone *R* under consideration and then to listen on a receiver *R′*, near to *T*, for the eventual echo return—the delay indicating, naturally, the total duration of the double trajectory *TIRIR′*. If the departure angle is varied, probings can be made at different distances; if the directional aerial is rotated about a vertical axis, we can probe at different azimuths; and finally if the frequency is varied, we can draw a map of the possible reflections and ranges all round the station at a given moment.

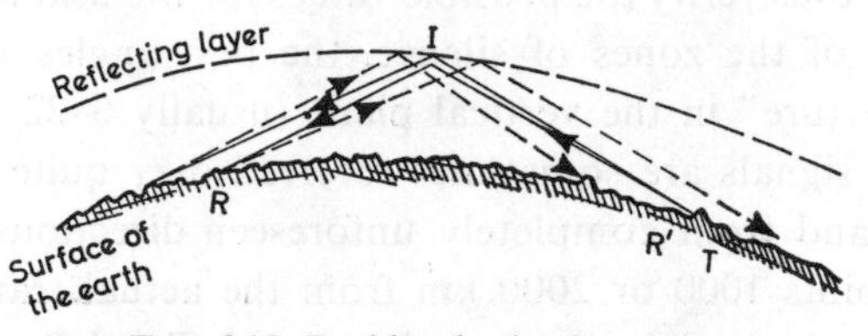

FIG. 6.10. Probing by back-scatter

This new technique is thus full of promise for enhancing—and perhaps partly replacing—the technique of vertical probing, especially in desert or polar regions and oceans where the network of vertical probing stations is necessarily too sparse to permit of an efficient survey.†

6.3. Constitution and "normal" variations of the ionosphere

The combined use of all these methods of research over the last 30 years has led to a substantial—although still very imperfect—knowledge of the extremely complex phenomena which occur in the upper atmosphere.

† See, for example, Shearman, *Proc. I.E.E.*, part B, March 1956, pp. 210–23 and 232–5; Beckman, *Onde Électrique*, May 1957, pp. 416–20; Dieminger, A.G.A.R.D. Report, Copenhagen, Oct. 1958; Peterson *et al.*, *Proc. I.R.E.*, Feb. 1959, pp. 300–14.

It is impossible to summarize here† all that has been learned concerning its physical properties, its composition, the field strengths and radiations found there, etc. We shall confine ourselves to indicating first the general form of the variation of the three principal parameters—pressure, density and temperature—and then to discussing in a little more detail a fourth parameter which is essential for radiocommunication—the ionization density.

Figure 6.11 shows that the pressure and density decrease (curves 1 and 2) as the height increases but not according to simple laws; this is because the constitution of the atmosphere changes: nitrogen, molecular or dissociated oxygen, ozone, hydrogen, the rare gases, form a complicated sequence of different equilibriums.

As to the temperature, after first dropping to about 170°K (i.e. –100 °C) at a height of about 75 km, it rises rapidly to the neighbourhood of 1000 °K at $h = 200$ km and then, perhaps, as far as 100,000 °K beyond that.

All of these indications are, however, only provisional and are therefore liable to considerable alteration as research work progresses.

Finally, the ionization density exhibits the remarkable property of having a profile with several maxima, forming the successive *D*, *E*, *F* "layers" or "regions" already mentioned.

The main cause of this ionization is obviously the light of the Sun—hence the diurnal maximum and nocturnal decrease—either by "recombination" of the molecules or by electrons becoming attached to neutral molecules so as to form negative ions. The existence of several maxima is the result of variations in the gaseous composition of the atmosphere; on the other hand, the persistence of a certain reduced amount of ionization during the night and various other features also suggest other secondary causes, for example, *solid particles* emanating from the Sun,

† There are a number of good general reviews, including: Waynick, *Proc. I.R.E.*, June 1957, pp. 741–9; Rawer, *Die Ionosphäre*, Groninger, 1953; *Proc. I.R.E.*, special issue, Feb. 1959.

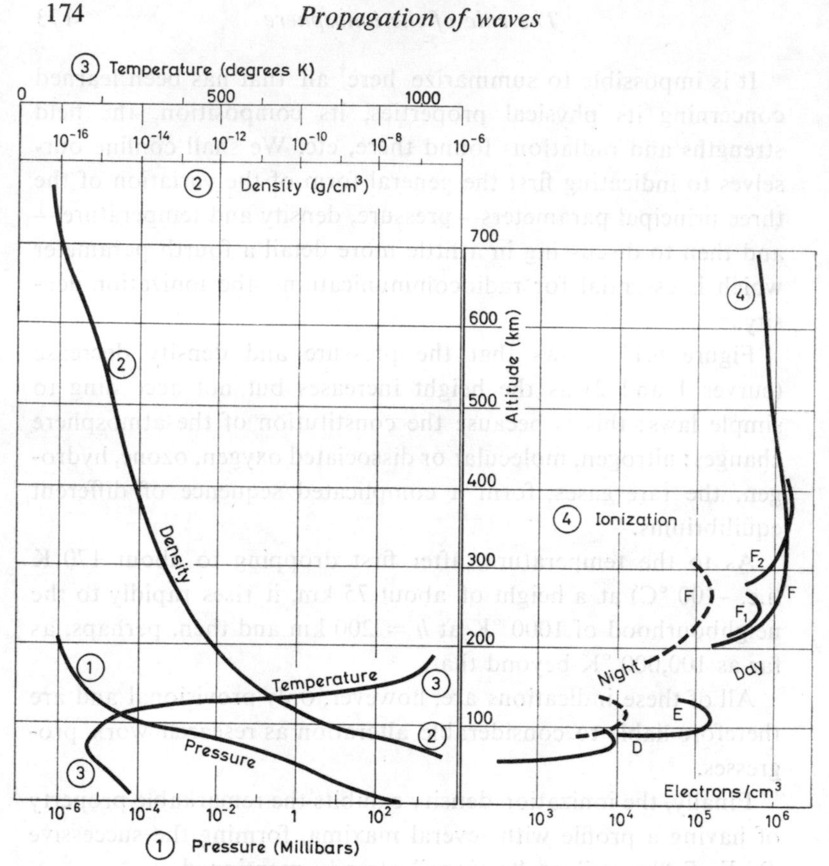

FIG. 6.11. Properties of the ionosphere

meteorites (we shall be discussing them), and perhaps also cosmic radiation.

First we shall examine various features of these regions.

6.3.1. *D* REGION†

The lower region (the *D* region) is usually defined as lying between about 60 and 90 km in altitude; it seems to be caused by ionization of nitric oxide (NO); the profile of this ionization

† See Gibbons and Waynick, *Proc. I.R.E.*, Feb. 1959, pp. 160–1.

(i.e. its variation with height) is rather uncertain; figures ranging from 10 electrons/cm^3 at height $h = 65$ km to 10^4 electrons/cm^3 at $h = 85$–90 km are found; it depends, moreover, on the time of day, the season and the latitude. But periodic maps have not been drawn for this region since reflections from it are produced only at very long wavelengths, well outside the range of normal probings.

By virtue of the number of molecules of air (10^6–10^8 per cm^3) and the consequent collisions, this region is characterized especially by the strong *absorption* which it produces, especially by day, at long waves, or (as will be seen later) at certain periods in the polar regions.

6.3.2. *E* Region

There exists a "normal", thin, *E* region, at a very stable height of about 100–110 km, in the molecular oxygen dissociation zone. The electron density has a maximum (order of 10^5) about midday and falls during the night to a much lower value (10^3–10^4) which is, however, still sufficient to ensure oblique reflection of medium waves; it appears that this diurnal variation can be approximated as a function of the zenith angle of the Sun, χ, by laws of the form:

$$f_c = k(\cos \chi)^n$$

but the exponent n is found to vary greatly from 0·25 to 1·5.

The seasonal variation is weak (see Figs. 6.12 and 6.13), about 50% more in summer than in winter.

We can add some supplementary remarks on the constitution of this layer:

(a) its lower part is more or less "turbulent" or "stratified", so that a certain amount of *diffusion* of the waves takes place (in addition to the reflection proper);

(b) the localization of echoes and meteor trails (which start at precisely this height), and the measurement of the "Doppler" effect in the received echoes confirm that this region is subjected to rapid horizontal displacements, as if pushed by regular "winds" at up to 60–80 m.p.h.

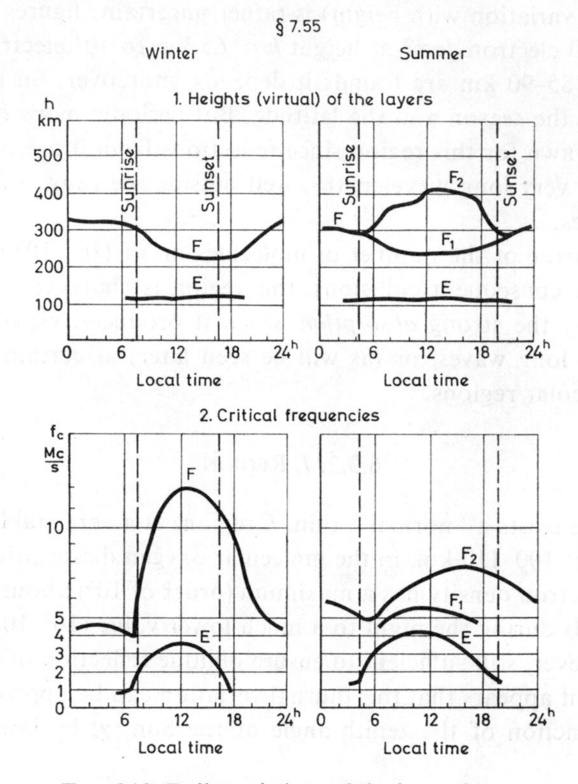

FIG. 6.12. Daily variations of the ionosphere

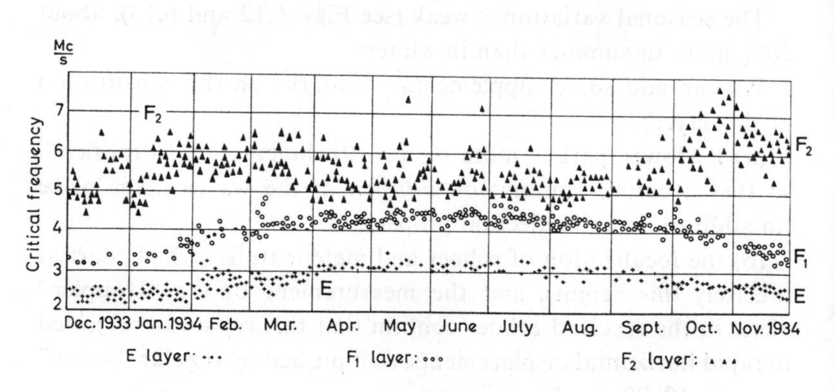

FIG. 6.13. Annual seasonal variation of the ionosphere

In addition to the "normal" E region there exists a much more ionized region, originally designated "abnormal" but later changed to "sporadic" (because of its frequency); this region should not be thought of as continuous but rather as being formed by localized *clouds;* these clouds, much more ionized than the rest, are capable of reflecting very high frequencies (up to 50, 60, 80 Mc/s); but relatively less dense spaces between them introduce a certain "transparency".

The sporadic E region depends very much on the latitude and is most often found at high latitudes; in fact, in the polar regions, it assumes such an importance that it is sometimes split into several components:† the "Thule" layer, the "auroral belt" and the "mixed layer"; the development of aerial navigation in this zone at present justifies special research work whose practical difficulty is obvious.‡

6.3.3. *F* Region

As already indicated, the F region is more variable than the E layer.||

It is at its minimum height of about 250 km about noon; in summer it splits and F_1 rises slightly whilst F_2 rises to about 400–500 km; there is also a variation with latitude, with a maximum at the equator.

The ionization is still a maximum (about 10^6 electrons/cm^3) about midday, as might be expected. But there are several surprising and, as yet, more or less unexplained peculiarities or variations, for example:

† See Penndorf and Coroniti, *J. Geophys. Res.*, Dec. 1958, pp. 789–802. The existence of an E_2 layer at a height of 150 km is also sometimes envisaged, see Whatman, *Proc. Phys. Soc.*, 1 May 1959, pp. 307–20.

‡ *E region general:* Appleton, *Proc. I.R.E.*, Feb. 1959, pp. 155–9.

Turbulence, Diffusion: Villars *et al.*, *Proc. I.R.E.*, Oct. 1955, pp. 1232–9; Gallet *et al.*, *Proc. I.R.E.*, Oct. 1955, pp. 1240–52.

Winds: Manning *et al.*, *Proc. I.R.E.*, Aug. 1950, pp. 877–83; Beynon *et al.*, *J. Atm. Terr. Phys.*, Dec. 1958, pp. 180–2.

Sporadic E: Dieminser, *Ann. Géophys.*, Jan.–March, 1959, pp. 23–30.

|| Regarding the F region, see Martyn, *Proc. I.R.E.*, Feb. 1959, pp. 147–55.

(a) in the equatorial regions, the maximum can occur before noon (local time) or there can even be two maxima, one before and one after noon;

(b) also in the equatorial regions, there can be two maxima separated in latitude, one above and one below the magnetic equator;

(c) there may be secondary maxima after sunset or even during the night;

(d) there are also indications of a tidal movement with a period twice that of the Moon.

6.3.4. Other variations. Ionization maps

The diurnal and annual (or seasonal) variations are not the only ones.

There is also an 11-year variation, synchronous with the solar cycle, which caused radio engineers a great deal of trouble before it was tracked down (Fig. 6.14). Between 1933 and 1950 two successive minima and maxima were observed and the variation in critical frequency f_c between them was roughly as shown in Table 6.1.

This variation was found again during the following cycle, which has just finished.

It is particularly large for the F_2 layer: the ratio of 1–4 (for N) is the biggest known variation in geophysical parameters.

There are also variations with longitude, related to the Earth's magnetism (and independent of the local position of the Sun).

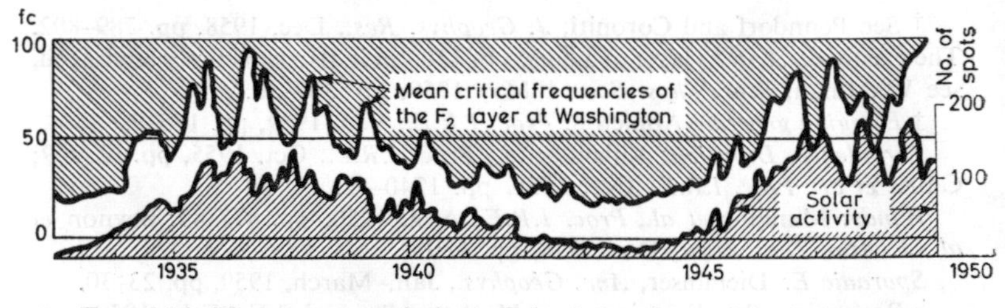

Fig. 6.14. Eleven-year variation of the ionosphere. Correlation with number of sunspots. (From Rawer)

TABLE 6.1. *Approximate variation in critical frequency during the 11-year solar cycle*

F_2	noon	+100%
	midnight	+ 75%
E		+ 30%

In order to represent all these variations conveniently, prolonged observations and forecasts extrapolated therefrom are used to draw maps of the *E*, sporadic *E* and *F* regions. On them are plotted curves of equal ionization, with either critical frequency or a coefficient proportional to f_0 as parameters.

These curves are centred on the "noon" meridian and the equator; the abscissa is local time, the ordinate is latitude; one can thus read at a glance the properties of the layer at any point of given latitude at a given time.

For the *D* and *E* layers, which vary little with solar activity, it is possible to prepare a series of maps, valid for each month, a long time in advance. Examples will be found later, Figs. 8.4 and 8.10.

For the F_2 region, which is much more sensitive to periodic and accidental variations, of the sun, it is at present impossible to make forecasts more than 3 months in advance, and the specialist departments publish in their periodic bulletins forecast maps similar to that of Figs. 8.5 and 8.6, which are valid for a given month of a given year. There are great differences from one month and one year to the next.

We shall deal at length with the methods of using these various maps in § 8.4.1.

We should add that on top of these *main* periodic variations there are other *secondary* variations which are only of academic interest and scarcely affect radio communication:

(a) a slight variation in height of the regions with the lunar period (ionospheric tide, etc.);†

† This is the great speciality of the Australians. A whole section of the International Radio Scientific Union is devoted to it (U.R.S.I. special report no. 2).

(b) a periodicity of 116 sec, also found in the Earth's magnetism;†

(c) a connection with the atmospheric phenomena of lower regions (meteorological, etc.).‡

6.3.5. Turbulence. Diffusion

In the above we have in no way defined to what extent the various regions are homogeneous and their lower surfaces smooth. The simplest hypothesis is obviously to assume that (except for a certain penetration which is taken into account in the "virtual height" found) reflection of the waves occurs at the lower surface of a layer, almost as if it were a mirror, hence the application, in § 6.2.3, of the formula from § 2.1.

In fact, as will be seen later, this simplification is acceptable as a primary basis of calculation for determining usable frequencies and ranges.

However, a number of experiments show that this is not always the case: signals originating from directional aerials are sometimes received in azimuths or at distances which are absolutely incompatible with the laws of regular reflection. The explanation is simple: it is only necessary to assume that the lower surface of a layer exhibits irregularities or turbulences (which is quite natural when it is compared with the lower surface of a cloud), and to invoke therefore *diffusion* phenomena (§ 3.2.6) superposed on the classical reflection. The same assumption explains certain intensity fluctuations (fading), the "scintillation" of radio noise originating in the stars, and other accidental peculairities.

Although it was pointed out by Eckersley|| in 1939, this turbulence has not been studied systematically until quite recently—as a result, it seems, of tropospheric turbulence;* extrapolating the

† Harang *et al.*, *Funktechnische Monatshefte*, Jan. 1940, p. 14.

‡ See Bannen *et al.*, *Proc. Roy. Soc.*, **174**, 21 Feb. 1940, p. 958.

|| *Nature*, **143**, 1939, p. 33.

* See, for example, Booker, *J. Geophys. Res.*, Dec. 1956, pp. 673–705; Gallet, *Proc. I.R.E.*, Oct. 1955, pp. 1240–52; Lepechinsky, *Onde Électrique*, July 1958, pp. 519–21; Hines, *Proc. I.R.E.*, Feb. 1959, pp. 183–6.

formulae of meteorology, we can predict, at an altitude of about 100 km, "vortices" due to ascending winds or gradients in horizontal winds. These vortices can first become established with dimensions of the order of a kilometre and then subdivide into smaller vortices whose energy is finally converted into heat by the viscosity of the air. The maximum dimension can easily explain the forward-diffusion of waves of 30–60 Mc/s; smaller dimensions can explain back-diffusion.

Furthermore, charged particles can behave differently from neutral particles and produce irregularities distributed along the lines of force of the terrestrial magnetic field, which explains why there is some lack of symmetry (effectively observed) between north–south and east–west trajectories.

We give now a few supplementary details about transmission by "ionospheric diffusion", on account of its growing importance in recent years.

During the year 1951, a new mode of propagation of metric waves was discovered in the United States. A signal transmitted on a wavelength of 6 m was received continuously at a distance of about 800 miles. The level of the received signal was weak, but the use of a powerful transmitter (25 kW), highly directional aerials ("diamond" aerials with a gain of the order of 20 dB more than an omnidirectional aerial) and a sensitive receiver ensured permanent transmission: the perturbations in the ionosphere which we shall be studying in §§ 6.4.1 (ionospheric storms) and 6.4.2 (black-outs) were not, in general, accompanied by any lowering in the signal level and frequently even led to a reinforcement of the signal, whereas at decametric waves, on the other hand, communication might be completely paralysed. Numerous experiments have been carried out since 1951. These studies have tended to attribute the observed propagation, as indicated above, largely to a slight diffusion of the transmitted radiation in a zone of strong ionospheric turbulence at the base of E region of the ionosphere, at a height of about 90 km, such that a fraction of the energy is reflected towards the receiver. But it was apparent that at certain times, especially at night and early daylight, reflections from the ionization trails due to meteorites which are always passing through the upper atmosphere and produce maximum ionization at about 90–95 km, too, make an important, or even dominant, contribution to the received field strength. An appreciable part is also played by reflections from more or less marked discontinuity surfaces in the ionosphere, from "islands" of strong ionization which constitute what we call the sporadic E region around 110 km in altitude, and, in the polar regions, from equally strongly ionized zones of a type peculiar to these regions.

Transmissions are possible over a whole range of wavelengths which extends in practice from 4 to 10 m approximately; the mean received field decreases rapidly, as λ^3 or λ^4, as a function of wavelength λ. The variation of field strength with distance depends especially on the angle of diffusion θ which the incident rays make with the diffused rays. Theory leads to a $(\sin \theta/2)^{-n}$ variation in mean received power, where n is a coefficient which, depending on the hypothesis made, lies between 5 and 12, experimental values being distributed evenly throughout this range. Links of from 600 to about 1500 miles are possible, the lower limit being due to the fact that as θ becomes too large, the received field becomes too weak, and the upper limit due to the fact that the Earth's curvature leads to a maximum realizable range, for a fixed diffusion altitude (about 90 km). Certain observations lead us to believe that in the equatorial regions diffusion would also be considerable in the F_2 region of the ionosphere, at a height of 300 or 400 km: the limiting range could then be from 2000 to 3000 miles.

The received field strength, like that of "tropospheric diffusion" studied in § 5.3.3, exhibits rapid fluctuations, the periods of the fluctuations sometimes being as little as a few tenths of a second. Slower variations of the horary median levels are also observed. A noticeable effect is seen on the received signal, with a minimum around 7–10 p.m. and a maximum at noon, very marked in temperature regions, much more subtle in the polar regions where a secondary maximum sometimes appears in the first hours of daylight. The distribution of the maximum values at noon reveals a seasonal effect, with a maximum in June and December and a minimum in March–April and September; but there is no well-defined law for the variation of the night field strength; nor is there any well-defined variation from year to year.

One feature of propagation by diffusion—already pointed out in connection with "tropospheric" diffusion—is that it leads to aerial gains which are much lower than those which they would have on the other types of link. The received field is, in effect, formed by summing elementary fields arriving by diffusion at different site and azimuth angles; the apparent gain of the aerials is then a kind of average gain corresponding to these different arrival angles and not the maximum in the axis of the beam. In practice, the aerials used have a maximum gain of the order of 100 (compared with an omnidirectional aerial): for example, "lozenges" with large dimensions (300 m side), networks of "yagi" aerials, consisting of arrays of doublets (4 by 15 or 4 by 17 in some American experiments), and, especially nowadays, corner reflector aerials and cylindro-parabolic reflector aerials. These reflector aerials give the best results because their polar diagram has practically no parasitic lobes outside the main beam. Parasitic lobes produce parasitic trajectories which cause distortion of the transmitted signals.

The rapid fluctuations in the signals received by two aerials at least 12–15 wavelengths apart, perpendicular to the propagation trajectory, are not correlated, so that fadings are not produced simultaneously. We can profit from this fact to improve the stability of the received signal by combining the signals from two such aerials by the process of space diversity. Another type of diversity reception can be employed. It has been observed, and moreover, it can be shown theoretically that at times when meteoric reflection predominates, the regions of maximum meteoric reflection, at a height of 90 km, are often displaced with respect to the vertical transmitter–receiver plane (Fig. 6.15). The problem is then to swing the beam axes of the aerial in azimuth with respect to the axis of the link, by an angle which can vary from 0 to about 10°, according to circumstances. Experiments have been carried out on these lines (apparently with some success) using a two receiving-aerial diversity, the beam axes deviating symmetrically with respect to the axis of the link (the devia-

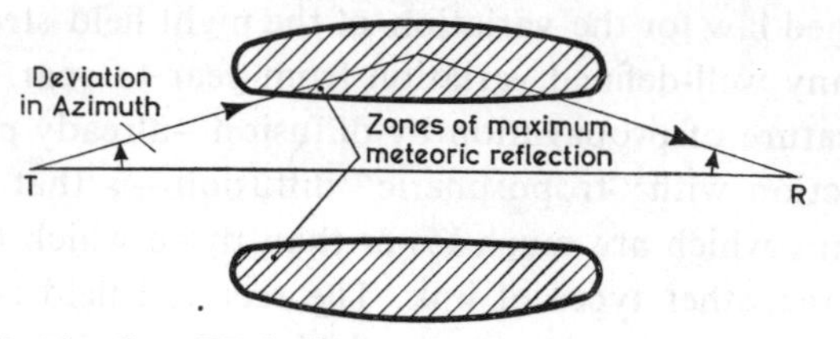

FIG. 6.15. The line *TR* indicates the vertical plane containing the transmitter and receiver

tion was $\pm 2{\cdot}5°$ on one trajectory studied in Britain, for instance).

Several links using "ionospheric diffusion" are already in existence, for example, there is a Gibraltar–England link.† Another link with 4 sections transmitting from four to eight telegraphic channels and one telephone channel has been set up by the U.S. Air Force between the U.S.A. and Britain via Labrador, Greenland and Iceland.

The transmitters have powers of several tens of kilowatts and are in general single-sideband amplitude modulated. The wavelengths used are of the order of 8–10 m.

6.3.6. Splitting of the reflected ray. Role of the Earth's magnetic field

1. It frequently happens that the trace of experimental curves of $h(f)$ on the oscilloscope exhibits a "splitting" or "bifurcation" (Fig. 6.16) with two similar branches O and X, separated by the order of a megacycle.

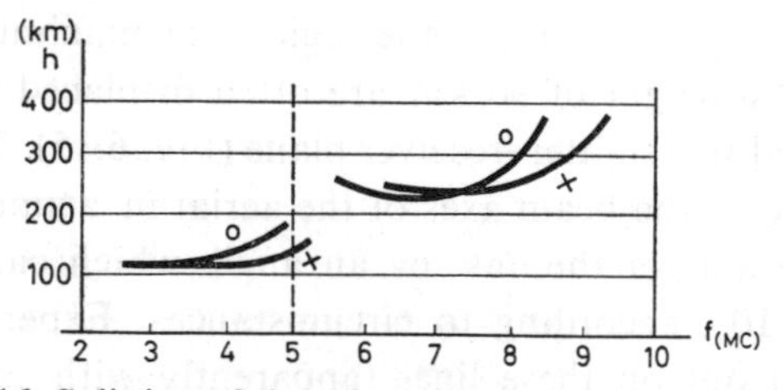

FIG. 6.16. Splitting of the echoes by the Earth's magnetic field

† Bain, *Proc. I.E.E.*, part B, **108**, 39, May 1961, pp. 241–56.

2. In parts of the world where the Earth's magnetic field H is *horizontal* (e.g. in Huancayo, Peru), either component can be separated and obtained at will by changing the orientation of the aerials:

(a) If one transmits and received on horizontal north–south aerials (E field parallel to H), one obtains the O-component only.

(b) If one suddenly switches to an east–west aerial (E field perpendicular to H), one finds the X-component only.

3. For an aerial pointing in any other direction, or in any other part of the world, the two components cannot be separated; but the O-component is often more intense than the X-component.

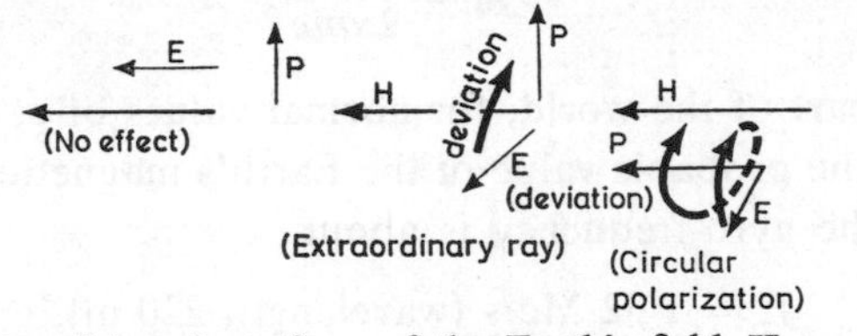

FIG. 6.17. Possible orientations of the Earth's field H, propagation direction P and electric field E

4. It is sometimes possible, by means of improved, complex aerials (frame–aerial combinations, etc.), to analyse the polarization of the received ray.

For a fair-sized trajectory in the north–south direction, the polarization is found to be *circular* and in opposite directions in the two hemispheres.

5. The influence of the poles and the magnetic equator on the distribution of ionization was discovered in 1943; they are the cause of the appearance of *two* maximum zones on the maps (Figs. 8.9 and 8.10); they make it necessary to split the surface of the globe into three zones (west W, intermediate I and east E) (Fig. 8.5) for which different maps are required.

Here we recognize the phenomenon of "double refraction" due to magnetic fields, which is well known in optics and explained by taking account of the Earth's magnetic field (vector H). The results of the calculation depend on the relative orientations of this field H, the direction of propagation P and the electric field E (Fig. 6.17).

If E is *parallel* to H (and therefore both perpendicular to P), the magnetic field has no effect on the motion of the electrons in its direction. The preceding theory applies as if H did not exist, giving the "ordinary" ray O.

If E is *perpendicular* to H, the magnetic field exerts on the moving electrons a force proportional and perpendicular to their speed, i.e. a *deviating* force.

The trajectories become curved and they have a certain "characteristic frequency" or "gyro-frequency", which is shown by calculation to have the value:

$$f_0 = \frac{eH}{2\pi mc}.$$

In our part of the world, for normal values of e, m, c and for $H = 0{\cdot}5$, the probable value of the Earth's magnetic field at high altitude, the gyro-frequency is about

$$f_g = 1{\cdot}32 \text{ Mc/s (wavelength 220 m).}^{\dagger}$$

The effect of this curvature of the electron trajectories varies, of course, according to the relative directions of the vectors P and H.

For a direction of propagation *P perpendicular to the field H*, the deviation is in the plane of propagation; it does not destroy the symmetry of the propagation but merely modifies the intensity of the electronic current. It can be demonstrated that the apparent decrease in the dielectric constant is given, not by formula (2.9), but by the new value

$$\eta' = \left[\varepsilon - \frac{Nq^2}{4\pi^2 mf(f-f_g)}\right],$$

as if the critical frequency (2.9a) were replaced by the corrected value

$$f_x' = \sqrt{f(f-f_g)}.$$

Reflection thus occurs for a smaller electronic density N' and therefore *lower down*, and continues up to a *higher* frequency.

† This figure varies with H according to place, from 0·7 Mc/s at the equator to 1·6 Mc/s at the poles; a map will be found in the *Radio Propagation Handbook*, Washington, 1943.

The ray thus modified is called "extraordinary"; it is sometimes noticeable from the F region, but not from the E region.

For a propagation P *parallel* to H, the electrons are deviated outside their normal plane of propagation and their trajectories become *helical.* It can be shown that the field is divided into two components, polarized circularly and in opposite directions and with very different attenuations: in fact, in the formula for the conductivity σ', § 6.2.2, the ω^2 term in the denominator must be replaced by $(\omega \pm \omega_g)^2$, where ω_g is 2π times the gyro-frequency; the + sign applies to the ordinary ray, the − sign to the extraordinary ray; the former is therefore *attenuated less* than the latter and can practically exist alone in certain cases.

Finally, of course, in the general case of a wave describing an *oblique trajectory* with *oblique polarization*, a *mixture* of various effects, more or less distinct, will be observed. In particular, if the ionization gradient of the reflecting layer is high, the O and X rays will be reflected to approximately the same point and will remain comparable and indistinguishable (they can, however, interfere and cause the amplitude of the echoes to vary, § 6.3). On the other hand, if the ionization gradient is small, the two rays will be reflected with an appreciable path difference and will yield two distinct echoes.

Two closing remarks:

1. The existence of a "gyro-frequency" of about 200 m wavelength explains certain odd effects and poor reception in this range.
2. Anomalies sometimes exist: the probing layer bifurcates into, not two, but *three* or *four* branches.

Explanations (impossible to summarize here) have been proposed by Haubert and Rydbeck.†

The problem of propagation in a rarefied ionized medium in the presence of a magnetic field is connected with that of "plasmas" and "magnetohydrodynamics".‡

† Note prélim. Lab. Nat. Radio., no. 127, 1948; *Onde Électrique*, Feb.–March 1951, pp. 70–81 and 153–6.

‡ See *Proc. I.R.E.*, special issue, Dec. 1961, and *Théorie des Ondes dans les Plasmas*, Denisse and Delcroix, Dunod, 1961.

6.3.7. Ionization by meteorites

Independently of the D and E layers, but in the same region (i.e. at a height of about 100 km) there is an additional ionization band due to meteorites, already mentioned in § 6.3.5. It is surprisingly permanent, for there are few meteorites big enough to be visible in the form of "shooting stars" (perhaps a hundred a day, weighing over a kilogram, except in the well-known periods when they are more abundant). But it has been discovered that much smaller meteorites (upwards of 0·01 mg, i.e. mere specks of dust) leave behind them *ionized trails* which (although invisible to the eye) are capable of reflecting radio and radar waves; indeed, the number of such specks is estimated at about 10^{10} per day; that is to say, it is only necessary to wait a few minutes—sometimes a few seconds—before a signal transmitted from any point of the globe meets a "trail" which will return it to some other point 500–2000 km away. Naturally, this intermittent transmission does not suit all types of signals; but in certain difficult telegraphic links, one can have recourse to it if one confines oneself to sending a brief message, at high speed, during the period of existence of a trail (order of a second). We shall discuss this again later.†

6.3.8. The magnetosphere and interplanetary space

It was remarked in § 6.3.3 that the ionization level of the F-region passed through a maximum (about 10^6 electrons/cm^3) during the day at an altitude of 200–500 km.

Above this height, recent measurements by rockets and satellites have indicated a relatively slow decrease in ionization with altitude: electron densities of some 10^4 per cubic centimetre at 1000 km, 10^3 at 10,000 km and about 10 electrons per cubic centimetre even at heights of 50,000 or 100,000 km and beyond, in the interplanetary plasma surrounding our planet.

† See *Meteor Astronomy*, A. C. B. Lovell, Oxford, 1954; Villard *et al.*, *Proc. I.R.E.*, Oct. 1955, p. 1473; *Proc. I.R.E.*, special issue, Dec. 1957.

Now, above 500 or 1000 km altitude we leave the ionosphere proper, which forms part of the terrestrial atmosphere, and enter the *magnetosphere*, a zone where the Earth's magnetic field, whose lines of force are shown in Fig. 6.18, plays a dominant part in the movement of the electronic particles of the medium (which is completely ionized): electrons and protons of solar origin.

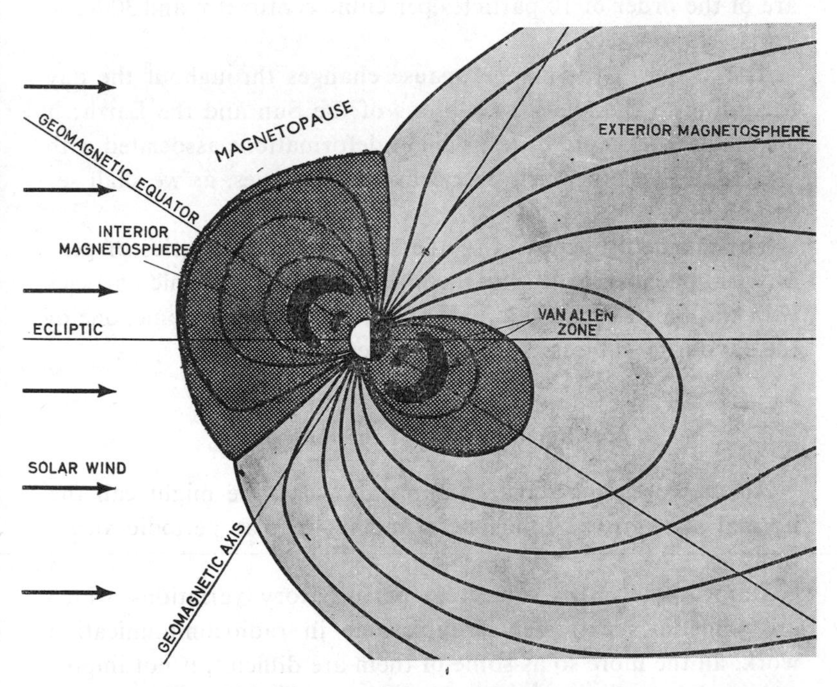

FIG. 6.18. Diagram of the Earth's magnetosphere

In particular, this is the region where the magnetic field traps the high-energy particles which form the "radiation belts" or "Van Allen belts" which are such a menace to cosmonauts and to electronic equipment alike.

The magnetosphere is bounded on the outside by a fictitious surface, the "magnetopause", of asymmetrical shape, at a distance of the order of 6–12 radii (40,000–80,000 km) from the centre of the Earth in the direction of the Sun and extending very far in the opposite direction.

Beyond the magnetopause, the Earth's magnetic field—which decreases inversely as the cube of the distance from the centre of the earth—is masked by interplanetary magnetic fields associated with a permanent flux of electronic particles ejected by the sun (the "solar wind", consisting mainly of protons), whose density and velocity in the neighbourhood of the magnetopause are of the order of 10 particles per cubic centimetre and 300–500 km/s, respectively.

The shape of the magnetopause changes throughout the day according to the relative positions of the Sun and the Earth; it also undergoes more or less violent deformations associated with solar flares accompanied by ejection of particles, as we shall see in § 7.3.

The magnetosphere is also the seat of a whole series of geophysical phenomena due to the interaction of electronic particles with the Earth's magnetic field; it is, at the present time, one of the favourite subjects for space research.

6.4. Perturbations of the ionosphere

All the foregoing remarks apply to what one might call the normal state of the ionosphere and its regular periodic variations.

Unfortunately it is subject to perturbatory variations which are sometimes extremely troublesome in radiocommunication work, all the more so as some of them are difficult, if not impossible to foresee.

6.4.1. "Black-outs"†

It sometimes happens that all short-wave long distance links over a certain area of the world are suddenly cut off so sharply that operators might suspect a power cut.

† The phenomenon was first reported in 1930 by Mogel but remained unnoticed until the French Scientific Radiotelegraphy Committee showed its interest in 1933. In 1937 it was observed by Dellinger and it is sometimes called the "Mogel–Dellinger" or "M–D" effect.

The normal state of affairs then gradually returns, over a period of half an hour, for example.

Methodical study has shown:

(a) that these sudden disappearances of signals involve also all the echoes in the probing stations, the initial and final transit periods revealing an enormous increase in ionization (Fig. 6.19) and the *E* and *F* regions remaining exactly in their normal places;

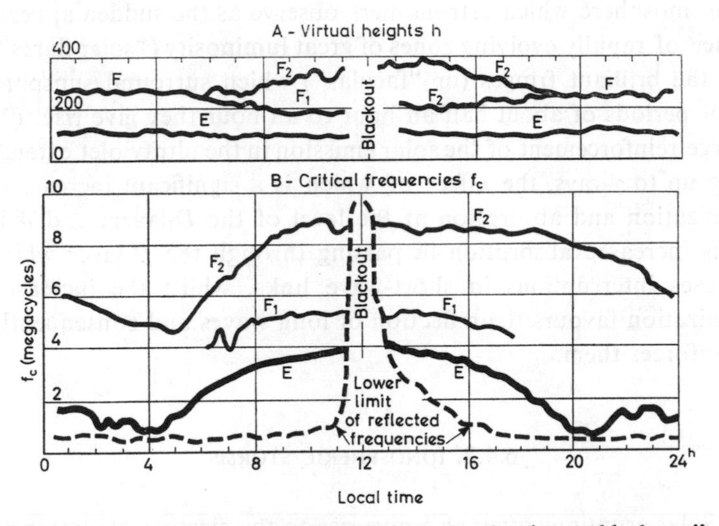

FIG. 6.19. Observations by ionospheric probe showing a "black-out" (Bernes, Wells, 31 July 1937)

(b) that they affect the entire short-wave band simultaneously, but that they have no effect on medium waves and, on the contrary, actually produce a *reinforcement of signals and atmospherics at long waves*;‡

(c) that they affect only *daytime trajectories* and never links operating in darkness;

See Lepechinsky, *Onde Électrique*, March 1953, pp. 157–64, and July 1958, pp. 521–4, and several articles in *Proc. I.R.E.*, Feb. 1959 and *Proc. I.E.E.E.*, Nov. 1963.

‡ See, for example, Bureau, *Onde Électrique*, Feb. 1947, pp. 45–56.

(d) that their number is, on average, affected by normal seasonal and decennial variations (Fig. 6.20);

(e) that they often coincide with sunspots or solar flares and *small* variations in the Earth's magnetism (*not* magnetic storms);

(f) that it is sometimes possible to re-establish contacts interrupted by the perturbation, by raising the operating frequency, sometimes above 30 Mc/s.

These perturbations are associated with eruptions in the Sun's chromosphere which astronomers observe as the sudden appearance of rapidly evolving zones of great luminosity ("solar flares") in the brilliant fringes (or "faculas") which surround sunspots. For periods of about half an hour to an hour they give rise to a large reinforcement of the solar emission in the ultraviolet extending up to γ-rays, the effect of which is a significant increase in ionization and absorption at the level of the D-layer; and it is this increased absorption in passing through the D-layer which causes interruptions in short-wave links, whilst the increased ionization favours the reflection of long waves and consequently reinforces them.

6.4.2. Ionospheric storms

Solar eruptions often also give rise to the ejection of electronic particles (essentially protons) corresponding to fluxes with an intensity and velocity (1000–2000 km/s) much greater than those of the "solar wind" of the calm sun (§ 6.3.8). When the sunspot responsible for the eruption is near the central meridian plane of the sun as seen by an observer on Earth, the ejected particles are directed towards the earth and, after travelling for 24–48 hr, start to bombard the Earth's magnetosphere; the magnetic field "guides" most of them into the auroral regions, latitudes 65°N. and S.

A whole series of geophysical phenomena are then observed: magnetic field perturbations (magnetic storms), atmospheric pressure waves detectable with infrasound equipment, deformation of the magnetopause (the boundary of the magnetosphere

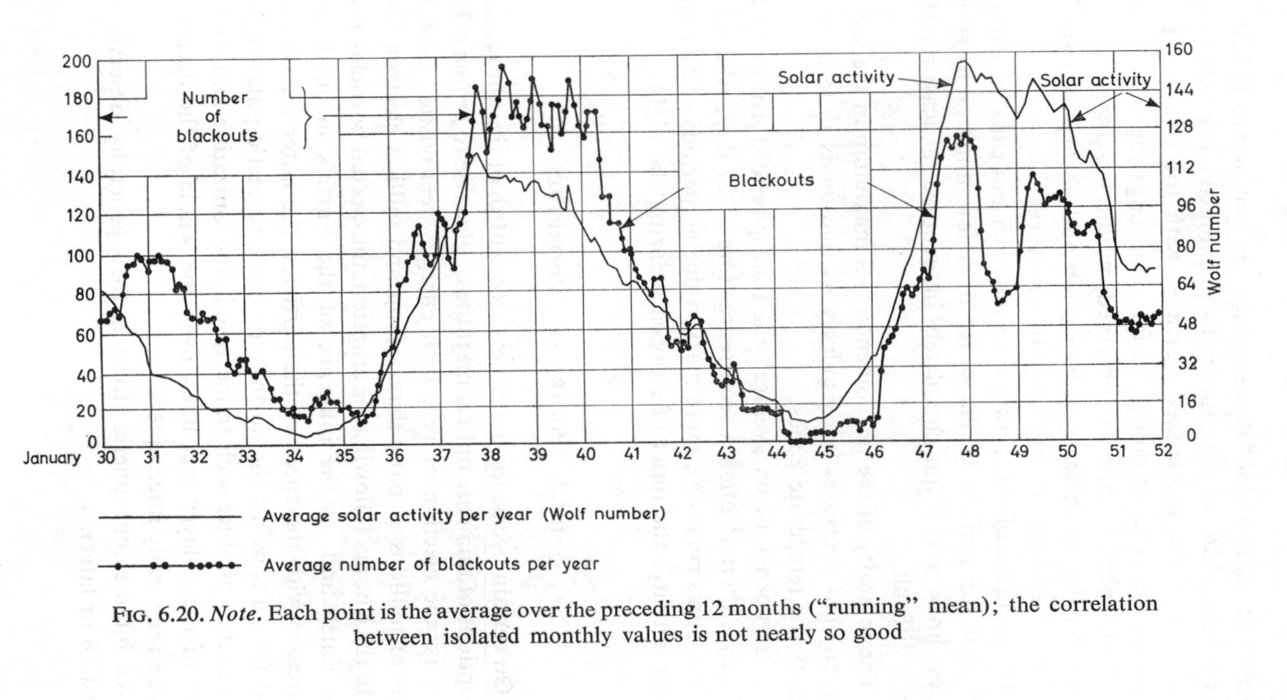

FIG. 6.20. *Note.* Each point is the average over the preceding 12 months ("running" mean); the correlation between isolated monthly values is not nearly so good

"cavity"), emission of "atmospherics" at long waves (§ 7.5), and sometimes *polar auroras*, luminous phenomena attributed to the collision of the accelerated particles with the particles of the atmosphere. And, of course, as far as propagation is concerned, an "ionospheric storm" is observed, modifying appreciably the *D*-region (by an increase in ionization and absorption) and the *E*- and *F*-regions (by a change in structure).

Propagation of short waves (decametric) can thus be seriously affected for periods of the order of a few minutes to several days; links, are, if not completely interrupted, at least rendered very difficult.

The intensity of the phenomenon is a maximum in the auroral regions but decreases fairly quickly as one tends in latitude towards the temperate zones.

Its frequency is, on average, one ionospheric storm every 10–30 days; there is often a recurrence time of 27 days (the period of the Sun's rotation), more marked during periods of minimum solar activity; the onset often occurs during the night.

6.4.3. "Polar cap" absorption

On certain occasions, much less frequent than the above, solar eruptions of immense optical brightness, give rise to *proton events*, that is, the ejection of very high energy "relativistic" protons: tens of millions at more than a thousand million electron volts.

In the 6 hours following the eruption, the ejected particles reach the Earth and, as the influence of the Earth's magnetic field increases with the speed of the particles, the latter are deflected and forced towards the north and south magnetic polar regions where they produce such an increase in the ionization and absorption of the *D*-layer that links over the whole of the decametre range are totally interrupted.

An intense ionospheric storm is then generally observed (at least 18 hr later).

6.4.4. Effect of nuclear explosions

Although few reports have been published on the subject of nuclear explosions, it appears that these experiments are likely to produce very serious perturbations in the ionospheric regions and, consequently, prolonged interruptions in certain radiocommunications. According to a rough evaluation† a one kiloton bomb can produce $1 \cdot 3 \times 10^{23}$ fissions, each of which can release 6×10^{6} electrons and ions, making a total of 8×10^{29} charged particles; distributed uniformly through 500 km^3, these charges therefore represent a density of 10^{12} electrons/cm^3, i.e. 10^6 times the normal for the *F*-region, thus multiplying the critical frequencies by 1000; this means that all normal links would be out of action. And although in the lower reaches of the atmosphere these free electrons can disappear quite quickly by recombination, at heights greater than 200 km rarefaction slows this process down considerably and the perturbation can persist for a very long time (several hours or even days?). Experience of high altitude explosions has shown appreciable increases in ionization level up to 1250 miles and variations in the Earth's magnetic field at even greater distances.

In an analysis of a Russian explosion in 1961, Arendt *et al.*‡ found:

a stationary "hole" stretching far into the ionosphere;

a mobile perturbation travelling along the lines of force of the earth's magnetic field.

Other authors|| observed that the explosion produced a momentary "flash of lighting" which was likely to damage receivers.

† *Missiles and Rockets*, 13 April 1959, pp. 21–24, and 20 April, pp. 26–30.

‡ Arendt *et al.*, *Proc. I.E.E.E.*, June 1964, pp. 672–6.

|| Hays, *I.E.E.E. Spectrum*, May 1964, pp. 115–22.

6.4.5. Echoes with very long delay times

Another odd phenomenon is that in which echoes are received with delay times of 10, 20 or even 30 sec.

Although exceptional at our latitudes, they were observed in 1929 in Tonkin by a French mission (Commandant Talon), with incredible signal strength and regularity, except for total disappearance during a solar eclipse.

There are two possible theories:

Stormer assumes that the signal escapes completely from the Earth and is reflected in space from charged corpuscles which are brought together, outside the Moon's orbit, by the Earth's magnetic field and which, moreover, already account for various other astronomical phenomena.

Others (Van der Pol, Pedersen, Fuchs) attribute the long delay time not to an effective path outside the atmosphere but to an exceptional reduction in the group velocity.

However, it does seem curious that such a reduction in velocity can be obtained in a stable manner and without prohibitive absorption.

6.4.6. Wave interaction ("Luxembourg effect") by non-linearity of the ionosphere

Finally, we mention a phenomenon discovered in 1933–4 on the occasion of the commissioning of the high-power Radio Luxembourg transmitter (120 kW; 1300 m); other analogous cases appear from time to time.

Waves which, during their trajectory, happen to be reflected from the ionosphere directly over a very powerful transmitter are affected and lightly modulated by the transmitter.

In a special series of measurements it was found that the interference could be appreciable in an extended ray (several hundred kilometres).†

† See, for example, Doc. 19 of the VIth Commission of the C.C.I.R. (Stockholm, 1952) and Doc. 91 of the C.C.I.R. (London, 1953); Doc. 118 of the C.C.I.R. (Los Angeles, 1959).

After being surprised by this "interaction" and even doubting it, radio engineers now think of it as quite natural. An interaction of this type is characteristic of non-linearity somewhere. Now, it is quite possible that the motion of electrons in the *K–H* layer brings into play non-linear factors omitted from our analysis in §§ 2.1 *et seq.*: for instance, the increase in the number of collisions as a result of the variation in average velocity and hence distortion of an intense wave and perturbation of weaker waves arriving simultaneously at the reflecting region.

If this phenomenon is really general and the power of radio stations continues to increase in the future, it could play an ominous part in their mutual interaction.† In the meantime, it can be utilized to calculate electron densities and collision frequencies.

6.4.7. Eclipses

These provide opportunities to carry out experiments in the course of which one invariably observes the transition from "day" to "night" propagation conditions in the space of a few minutes.‡

While confirming the action of the ultraviolet light from the Sun (or extremly fast propagation corpuscles), these observations also furnish proof that the ionization is dissipated very quickly when its cause disappears.

† Bailey, in a theoretical study of this phenomenon, found that a vertically radiated power of 500 kW would be sufficient to produce visible luminosity in the ionosphere and that a power of a million kilowatts would produce an "artifical aurora" with a luminosity comparable with that of the full Moon.

‡ A review of the observations made during the eclipses of 1949–52 will be found in articles by Steinberg, *Onde Électrique*, June 1953, pp. 281–3 and Aug.–Sept. 1953, pp. 531–2—as well as in the *C.R. de l'U.R.S.I.* (Joint Commission on the Ionosphere, 1951, pp. 66–75).

CHAPTER 7

INTERFERENCE

7.1. General

No book on wave propagation would be complete without a mention of certain perturbatory phenomena which affect radio-communications, not by direct action on the signal but indirectly by the addition of *noise* to it (the word *noise* is an exact description of what happens in audio receivers; it has been generalized to include all other types of receiver: in telegraphy it denotes random movements of a relay; on a picture it implies the presence of small dots which reduce the definition, etc.).

Noise has been observed since the early days of wireless; even with receivers of poor sensitivity, "crackling", apparently associated with local storms, was often a source of trouble.

With the advent of high gain receivers the effect became much more common and the range was very often limited not by the absolute value of the signal but by the degradation of the "signal/noise" ratio.

A great deal of effort, both theoretical and experimental, was put into improving this ratio and observations of various types of traffic, over the whole frequency range, soon showed that there are not *one* but *several* types of noise, of different origin, their effects being additive; thus the first thing to do is to determine which type is dominant in each case and to concentrate the effort on reducing it, since reducing the other weaker types would not yield a noticeable improvement.

The various types of noise can be classified as follows:

1. A certain amount of *internal noise* is inevitable in a receiver as a result of thermal agitation of electrons in the circuits and

the early amplifying stages; an analogous noise also arises in the aerial, even if it is enclosed in a perfectly efficient "Faraday cage".

2. However, the receiver aerial also captures, in addition to the signal, various types of *external noise* which can arise from three sources:

The first, known as *atmospherics* or *natural* or *terrestrial* noise, has its origin in storms or electrical discharges between more or less distant clouds; the noise level depends largely on place, season, time and frequency: it is most troublesome at long waves.

The second, known as *artificial* or *industrial* noise, is produced by any electrical apparatus with a transient (switching) regime: contactors, motors with commutators, luminous discharge tubes, motor-car ignition systems; it is encountered over almost the whole range of frequencies used, from long waves to metric waves; the noise level can be high, especially in towns; but it can be eliminated at source, which poses legal as well as technical problems.

A third category was discovered about 1935, namely *extra-terrestrial* noise; this may be subdivided into *solar* noise, originating in the Sun, and *cosmic* or *galactic* noise, originating in other stars or galaxies, visible or otherwise. The astral radiation is weak but it is perceptible and can sometimes be troublesome in the most sensitive receivers at metric and shorter wavelengths; and it has the advantage of providing the astronomer with new data on the structure and dimensions of the universe: this has led to the creation of a specialist branch —Radio-astronomy— which is now much to the fore.

The fundamental observation is that all these types of noise are, to a first approximation, random phenomena analogous to "thermal noise".†

It is well known that the motion of electrons creates in any

† Although this analogy has become an "article of faith" for the majority of radio engineers, we would warn the reader that it is only an approximation which can lead to erroneous conclusions when applied to certain "quasi-impulsive" interference (atmospherics at long wavelengths, car ignition sparks, etc.) (see § 7.5.2).

system with one degree of freedom—for example, a passive circuit—at absolute temperature T and in bandwidth B, a noise power

$$w = 4k \times T \times B \quad (k = \text{Boltzmann's constant})$$

or, in practical units,

$$w\,(\text{W}) = 4 \times 1{\cdot}38 \times 10^{-23} \times T\,(^\circ\text{K}) \times B\,(\text{c/s}). \qquad (7.1)$$

As the noise power is proportional to the bandwidth B, it is important to reduce B to a minimum; but this minimum value is set by the nature and speed of the traffic. For a given value, therefore, comparing noise powers becomes a matter of comparing noise temperatures.

Now, the theory of functions of random variables shows that noise powers arising from different sources are purely and simply additive: if the powers from two sources are w_1 and w_2, the resultant is the same as if there were a single source of power $w_1 + w_2$.

Thus if interference is analogous to thermal noise and if any source introduces into the circuit an additional noise power w_p (in addition to the inevitable thermal noise w which is already there), the effect can be represented by attributing to the circuit, instead of its real temperature T, a fictitious *noise temperature* $T' > T$, defined by

$$T' = \frac{w + w_p}{4kB} = T + \frac{w_p}{4kB}. \qquad (7.2)$$

This enables us to compare the effects of various types of noise by quoting the equivalent noise at some given point in the circuit, in the form of noise temperatures or factors bearing a simple relationship to these temperatures.

The point chosen for comparison is usually the input of the receiver.

7.2. Internal noise in the receiver

It is well-known that internal receiver noise arises first and foremost in the head amplifier and also in the later circuits and amplification stages (valves, transistors, etc.), which make a rapidly less significant contribution. If we take account of the gains of the successive stages, we can reduce all these noise sources to a single source at the input terminals: let this noise power be w_r.

In addition, the input circuit is coupled to the receiving aerial and receives from it, not only the signal, but also an amount of noise which is the sum of the aerial noise itself and any external noise captured by it.

The aerial noise, w_a (ignoring the external noise) is the noise corresponding to its temperature of equilibrium with its surroundings, that is to say, if it were enclosed in a perfectly efficient Faraday cage at temperature T, the aerial noise would be that given by eq. (7.1) for the temperature T.† If it is slightly directional and located near the ground, it can be assumed that the ground acts as "surroundings" at its mean temperature which will then be the noise temperature. If it is highly directional and pointing at a region of temperature T_1, the exchange with the other directions can be neglected and it can be assumed that the aerial noise is that corresponding to temperature T_1, i.e. $w_a = w_1$.

The above, we repeat, ignores external noise sources. But the latter will, in general, introduce additional noise w_p (possibly much greater than the intrinsic noise of the aerial) which will ultimately determine the total noise level in the aerial, or, to express it another way, the "equivalent noise temperature"

$$T_A = \frac{w_p + w_a}{4kB}.$$

Since the aerial behaves like a generator feeding the input circuit of the receiver, it will feed in (assuming perfect matching) the "usable" fraction (i.e. one quarter) of this power:

$$w_A\,(\mathrm{W}) = kT_A \times B = 1{\cdot}38 \times 10^{-23} \times T_A(^\circ\mathrm{K}) \times B\,(\mathrm{c/s}). \qquad (7.3)$$

This, then, is the power which is added to that of the intrinsic noise of the receiver w_r, and with which it will have to be compared.

† If the aerial is in thermal equilibrium with the surroundings, the thermal power received is equal to the power radiated by its "radiation resistance" at temperature T.

The ideal of the receiver designer is to reduce the intrinsic noise w_r to a value low compared with the probable minimum aerial noise w_A, so as to be able to use the weakest signals which can be resolved from the noise w_A.

Receiver sensitivities could thus be classified by giving their "internal noise power at the input", w_r. But this is not very convenient because this power is a function of the bandwidth B which depends on the tuning and type of traffic; it is therefore more impressive to make comparisons in terms of corresponding "noise temperatures" according to formula (7.1).

This temperature is usually above ambient temperature (about 300°K, say) since, not only are all the components of the receiver at this temperature, but also the valves and transistors introduce additional noise.

However, the recent invention of "parametric amplifiers" and "masers" has led to the design of preamplifier stages working at very low temperatures—almost down to absolute zero. With these, the intrinsic receiver noise temperature can drop to 30°K, 10°K or even lower. The total noise is then almost entirely aerial noise and if a highly directional aerial is used, on a site free from interference, and the beam is pointed towards different parts of the sky, certain "dark" or "cold" regions are discovered in which the noise temperature collected by the aerial is very low—even lower than the "reference" temperature—whereas in other areas the temperature is quite high: the narrower the exploratory beam, the better the "hot spots" can be located. In particular, looking at the Sun, one can distinguish the temperature of the centre from that of the edges.

The same apparatus can, of course, be used to receive very weak signals from space craft and the "darker" the background, the greater the range at which they can be detected: present records are of the order of 125 million miles.

7.2.1. "NOISE FACTOR" OF A RECEIVER

Instead of giving the "noise temperature" of a receiver, we can define a factor which bears a simple relation to it, in particular by comparison with a "reference temperature".

Before the days of radio-astronomy or satellites or maser amplifiers, it was thought that the minimum noise temperature of an aerial or a receiver could not be lower than the ambient temperature (around 300°K). A value of $T_0 = 288°K$ was therefore taken as the reference temperature, giving $k \times T_0 = 4 \times 10^{-21}$. Assuming the aerial to be at this temperature, the noise power into the input circuit is $w_{A0} = k \times T_0 \times B$, and if the intrinsic noise of the receiver is equivalent to w_r in the input circuit, the *noise factor* of the receiver is defined as the ratio of the total noise power $(w_{A0} + w_r)$ to the reference aerial value:

$$F = \frac{w_{A0} + w_r}{w_{A0}} = 1 + \frac{w_r}{w_{A0}},$$

whence

$$w_r = (F-1)\, kT_0 B \tag{7.4}$$

which can be expressed in decibels:

$$F\,(\text{dB}) = 10 \log \left(1 + \frac{w_r}{w_{A0}}\right). \tag{7.5}$$

On the initial assumptions, this factor was always well above unity—1·5–4 in the best decametric wave receivers, 10–40 at centimetric waves—and figures of this order described quite well the effective variation in sensitivity of the receiver.

The advent of "masers" and the importance attached to reception of signals from space craft and radio stars using highly directional aerials have produced systems in which the noise power fed in by the aerial w_A and the intrinsic noise of the receiver w_r are both considerably less than the power w_{A0} taken as the reference minimum.

Consequently, the noise factor F is seldom greater than unity, its logarithm a small fraction of a decibel and large variations in sensitivity now appear as quite paltry variations in F, which are not sufficiently convincing.

We therefore use a noise temperature which we shall define as

$$T_r = \frac{w_r}{kB} = (F-1)T_0 \text{ (°K)} \tag{7.6}$$

by analogy with the usable power due to external noise.

For example, if $w_r = 0{\cdot}14\ w_{A0}$, we obtain $T_r = 40$°K and $F = 1{\cdot}14$ (or 0·6 dB).

Let us suppose that we could manage to reduce w_r to a quarter of this value, which would be a remarkable achievement: the noise temperature falls to 10°K, which is highly indicative; but the noise factor becomes $F = 1{\cdot}035$, or 0·14 dB, which gives the deceptive impression of having gained less than half a decibel. (There is no point in discussing the relative merits of this definition compared with that of the reference temperature; the essential rule is always to use the same definition in making comparisons.)

Notes

1. If the temperature in the area towards which the aerial is pointing is not uniform, it will be necessary to introduce the variation in gain G and temperature T in each elementary cone of solid angle $\mathrm{d}\Omega$ and integrate over all space to obtain the equivalent temperature T_A of the aerial:

$$T_A = \frac{1}{4\pi}\int G(\Omega)\cdot T(\Omega)\cdot \mathrm{d}\Omega. \tag{7.7}$$

For the majority of discrete sources (radio stars, planets) the angular width θ as seen from the Earth is small compared with the angular beam-width β; formula (7.7) then gives an equivalent temperature T_A which is much lower than the true temperature T:

$$T_A = T \times \frac{\theta^2}{\beta^2}$$

and the galactic background noise is predominant.

On the other hand, if $\theta > \beta$, as is the case for the Sun, we have

$$T_A = T.$$

If there are several sources of noise acting on the aerial, formula (7.7) will have to be applied to each one (not forgetting the ground) and the resulting equivalent temperatures added together.

2. If a non-transparent medium, of absorption coefficient (power) equal to α and temperature T', is interposed between the aerial and the region being studied (temperature T_A), the working temperature must be taken as

$$T_A'' = (1-\alpha)\,T_A + \alpha T'. \tag{7.8}$$

Thus, at centimetre and millimetre wavelengths, it is necessary to take account of the passage through the troposphere, for which the mean temperature can be taken as of order 300°K.

Taking account of absorption by oxygen and water vapour, it is found that the absorption, and therefore the equivalent temperature, increase as the angle of elevation decreases, in other words, as the path length through the atmosphere increases.

Absorption by rain and clouds must also be taken into consideration if necessary; in periods of heavy rain the equivalent temperature of the sky, at centimetre and millimetre wavelengths, approaches 300°K. For an angle of elevation of the order of 10°, it has been observed that at a frequency of 6000 Mc/s the sky temperature increased from 3 to 70°K during rain and to 120°K during a storm.

Finally, it should be noted that absorption in the aerial "feeder", like tropospheric absorption, leads to an apparent increase in the aerial temperature:

$$\Delta T_A = \alpha T_f,$$

where α is the absorption coefficient and T_f is the actual temperature of the feeder. For example, at room temperature (T_f = 300°K), each tenth of a decibel of absorption raises the temperature by 7°K.

3. If the source of external noise is not known by its temperature but by the field E_p which it produces in the aerial (as in the case of atmospherics), we obtain the captured power and the usable power as a function of the parameters of the aerial, just as if it were activated by a signal (§ 1.1.2); in particular, we have

the "usable power" applied to the receiver by formulae (1.17), (1.19) and (1.21), depending on whether the aerial is a small doublet, a half-wave dipole or a directional aerial of known gain G_i.

For example, with a small doublet connected to earth (and whose radiation resistance is therefore doubled) we obtain

$$w_A\,(\text{W}) = \left[\frac{300}{f\,(\text{Mc/s})}\right]^2 \frac{E_p^2\,(\text{V/m/kc/s}) \times B\,(\text{kc/s})}{4 \times 1580} = 14{\cdot}25\,\frac{E_p^2 \times B}{f^2} \tag{7.9}$$

and, in the evaluation of the noise factor, the bandwidth B vanishes so that we are left with the "noise factor due to the interference level E_p":

$$F_p = \frac{14{\cdot}25}{4 \times 10^{-18}} \times \frac{E_p^2\,(\text{V/m/kc/s})}{f^2\,(\text{Mc/s})}\,. \tag{7.10}$$

If E_p is expressed (as sometimes happens) in "decibels above 1 μV/m" (for a 1 kc/s bandwidth), we can write

$$F_p\,(\text{dB}) = E_p\,(\text{dB}) - 20 \log f\,(\text{Mc/s}) + 65{\cdot}5. \tag{7.11}$$

It will be observed that, in this case, the relation between the field strength E_p and the noise factor F_p [and therefore the noise temperature T_r, from (7.6)] contains the term f^2; this is because, for a doublet of effective height h, the e.m.f. picked up in the field E_p is $h \times E_p$, independent of frequency; but when we are considering power we have to take account of the radiation resistance which is itself proportional to $(h/\lambda)^2$. If the doublet is used in conjunction with a reflector of area S, the gain is proportional to S/λ^2 and the power picked up is now independent of frequency.

In spite of this inconvenience, it is often desirable, when comparing thermal noise with atmospheric and industrial interference (usually measured in terms of their field strength), to draw curves of noise level against field strength. Thus, in Fig. 7.1, curves 3 and 3′ give the "thermal noise field strength" for total noise temperatures, $\theta = T_A + T_r$, of 300°K and 10°K respectively; the first would correspond, for example, to the case of an omnidirectional aerial near the ground, or a directional aerial pointing towards the ground, and a receiver with negligible internal noise, i.e. $T_A = 300°K$ and $T_r \simeq 0$; the second would correspond to the limiting case of a directional aerial pointing towards the black regions of the sky and, again, a receiver with negligible internal noise, i.e. $T_A = 10°$, $T_r \simeq 0$; we shall see in § 7.3 that we can

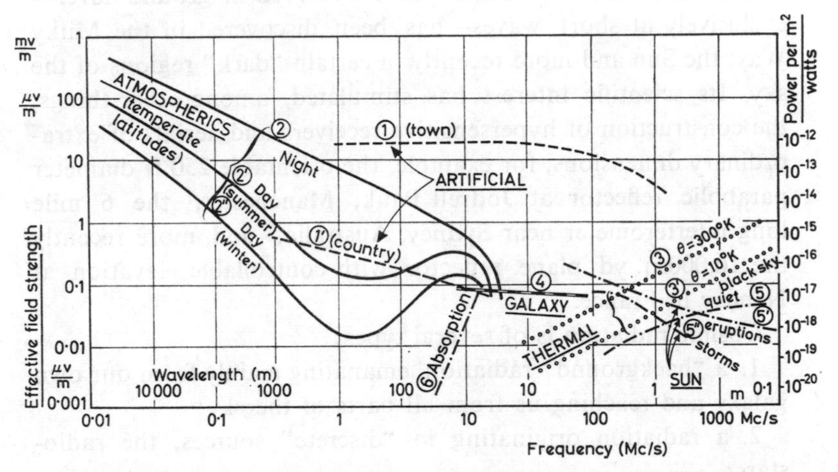

Fig. 7.1. Comparison of noise levels (approximate) in a 1 kc/s bandwidth

approach these conditions with frequencies of the order of gigacycles per second.

4. Finally, it may also happen that the interference level is not measured directly but in terms of the "minimum signal required to dominate it".

In this case it is obviously necessary to subtract the number of decibels corresponding to the required signal/noise ratio (usually 10–30 dB) from this "minimum level".

Conversely, if, after measuring or calculating the interference level, it is desired to derive the minimum usable signal level, it will be necessary to remember to add the signal/noise ratio.

7.3. Extra-terrestrial interference. Noise radiated by the atmosphere

The "sky" can be characterized, according to the observation direction, by an "equivalent temperature" which takes account of sources of cosmic radiation and, at some frequencies, absorption due to passage through the lower atmosphere.

Extra-terrestrial or cosmic noise observed at ground level—exclusively at short waves—has been discovered in the Milky Way, the Sun and more recently in certain "dark" regions of the sky. Its scientific interest has stimulated, among other things, the construction of hypersensitive receivers and aerials of extraordinary dimensions, for example, the orientable 250 ft diameter parabolic reflector at Jodrell Bank, Manchester,† the 6 mile long interferometer near Sydney, Australia,‡ and, more recently the 10,000sq. yd plane reflector with controllable elevation at Nançay in France.

Cosmic radiation is of several types:||

1. a "background" radiation, emanating mainly from our own galaxy and reaching us from all parts of the sky;
2. a radiation originating in "discrete" sources, the radio-stars;
3. a radiation originating in the bodies of the solar system, the Sun, Moon and planets: Jupiter, Venus, Mars, etc.

The equivalent temperature of the galactic background noise level may be determined from curve *B* of Fig. 7.8 for frequencies below 100 Mc/s and from Fig. 7.2 for higher frequencies.

In the former case, the parameter given by the curves is an "equivalent noise factor" F_{am} (in decibels) which is converted into a power ratio to obtain the equivalent temperature by the simple formula

$$T = F\,(\text{power ratio}) \times 288°\text{K}.$$

† See *Wir. World*, June 1952, p. 238.

‡ See Australian report to U.R.S.I., 1952.

|| For information on extra-terrestrial noise and radio-astronomy see Nicolet, *Bruits solaires et cosmiques*, Brussels, 1949; Lowel, *Radioastronomy*, New York, 1952; Proceedings of Commission V of the U.R.S.I. and their summary; Laffineur, *Onde Électrique*, July, 1958, pp. 539–44; *Proc. I.R.E.* (special issue), Jan. 1958.

(This formula does not allow for absorption due to passage through the ionosphere at frequencies below 30 Mc/s; the indicated values therefore correspond, in this range, to an upper limit of galactic noise.)

In Fig. 7.2, for frequencies above 100 Mc/s, in order to allow for the fact that the distribution of radiation throughout the sky

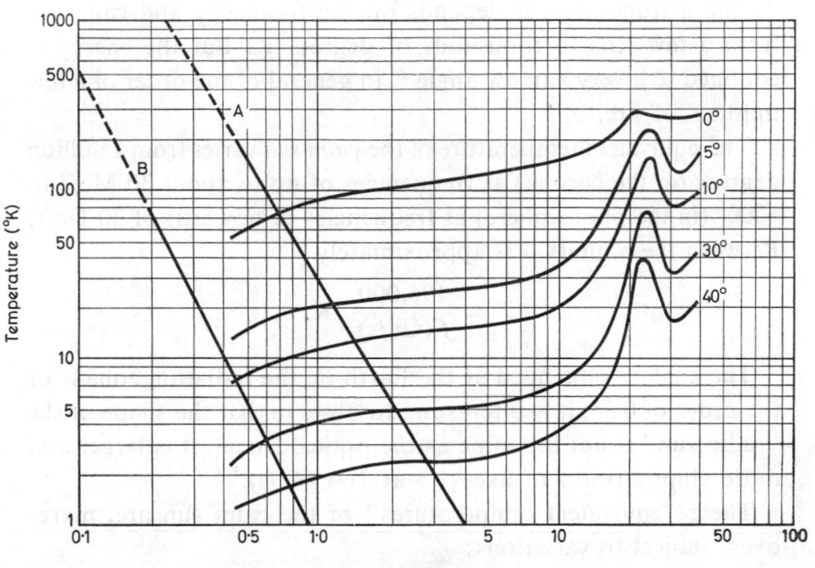

FIG. 7.2. Equivalent temperatures: *left*, galactic noise level (A = maximum, B = minimum); *right*, atmospheric radiation level

is not uniform, two curves, A and B, are shown, corresponding to the two "limiting cases": one in which the receiving aerial is oriented towards the galactic noise *maximum* (near the centre of the galaxy) and the other at the noise minimum.

Curve 4 of Fig. 7.1 can also give an order of magnitude for galactic noise.

The known "radio stars" fall into several categories:

(a) Supernovae which have exploded, leaving gaseous residues (the Crab nebula, exploded in 1054 B.C.: Cassiopeia).

(b) Nebulae located in our Galaxy.

(c) Distant galaxies which radiate like our own (Archimedes' Spiral).

(d) Interaction of two colliding galaxies [the radio-star Cygnus which radiates a total power of 4×10^{35} W and whose radiation at a distance of 10 million light-years reaches us with an intensity greater than that of the Sun (at metric waves)].

The equivalent temperature of these radio-stars is determined by the astronomers; it depends on the frequency and can vary from a few tens to thousands of degrees K. But the energy is confined to a very narrow angle θ, in general of the order of a few minutes of arc.

The equivalent temperature of the *calm sun* varies from 1 million degrees (in the corona) at frequencies of up to about 30 Mc/s to 6000° (in the photosphere) at frequencies of upwards of 30 Gc/s. Between these limits it is approximately

$$T = \frac{196{,}000}{f\,(\text{Gc/s})}\ ^\circ\text{K}.$$

The angle θ subtended at the Earth by the radiation zone is of the order of 0·5°; it is interesting to observe that the shape of the "radio sun" is not the same as the optical shape: it is larger and more elliptic (ratio of axes 1·5 at 160 Mc/s).†

These "equivalent temperatures" of the calm sun are, moreover, subject to variations:

(a) During periods of intense solar activity, the temperature of the Sun can vary slowly, over a period of weeks or months, rising to values twice or three times greater than those indicated above—over the whole range from 500 Mc/s to 10 Gc/s—in correlation with the total area of sunspots covering the surface.

(b) During chromospheric eruptions which are accompanied by ejection of particles, *bursts* of solar noise are observed, particularly at metric wavelengths (20–300 Mc/s) where the peaks can be as much as several hundred times higher than normal. These bursts, which usually last a few seconds each, follow each other in sequences which may last anything from a fraction of a minute

† Blum *et al.*, *C.R.Ac.Sc.*, **233**, pp. 917–19; **234**, pp. 1597–9; Kraus, Ksiasek, *Electronics*, Sept. 1953, pp. 148–52.

to several hours; they constitute a warning of ionospheric storms to come.

Eruptions which cause severe storms and "proton events" give rise to bursts of a particular type ("type IV" in radio-astronomy terms) which are very intense and last several hours.

(c) Another type of solar noise storm which is sometimes observed in one which is not correlated with flares and lasts for some minutes, covering a very wide spectrum (20 Mc/s to 30 Gc/s) and having an equivalent temperature which reaches extraordinarily high values: 10^{10}–10^{13}°K.

Curve 5 of Fig. 7.1 gives the order of magnitude of the noise *field strength* of the calm sun as a function of frequency [it increases with f, whereas the noise temperature decreases due to the f^2 term in eqn. (7.9)].

Curves 5′ and 5″, on the other hand, correspond to the case of the sun perturbed by more or less intense chromospheric eruptions.

The interference level from the *planets* is quite appreciable: the noise temperature of Jupiter is of the order of 50,000°K at 400 Mc/s, 5000°K at 1 Gc/s, 180°K at 10 Gc/s.

Those of Venus, Mars, Saturn and the Moon are of the order of 600, 200, 100 and 200–300°K, respectively.

The general conclusion regarding noise of celestial origin is that at frequencies below 500 or 1000 Mc/s galactic background noise is dominant—except for highly directional aerials pointing towards the Sun or an intense radio-star. At higher frequencies, tropospheric noise is dominant: it is therefore only at angles of elevation greater than 5° that the "equivalent temperature of the sky" will be as low as 20 to 30°K for frequencies between 1000 and 5000–6000 Mc/s.

7.4. Artificial interference

Artificial interference is a nuisance not only in broadcasting but in all professional traffic, since the location of the receiver cannot always be chosen to be far from any source of perturbations.

It arises simply in the exponential decay régime which

exists in all circuits or parts of circuits in which a change of current occurs (switching off, change of direction, etc.). For example:

(a) any motor with a commutator in which the passage of a segment over a brush corresponds to a reversal of the current in one section of the winding, causing some degree of commutator sparking;

(b) all *switches*, contactors, commutators, relays, etc., where a current is broken and re-established, especially in highly self-inductive circuits;

(c) the electrical ignition parts of internal combustion motors where we find the elements of a small metre-wave spark transmitter (high tension magneto; sparking plug; low inductance and capacity resonant circuit);

(d) *luminescent neon, mercury, etc., tubes* (luminous signs, etc.) whose operation consists of more or less regular electronic discharges;

(e) various types of *medical apparatus* (diathermy, X-rays, etc.) combining several of the above elements.

The trouble which these perturbators cause is generally aggravated by the fact that there is a common mains supply line between them and the receiver, or that their supply lines, although separate, are strongly coupled (e.g. the case of a tramway or trolley-bus line and the mains supply to nearby houses); interference is propagated along these lines with only a small attenuation.

The principal objective in studying and measuring artificial interference is obviously its suppression, either at the source itself or during propagation or in the receiver.

At the source. The cause of the perturbations can sometimes be removed by changing the principle of the apparatus: for example, a commutator motor can be replaced by a squirrel-cage asynchronous motor, a bell with a vibrator by a buzzer worked by an a.c. supply, diathermal flash apparatus by a type using lamps, etc.

Sometimes, one can also modify the electrical constants of the offending circuit in such a way as to make its decay régime less troublesome: for example, by increasing its decay time (say, by putting a resistance of the order of 5000 ohms in series with the sparking plugs) or by changing its characteristic frequency† (putting a condenser of the order of 0·1 μF across the terminals of switches or commutators).

If it is impracticable to suppress the interference, one can *prevent* its being propagated.

For instance, direct propagation (by radiation) can be prevented by means of metallic screens, either solid or in the form of a

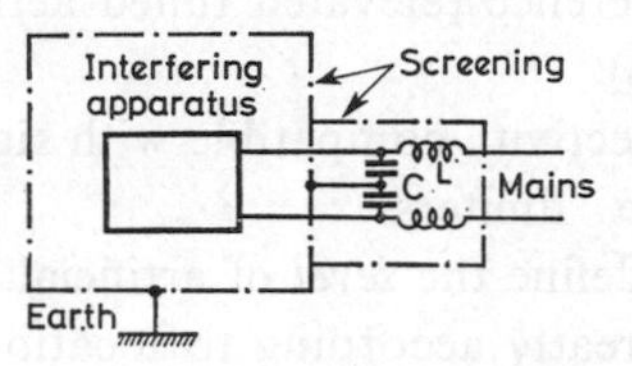

FIG. 7.3. A method of protection against artificial interference

very close mesh: the classical solution in the case of aircraft engine ignition, for example, but only effective if there is complete suppression.

Propagation along the mains line can be eliminated by interposing a "low-pass filter": inductance in series, capacitance in parallel. Theoretically, any desired degree of attenuation can be achieved by increasing the magnitude of the components or the number of successive units. But in practice, one is often limited by considerations of size and cost (as regards the construction of inductances which will carry the high currents used, for instance, on tramways) or the safety of operators (maximum imposed on the capacitances in domestic apparatus with no earth connection). This solution (whose efficiency, in spite of all the promises, can never be strictly guaranteed) is very widespread and most of the "anti-interference devices" on the market are very simple combinations of this type.

The system is shown schematically in Fig. 7.3; the interference

† This artifice, although advantageous for reception in a narrow band, is sometimes illusory when one wants to receive on a very wide band, for it only serves to shift the trouble somewhere else and the interference, removed from one part of the spectrum, reappears in another.

level can normally be reduced by 40–60 dB: it is difficult to do better.

Finally, reasonable steps should be taken in the receivers themselves to reduce the possible effect of artificial interference; we list them below.

1. Maximum possible decoupling (at least 30 dB) between the receiver circuitry and the power unit (by filters, screens, screened transformers, etc.).

2. Use of aerials to pick up the maximum of signal and the minimum of interference (elevated tuned aerials, with screened leads; frame aerials).

3. Maximum selectivity compatible with signal distortion.

4. Connection to "limiters".

It is difficult to define the *level* of artificial interference, which obviously varies greatly according to location.

In principle, it is negligible *in the country*; but it must be remembered that power line pick-up can be appreciable at a hundred yards; with an elevated aerial, interference due to a car, motorcycle or tractor a mile away can cause trouble. Consequently, in the absence of special precautions and aids around the receiver, the level seldom drops below that indicated by curve 1′ in Fig. 7.1 (according to Report 65 of the C.C.I.R.).

In towns, the artificial interference level is naturally much higher and very variable.†

† A bibliography of some recent papers on artificial interference includes:

New French legislation: *Journal Officiel*, 26 June 1951, pp. 6638–50 and *Toute la Radio*, Oct. 1951, pp. 271–4.

British measurements and statistics: Whitehead, *J.I.E.E.*, III, Dec. 1943, and Aug. 1944, pp. 324–30; Maurice *et al.*, *Wir. Eng.*, Jan. 1950, 27, 316, pp. 2–12; Ball, *Proc. I.E.E.* part B, May 1961, pp. 273–8.

Interference due to high voltage lines: Cahen, Pelissier *et al.*, *Bull. Soc. Française des Electriciens*, Dec. 1949, pp. 693–709, and the entire issue of July 1953.

General methods and devices for eliminating industrial interference: Veyssière, Tech. note 1087, Centre National d'Etudes des Télécommunications, 1948.

Spark interference: *Wir. World*, Jan. 1949, pp. 31–6; August 1959, pp. 251–5; April 1953, pp. 189–90.

Interference suppression in motor vehicles and electrogenic groups: *Manuel Technique des Transmissions* (Guerre), no. 129, 1948.

In certain cases it can even affect the reception of relatively very strong signals, for example 10 mV/m: this is the basis of broadcasting claims.

However, in the course of some work carried out in France to establish the first legislation (1933) on the subject, it was found that a large amount of urban interference could be reduced by the use of appropriate protective devices to such an extent that it would hardly affect field strengths of 1 mV/m, i.e. about 26 dB down. We have indicated this in curve 1 of Fig. 7.1 (normalized to a bandwidth of 1 kc/s).

The above applies to the normal range of amplitude-modulated sound broadcasting. The advent of frequency modulation and television at metric wavelengths has necessitated the extension of the curve towards higher frequencies; here, most of the sources of interference have certainly disappeared; but one very important one remains: motor-vehicle ignition systems (which, of course, are rapidly increasing in number). As long as this source is not eliminated by effective legislation, it is a fact that the average interference level does not fall appreciably with wavelength: perhaps by 10–20 dB between 50 and 500 Mc/s, i.e. 6 m and 60 cm (and it must be remembered that in television, the bandwidth is multiplied by a factor of the order of 1000, i.e. the sensitivity to noise is increased by 30 dB).

7.5. Atmospherics

Atmospherics are even more difficult to study than artificial interference, for several reasons.

Firstly, their source is beyond our reach and we cannot switch them on and off at will.

Next, their exact nature is almost undetectable. Although the

---

General conditions for suppression in aircraft: Direction technique et industrielle de l'Air, doc. 2025 MC, 1 July 1949.

Report of the Special International Committee on Radiophonic Perturbations (C.I.S.P.R.), London, 1953; *Onde Électrique*, June 1954, pp. 536–8.

Radio Interference Suppression Techniques (U.S. Dept. of Commerce, P.B. 11, 611); *I.R.E. Trans.*, Vol. CS–3, no. 1, March 1955, pp. 8–13.

slowest and most powerful of them can be recorded directly by an oscilloscope on an aerial, on the other hand, most of them, the fastest—and therefore the most troublesome—escape from this analysis and are only recognized by their effects on receivers, i.e. through the complicated phenomena of tuning in the transition regime, amplification, non-linear detection, etc., which are always tricky to interpret.

Thirdly, atmospherics reaching our aerials have, since their origination, described a more or less lengthy path, with absorption, reflection by the ground and the ionosphere, etc., all of which have reacted on their amplitude and form; to the uncertainties regarding their origin, therefore, we must add all those concerned in their propagation.

This complexity leads to two series of investigations with different mental approaches:

We can try to localize the interference and analyse each occurrence, considered as a scientific phenomenon capable of improving our knowledge of geophysics, meteorology, storms, wave propagation, and so on; atmospherics are thus permanent, gratuitous emissions, notably at frequencies below 15 kc/s not used by radio stations. We can seek to pick them up for their own sakes, without being troubled by the reception of any signal. We shall say no more about these investigations here.†

The other point of view, is naturally, that of the radio engineer, for whom atmospherics are enemies, annoying perturbators which limit the sensitivity and range of this receiver. He is not therefore interested in their actual form but simply wants to define the signal level which he can receive in spite of them with a definite proportion of errors. As this level varies, not only with the type of signal envisaged, but with time, place, frequency and state of the equipment, the establishment of forecasts of "atmospheric noise" is, and will be, a long and tedious enterprise.

† See the reports of the International Radio Scientific Union, the Preliminary Notes of the Laboratoire National de Radioélectricité, as well as articles by Gardner, *Phil. Mag.*, Dec. 1950, pp. 1259–69; Bowe, *Phil. Mag.*, Feb. 1951, pp. 121–38; Rivault, *Onde Électrique*, July 1958, pp. 527–31; Foldès, *loc. cit.*, pp. 533–8.

Here we shall merely outline the essential features of the task. Let us first mention an international agreement on:

(a) the establishment of a primary "normalized" measuring equipment:† vertical 6·5 m aerial, 50 m feeder, 10 kc/s bandwith receiver, standard generator with automatic manipulator; a subjective observation by trained operators enables an estimate to be made of the e.m.f. in the aerial (and therefore the field strength) necessary to read a Morse message at slow speed (10 words/min) with 5% error;

(b) the establishment of a world network of a score of stations equipped in this way (see Fig. 7.5);

(c) the choice of normalized observation frequencies: 2·5, 5, 10 and 20 Mc/s, with eventual extensions below 1 Mc/s;

(d) the publication of data thus collected;

(e) various refinements, such as the gradual replacement by an automatic, objective measuring equipment, and the determination of the "parameters" of the noise field which ought to be measured in order to make possible the calculation of the interference with the various types of radiocommunication.‡

7.5.1. Form and duration. Microstructure. Variation with frequency

Here now are some of the principal results obtained.

First, of all, the shape of atmospherics and what is sometimes called their "microstructure": they are aperiodic or very rapidly decaying perturbations, rarely consisting of more than a few oscillations (Fig. 7.4). A minute analysis of these forms reveals various possible mechanisms of production and propagation.||

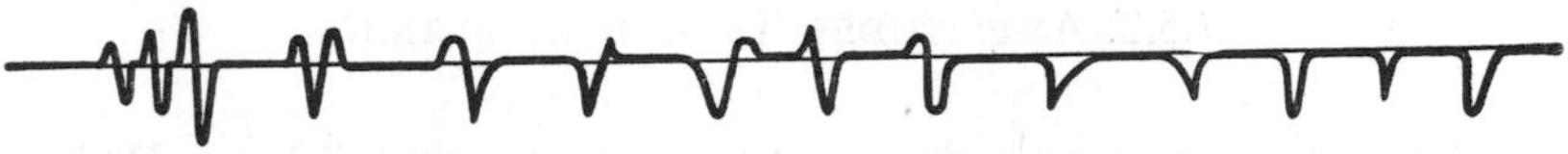

Fig. 7.4. The shape of atmospheric noise

† Thomas, *J. I.E.E.*, III, 97, Sept. 1950, pp. 329–34.

‡ Crichlow, *Proc. I.R.E.*, June 1957, pp. 778–82.

|| For example, Rivault, Notes prélim. Lab. Nat. Radio., no. 65, 1945; no. 32, 1942; *Onde Électrique*, June 1955, pp. 593–7.

The *duration* of the perturbations is extremely variable. It could be said that it was "randomly distributed", or that, on the graph which represents them, "the time scale is immaterial" (it would certainly be correct to say that all time scales are equally probable). This amounts to saying, in other words, that the "frequency spectrum" is, on average, uniform, which is the characteristic of "thermal" or "white" noise. This approximation is often used, although it is very rough, as we shall see.

In fact, if we consider extended intervals of frequency—and finally, the whole of the range occupied by radiocommunications—we see that the level of atmospherics decreases as frequency increases; this decrease can vary according to time and place, but it is incontestable, and radio engineers have found by long experience that atmospherics, although very troublesome at long waves, become rather less so at short waves and are practically exceptional at metric wavelengths and below. In France, for example, a decrease in noise level proportional to λ or $\lambda^{0.8}$ has been found, according to curves analogous to those of Figure 7.1, showing a regular decrease for the combined effect of night atmospherics (midnight) (curve 2), and others, less regular, with a minimum at about $f = 2$ Mc/s, for daytime atmospherics with the seasonal variation (curves 2′ and 2″).

These curves (mean of various authors) are very approximate and their object is simply to allow a comparison of orders of magnitude to be made with the other sources. Note that they represent the level of the interference itself and not that of the signal required to dominate it, which would, in most cases, be 10–30 dB higher.

7.5.2. Amplitudes. Their distribution

The above gave us a rough idea of the "average" level. How do the amplitudes vary on either side of this level? Certainly within wide limits.

A storm discharge releases colossal instantaneous power; at short distances, it certainly induces in our aerials voltages far exceeding any due to signals (this is what runs amplifier valves into their non-linear regimes and justifies the term "limiter"). But such power levels are exceptional: the "probability" of such amplitudes is low. Perturbations originating farther away are at the same time weaker and more numerous. We cannot therefore avoid introducing here the notion of "amplitude probability", defined in § 1.2.1.

It is important to define the law of this distribution: firstly, in order to relate the error probability to the signal level, assumed constant, and therefore to the emitter power, which will allow us to calculate the "net cost" of an increased safety factor in the link; and secondly, in order to extend these conclusions to the very general case in which the signal is not itself constant and its distribution probability has to be combined with that of the interference in order to determine the proportion of errors or the fraction of time during which the link will be correct.

Here again, a simple initial hypothesis is very tempting: namely, that the variation follows the Rayleigh or Gaussian laws (§ 1.2.1); the calculations are then very easy. For example, with thermal noise (Rayleigh's law) (Fig. 7.5), an increase in the field strength of the order of 10 decibels would be sufficient to reduce the percentage interference from 50% (median value), which is certainly unacceptable, to 0·1%, which is satisfactory for some traffic; with 2 or 3 dB more, it would fall to 0·01% or even 0·001%, which would satisfy the most exacting requirements.

Unfortunately, this approximation is optimistic; it was proved long ago† that far greater increases in the field strength do not produce such a decrease in the number of errors due to interference. It is this that makes the task of the radio engineer so difficult. At the 1959 conference of the C.C.I.R.,‡ the pooling of the results

† For example, according to Bellescize (*Les Communications radioélectriques*, 1935, pt. II, p. 191, Fig. 6) the probability that the signal is dominant goes from 10 to 50% for an increase in field strength of 25 dB (at short waves); according to Carbenay (note *Ac. Sc.*, 11 Aug. 1952) the proportion of errors goes from 2 to 15% for a change of 7 dB in the interference field strength (long waves).

‡ C.C.I.R. Los Angeles, 1959, Comm. II, Report 99. See also Linfield, Samson, *Proc. I.R.E.*, Aug. 1962, 50, 8, pp. 1841–2.

of several workers led to the curve of Fig. 7.5; it can be seen that in order to go from an error probability of 50% (median value) to a probability of 1%, an increase in the field of about 20 dB is necessary; over 30 dB is required for a reduction to 0·01%. Thus although Rayleigh's law may be approximately valid

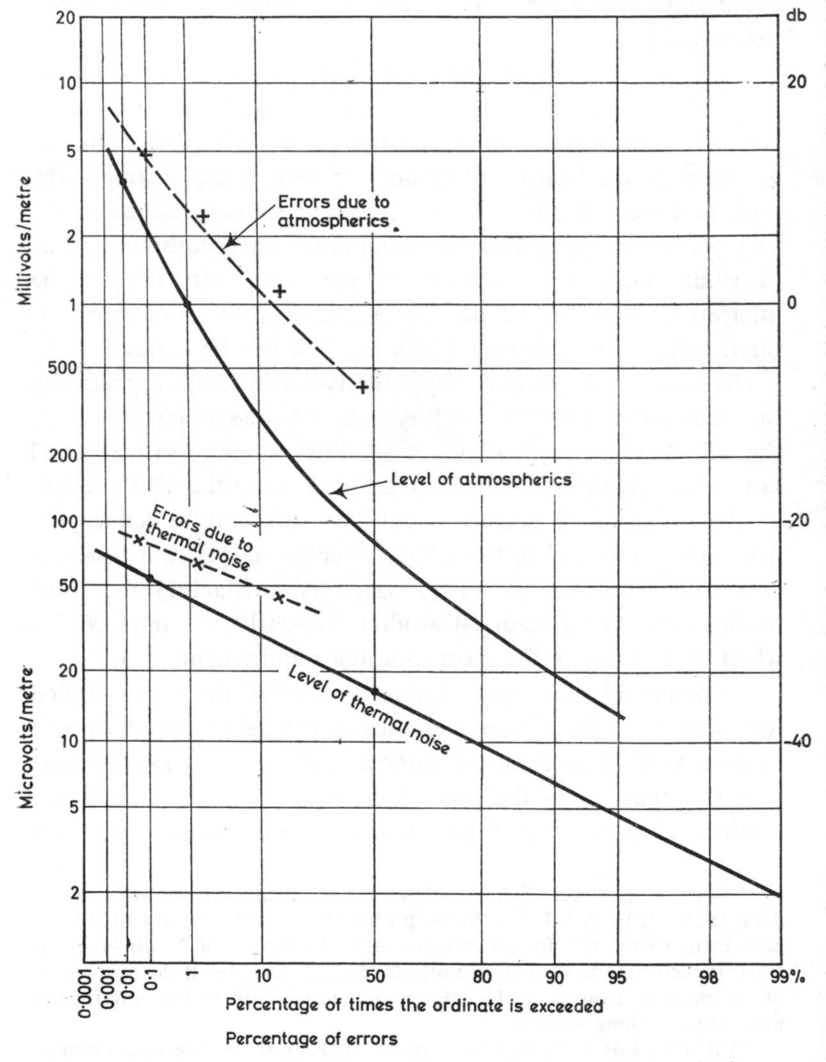

FIG. 7.5. Amplitude distributions of thermal noise and atmospherics

for natural interference of low level, and therefore high probability, it is on the other hand, completely false for interference of very high level which occurs only rarely or for short intervals of time; the probability of these very high levels decreases much more slowly (one might say that the "dynamics" of atmospheric noise is much greater than that of thermal noise). Thus if it is relatively easy to obtain a mediocre link with a high proportion of errors, it may be very costly, and sometimes even practically impossible, to bring the error rate down to the very small values (less than 0·01%) desired at present.

Another result of this difference between atmospheric noise and thermal noise is a difference in the behaviour of tuned systems and the effect of their bandwidth B. The energy in the system is always proportional to B, and therefore the *effective* voltage is proportional to $B^{0\cdot 5}$; but the *mean* voltage often increases with B more slowly than this; estimates from $B^{0\cdot 25}$ to $B^{0\cdot 42}$ are currently to be found. In receivers with a *limiter*, there may even be some point in increasing B in order to reduce the duration of the perturbation caused by short bursts of interference.

It should be noted that these considerations affect long waves in particular; at short waves, they scarcely apply except as regards fading, because of the decrease in the level of atmospherics and the possibility of realizing strong signals when propagation is favourable.

7.5.3. Localization. Atmospheric noise maps

Finally, the atmospheric noise level varies considerably from point to point of the globe. This is because atmospherics are due to storm manifestations which mostly arise in certain "preferred" regions and are then shifted by winds, hence the influence of mountains, coastlines, etc. It has been noticed over a number of years that the level of atmospherics increases significantly from the poles to the equator, but with violent irregularities. The recent installation of a network of measuring stations has enabled this distribution to be defined, as well as the diurnal and seasonal variations, in the form of "atmospheric noise maps", of which Figs. 7.6 and 7.7 are two specimens.

The curves are graduated in "atmospheric noise factors" (in decibels).

The production of these maps represented a huge task. A first edition had been published by the C.C.I.R. in 1956 (Report No. 65); at the next conference (1959), Recommendation No. 315 prescribed the "provisional use" of these curves "with a certain amount of prudence". A second, completely revised, edition was given in Report No. 322, 1963; it contains series of maps analogous to those of Figs. 7.6 and 7.7 for each season and each of the six 4-hour subdivisions of the day (in local time) and at a frequency of 1 Mc/s; the collection is completed by two other series, one of which is used for converting from 1 Mc/s to any other frequency in the range 10 kc/s to 20 Mc/s, approximately (an example is shown in Fig. 7.8); the other enables the statistical distribution of levels around the median value given by the curves to be determined.

This collection was deemed by the C.C.I.R. to be too big to be included in the normal published proceedings and forms the subject of a special issue;† for the same reason we cannot present it here.

A glance at Figs. 7.6 and 7.7 suffices to indicate the large variation in noise level with place, time of day and time of year: for example, in the morning, in winter, the noise level is low over the whole northern hemisphere (20–30 dB), whereas it is a maximum (80 dB) in S.E. Africa; by contrast, during the summer nights, it is Senegal and the Gulf of Mexico that have the maximum (100 dB level) and in southern Europe the level is 70–80 dB.

† World distribution and characteristics of atmospheric radio noise, Report No. 322 of the C.C.I.R., U.T.I., Geneva, 1964.

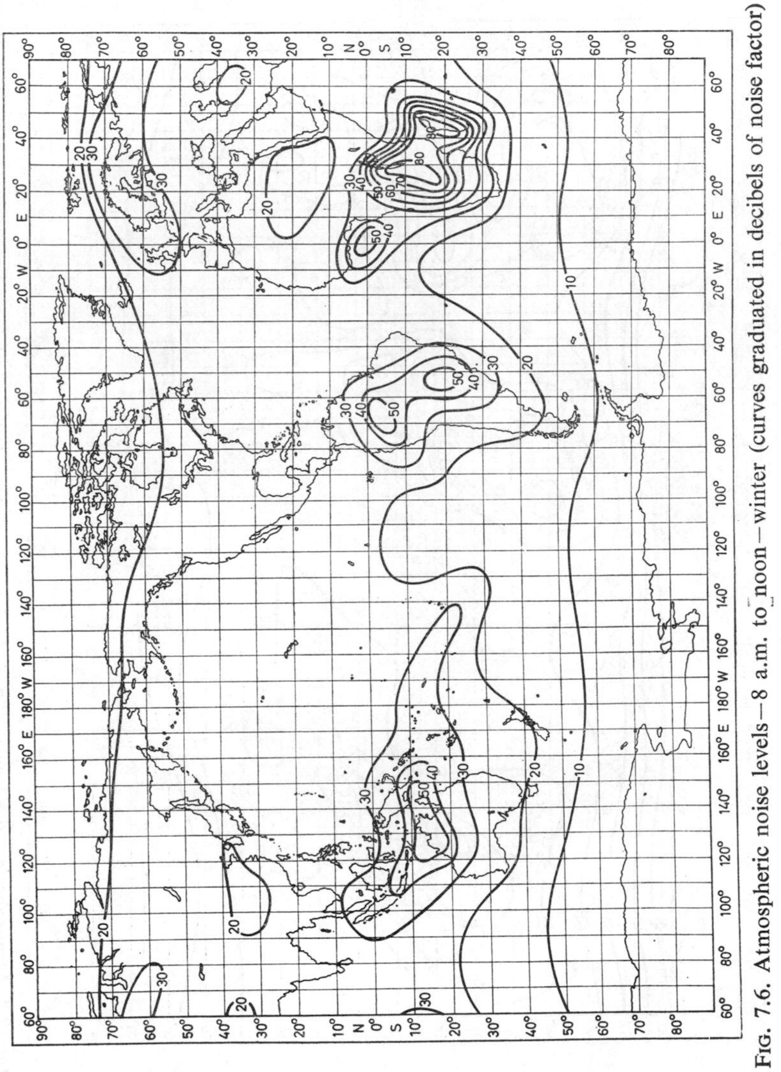

FIG. 7.6. Atmospheric noise levels—8 a.m. to noon—winter (curves graduated in decibels of noise factor)

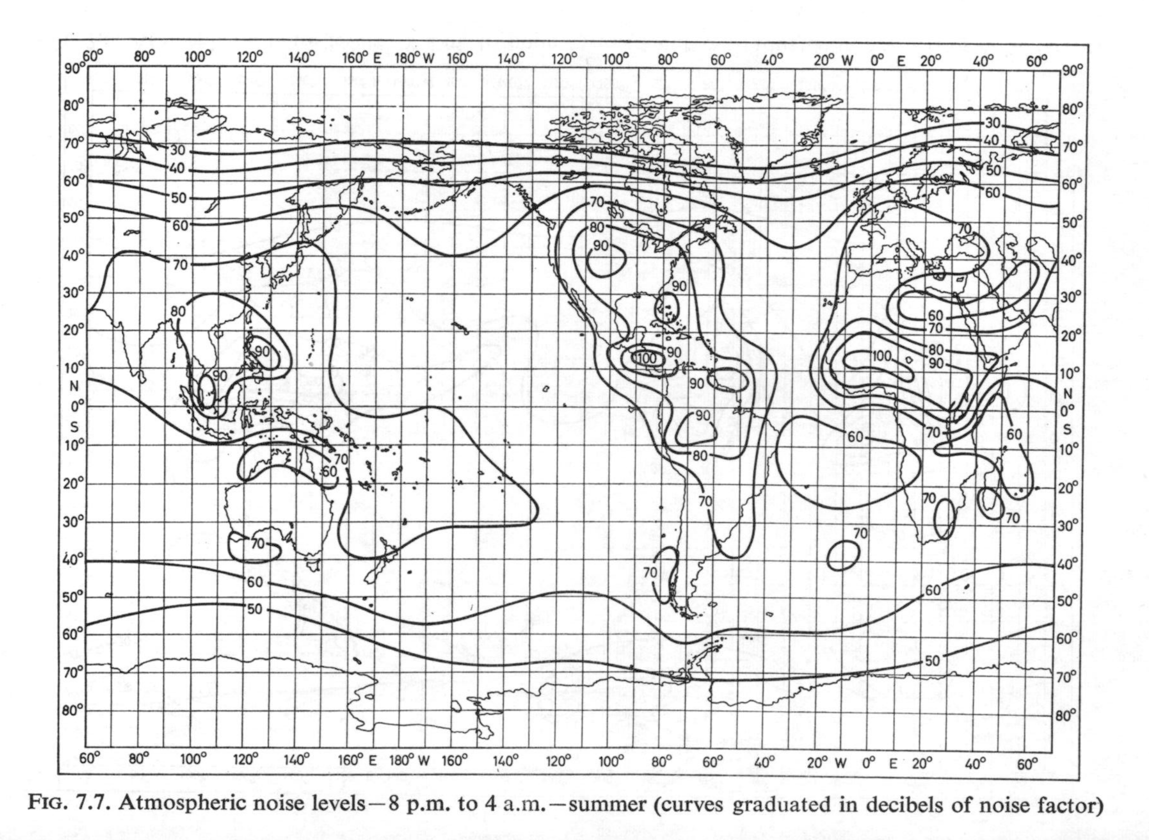

FIG. 7.7. Atmospheric noise levels—8 p.m. to 4 a.m.—summer (curves graduated in decibels of noise factor)

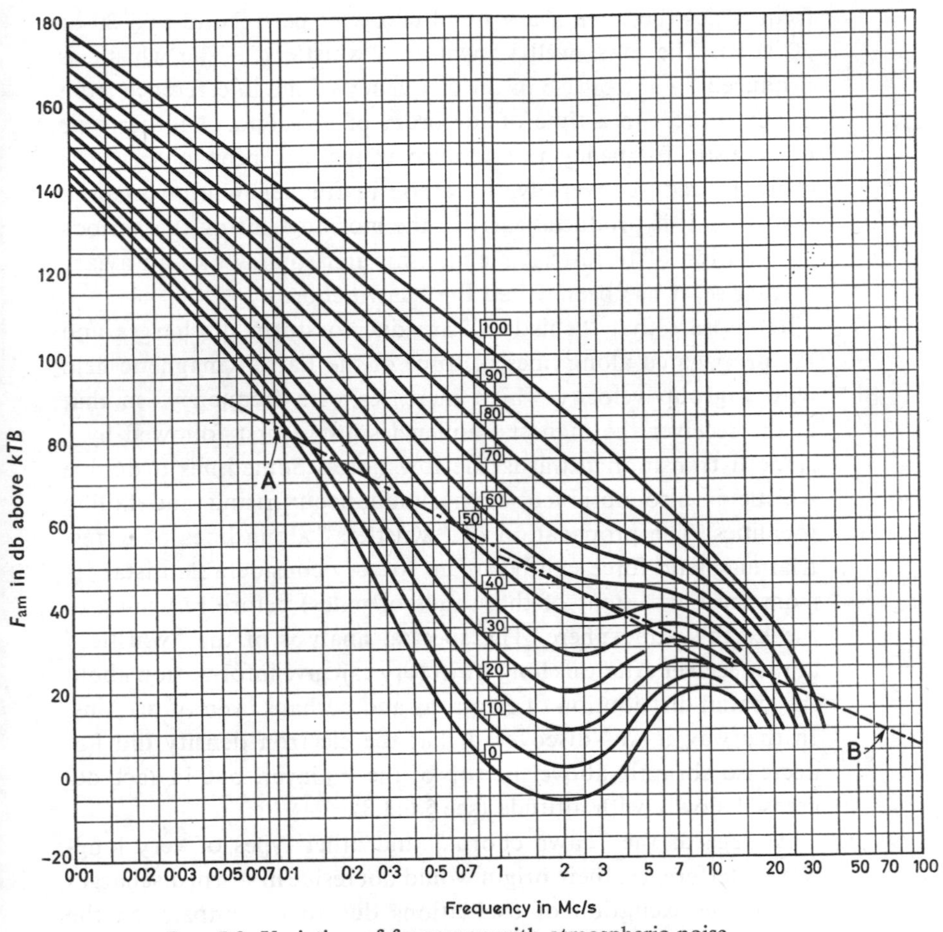

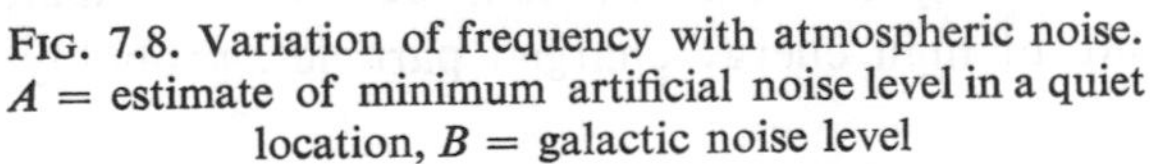

FIG. 7.8. Variation of frequency with atmospheric noise. A = estimate of minimum artificial noise level in a quiet location, B = galactic noise level

7.6. Other types of interference†

We must point out the discovery quite recently of a type of interference observed at very long wavelengths, at acoustic

† See, for example, Dinger, *Onde Électrique*, May 1957, pp. 526–34; Rivault, *loc. cit.*, pp. 539–40; *Journal U.T.I.*, No. 3, Mar. 1958, and *Bulletin U.R.S.I.*, No. 108, Mar.–April 1958, pp. 22–4; several papers in *Proc. I.R.E.*, Feb. 1959, pp. 200–9, 211–32, 328.

frequencies (between a few hundred cycles per second and 20 or 30 kc/s). The most well known are "whistlers". The whistle is produced by a decrease in the frequency of the interference from about 10 kc/s in a time of the order of a second. Another type of acoustic frequency interference which has been observed is characterized, conversely, by a rise in frequency above about 2 kc/s; as it is produced early in the morning during the periods of agitation of terrestrial magnetism, and sounds like an aviary at sunrise, it has been called the "dawn chorus".

It appears that "whistlers" are due to storm discharges and are propagated along lines of force of the Earth's magnetic field with a group velocity which increases with frequency, so that at the receiver the highest frequencies arrive first followed by a smooth transition towards the low notes. Sometimes a double whistle is also observed, with simultaneous rising and falling tonalities. The propagation of "whistlers" along lines of terrestrial magnetic force can occur up to very considerable distances from the Earth (tens of thousands of miles) before returning to the Earth's atmosphere. The spectral analysis of the "whistler" at the receiver after this long trajectory can give information about the spatial distribution of electrons and perhaps even of protons. In this way it was discovered that the electron density did not decrease abruptly above the top of the F_2 layer, but in fact decreased slowly with altitude (see § 6.3.8).

As regards the "dawn chorus" and other types of very long-wave interference, their origin would not reside in storm discharges but in the excitation of oscillations due to the impact on the ionosphere of high energy charged particles, probably of solar origin.

The list of "miscellaneous" types of interference is inexhaustible; we might include, for example, those produced in the aerials of aircraft either by dust or water droplets or by the *corona* effect;† we might also mention the "flash" of nuclear explosions,‡ etc.

† *Proc. I.E.E.E.*, Jan. 1964, pp. 44–52 and 53–63.
‡ See § 6.4.4.

7.7. Conclusions

A glance at Fig. 7.3 is enough to show the relative importance of interference of various origins, depending on the frequency considered:

At long waves (frequency below about 100 kc/s), atmospherics are the worst offenders; they are the features which normally limit the range, especially at night.

At medium waves, in towns or particularly perturbed places, artificial interference can become preponderant.

At short waves, atmospherics become far less troublesome; there is a preferred band about 30–100 Mc/s where the whole sensitivity of the receiver can be used up to the limit of thermal noise, provided it is situated far from sources of artificial interference (and a directional aerial is used, pointing towards the transmitter and not towards the Sun or the Milky Way!).

At metric waves and below, the above also applies, but the detection of solar or cosmic emission becomes much more frequent with high gain directional aerials—either deliberately for radio astronomy or as unwanted interference in the detection of aircraft and satellites at high altitude.

CHAPTER 8

APPLICATIONS. PROPAGATION IN VARIOUS WAVELENGTH RANGES

CHAPTERS 6 and 7 have shown us that the ionosphere and interference add more complications and unknowns to the calculation of the range of radio links and that their effect depends largely on wavelength.

We therefore propose to conclude by summarizing, *for each frequency range*, the various fundamental factors:

Ground wave (§§ 4.3 and 4.4);
Height of aerials (§§ 4.2–4.5);
Tropospheric refraction (Chapter 5);
The role of obstacles (§ 4.8);
Ionospheric reflection (Chapter 6);
Level of interference (Chapter 7);

grouping them together and combining their effects to obtain finally the rules and laws of propagation.

Such is the object of this last chapter.

For brevity, we shall use the terms:
short ranges, for those of order 5–50 miles;
medium ranges, for those of order 50–500 miles;
long ranges, for those of order 500–2000 miles;
very long ranges, for those over 3000 miles.

8.1. Propagation of very long waves (kilometric)

In the early days of wireless—around 1920—these wavelengths were used extensively for transoceanic or transcontinental communication as they yielded the maximum range, thanks to the appreciable diffraction of the direct ray round the Earth (§ 4.4) and the sizeable reflection of the indirect ray by the *E*-region of the ionosphere (§ 6.3.2).

However, such long ranges demanded high power and highly complex aerials, i.e. costly fixed stations.

Interest in them has therefore waned since the advent of short waves. Nevertheless, they still have several advantages:

(a) much better uniformity of propagation;

(b) relative independence of the nature of the terrain, obstacles, hills, etc.;

(c) quite deep penetration into the ground and for several feet into the sea.

These advantages are very valuable in certain applications, particularly goniometry, certain types of radio navigation, the transmission of very precise horary signals and, lastly, communication with submerged submarines.

As a result, the study of long-wave propagation, after being neglected for a time, has recently taken on a new urgency and prompted some important research.

8.1.1. Calculation of field strength. Formulae of Austin–Cohen, Zinke, Wait, etc.

At very long wavelengths, the received field strength at great distances is the resultant of two components: the direct ray, at ground level, and the indirect ray, reflected from the ionosphere. Although the latter component suffers considerable absorption in passing through the *D* layer (especially in day-time) and is affected by the diurnal and seasonal variations in the *E*-layer, it is relatively stable and soon becomes dominant with respect to the ground ray.

Thus, if we are content with an approximate value of the *mean* total field strength, we observe that:

(1) at *short* range, the direct field is dominant and decreases roughly as $1/d$;

(2) at *medium* range, interference occurs between the two components, which are of the same order of magnitude; the interference, which is quite stable, might appear, at a given point and a given time, as a reinforcement during the day and a decrease at night time—or vice versa at a different time or at a more distant point; this is due to the variable phase difference between the direct and indirect rays as a function of the height of the reflecting D or E layer:†

(3) at *long* range, the indirect field, after several reflections between the Earth and the D or E layer, becomes dominant; it is subject to large variations in which solar perturbations, diurnal, seasonal and cyclic influences can be distinguished; but, on average, the level falls exponentially as a function of distance;

(4) in the *antipodes*, a very clearly defined restricted zone is observed, which is due to the phase-coincident arrival of radiation emitted by the aerial in all directions over a trajectory of 12,500 miles (experiments carried out in the French sloop *Aldebaran* in 1923).

In the "long range" category which is usually the most interesting for the utilization of long waves, the mean field strength can be approximated by a semi-empirical formula of exponential type:

$$E\left(\frac{\mu\mathrm{V}}{\mathrm{m}}\right) = \frac{120\pi\cdot h\,(\mathrm{m})\cdot I\,(\mathrm{A})}{\lambda\,(\mathrm{km})\cdot d\,(\mathrm{km})}\times \mathrm{e}^{-\alpha d/\lambda^{x}}. \tag{8.1}$$

Austin and Cohen had originally deduced the following experimental values of the coefficients, for ranges below 10,000 km, from their measurements:

$$\alpha = 0{\cdot}0014 \quad \text{and} \quad x = 0{\cdot}6;$$

† It is even possible to deduce the height of the reflecting layer from this variation; we did so as far back as 1933 using measurements carried out at Algiers and Casablanca on a transmitter at Toulon. See also Irumata *et al.*, *J. Radio Research Lab.*, Japan, May 1964, pp. 147–51.

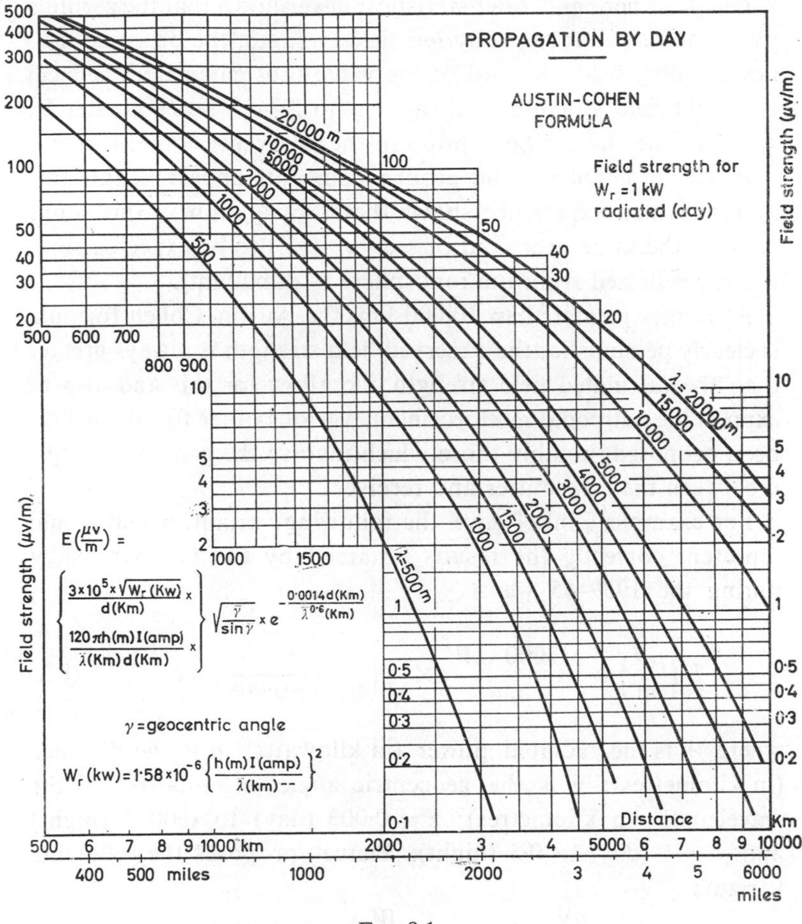

FIG. 8.1.

these were used to prepare Fig. 8.1; the curves agree quite well with those of the ground wave.

Various later experiments have led to a few adjustments; slightly different values, such as

$$\alpha = 0{\cdot}0015 \quad \text{and} \quad x = 0{\cdot}5$$

are often adopted.

On the other hand, later statistics† have shown that the formula works quite well for the *daytime* field strength, the observed field never falling below a third of the value thus calculated. Indeed, the night field strength is always up to 3 or 4 times greater by virtue of the smaller absorption in the D layer.

It will be noted that no ground constants appear in this formula; in fact, experience shows that propagation in this range is much the same over land or sea (which is to be expected since the ray reflected from the ionosphere is dominant).

At ranges greater than 10,000 km, the Austin–Cohen formula is clearly pessimistic: the observed field strength is always greater than the calculated field strength. To allow for this and also to express the antipodal reinforcement, various other formulae have been proposed, notably those which involve the geocentric angle γ between the transmitter and receiver.

For example, Zinke‡ gave the following variant, which would represent correctly the results obtained by the German Navy during the 1939–45 war:

$$E\left(\frac{\mu\text{V}}{\text{m}}\right) = \frac{15000\sqrt{W}}{\sqrt{d}} \times \sqrt{\left(\frac{\gamma}{\sin\gamma + 0{\cdot}008}\right)} \times \text{e}^{-\left(\frac{\alpha d}{\lambda^x}\right)} \tag{8.2}$$

where W is the radiated power (in kilowatts); d is the distance (in kilometres); γ is the geocentric angle (in radians); λ is the wavelength (in kilometres); $\alpha = 0{\cdot}003$ (day) to $0{\cdot}0005$ (night) and $x = 1$ (day) to $0{\cdot}5$ (night). Pierce|| proposes the following variant:

$$E\left(\frac{\mu\text{V}}{\text{m}}\right) = 210\sqrt{\left(\frac{W}{\sin\gamma}\right)} \times \text{e}^{-(0{\cdot}1 \times f)^{u} \cdot \gamma} \tag{8.3}$$

where W is the radiated power (in kilowatts); γ is the geocentric angle; f is the frequency (in kilocycles per second); u is an empirical coefficient whose value is 0·5 (sea) to 0·65 (land) at night, 0·9 (sea) to 1·05 (land) by day.

† Adams, Colin, *Electrical Communication*, 1946, **23**, 2.
‡ Zinke, *Frequenz*, Oct. 1947, pp. 16–23.
|| Pierce, Cruft Laboratory Report 158, 1952.

A number of recent authors have attempted to analyse long-wave propagation more deeply and the variations due to interference between the direct and indirect rays in particular; special attention has been given to the phase irregularity which arises and which seriously upsets certain measurements of time or distance.

These phase variations have been measured by various observers and found by some to be of the order of several tens of degrees† and by others sometimes as much as 200 or 300°.‡

Johler found that it was possible, using short pulses, to distinguish the direct and indirect waves up to about 3200 km.‖

But it is obviously more sound to revise the theory, not confining ourselves to "geometrical" optics but likening propagation between the earth and the ionosphere to that which takes place in a wave guide; it is then necessary to revise the theory of "modes" to include the additional complication of the Earth's curvature and magnetic field.

The complete calculation has been done by Wait:§ naturally, it abounds in very complicated formulae but a few simple practical conclusions can be drawn from it.

Thus, at intermediate ranges of 250–2500 miles, the mean field strength over sea decreases* approximately as $d^{-3/4}$ and, over land, an exponential attenuation is added (about 2·5 dB per 1000 miles at a frequency of 16 kc/s).

At ranges greater than 2000 km, one of the formulae simplifies and becomes[1]

$$E\left(\frac{\mu V}{m}\right) = 5{\cdot}2\times10^{6}\sqrt{\left(\frac{W}{R\sin\gamma}\right)}\times\frac{1}{\sqrt{f}}\times\frac{1}{h}\times10^{-\alpha d}, \quad (8.4)$$

where W is the radiated power (in kilowats), γ the geocentric angle (d/R), R the radius of the Earth (6400 km), d the distance

† Pressey, *Proc. I.E.E.*, part **B, 108,** Mar. 1961, pp. 214–26; Casselman, *Proc. I.R.E.* **47**, May 1959, pp. 829–39.

‡ Pierce, *Proc. I.R.E.* **43**, 1955, p. 584.

‖ Johler, *Proc. I.R.E.*, Apr. 1962, pp. 404–27.

§ Wait, *Proc. I.R.E.*, June 1957, pp. 760–7.

* Wait, *Proc. I.R.E.*, Jan. 1962, pp. 53–6.

1 C.C.I.R. Los Angeles, 1959, Doc. 195, Appendix A.

(in kilometres), f the frequency (in kilocycles per second), h the height of the reflecting ionospheric layer, i.e. 70 km by day and 90 km by night, α an exponential attenuation coefficient which, at about $f = 18$ kc/s, passes through a minimum value of 2·2 dB

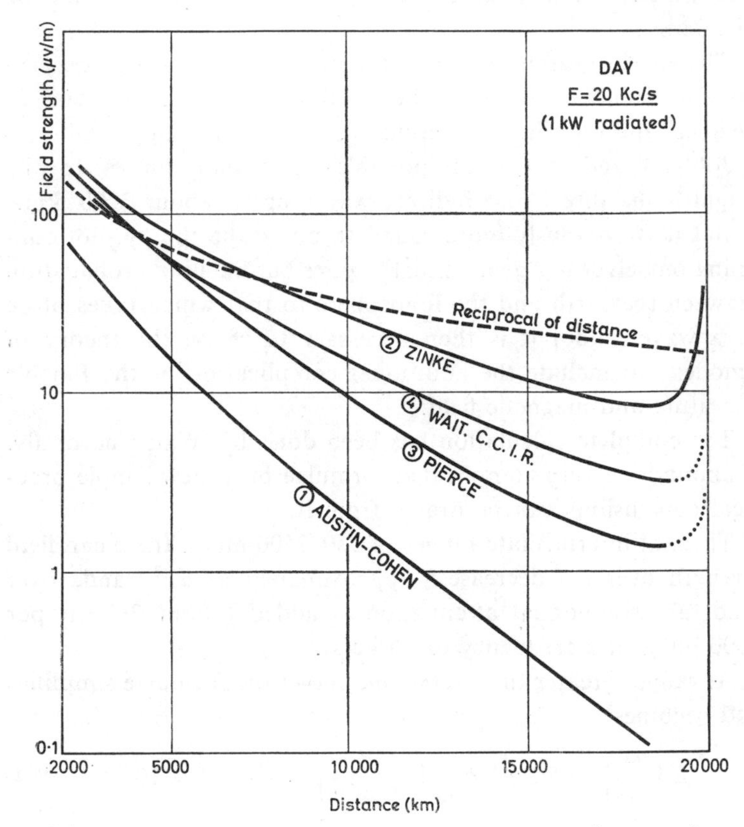

FIG. 8.2. Long range propagation of long waves, according to various formulae (day-time, $F = 20$ kc/s)

per 1000 km. (It is surprising that this formula gives a value which is smaller by night than by day.)

Figure 8.2 gives a comparison of the three formulae with that of Austin and Cohen for a frequency of 20 kc/s, assuming long distance, daytime propagation.

It can be seen that they are more optimistic than the Austin–

Cohen formula but still fairly divergent beyond about 6000 miles. Indeed, for $d = 20{,}000$ km, i.e. $\lambda = \pi$, formulae (8.3) and (8.4) give infinity which obviously is meaningless; but the addition of a small experimental correction to the $\sin \gamma$ term in the denominator, as in formula (8.2), would easily lead to a reasonable value for the reinforcement.

Finally, Wait's experiments enable him to report that it is practically impossible to suppress certain "higher modes" of propagation which are responsible for rapid variations which perturb the phase.†

Note

All the above theories are based on the hypothesis of a permanent and quasi-total reflection of very long waves by the *D* and *E* ionospheric layers.

However, a curious observation made with the Lofti satellite‡ showed that part of the energy passes through these layers so that reception is still possible above them.

The possibility of direct communication with satellites at long waves is, therefore, not ruled out.

8.1.2. Atmospherics

But the absolute value of the field strength is not everything.

The limit is set by the signal–noise ratio, because *atmospherics are important* at these wavelengths. Depending on time and place, *minimum* field strengths of 30–300 μV/m are necessary in telegraphy; application of formula (8.1) shows that for practical transmitter powers and with the poor performance of aerials at these wavelengths, the corresponding ranges are of the order of 3000–6000 miles at best. Formulae (8.2)–(8.4) are more optimistic, but experience confirms that ranges of more than 6000 miles are very risky.

† Wait, *Proc. I.R.E.*, Sept. 1961, pp. 1429–30. See also Technical Note 58 and Circular No. 77 of the National Bureau of Standards.

‡ *Missiles and Rockets*, 10 April, 1951, p. 35.

8.1.3. Obstacles

In addition to reliability of propagation, the "very long wave" range has two other advantages:

1. It is quite insensitive to obstacles since the latter are nearly always small compared with λ (except for mountain ranges) (see § 4.8); goniometry is therefore relatively accurate and errors are small; there are also only a very few variations in the duration of the trajectory, i.e. the apparent velocity or the phase; this is all very valuable in various systems of radio-guidance, as well as for high precision frequency measurements.†

2. Only very long waves are able to penetrate to a depth of several feet in the sea, and can therefore be received by submerged submarines; this is the chief reason for maintaining or building powerful very long wave stations since very short waves are capable of operating all surface links.

All this applies, of course, to *vertical* polarization only, the horizontal component being *practically zero* near the ground.‡

8.1.4. Examples

The transmission and propagation characteristics of very long waves mean that their use is manifestly confined to powerful stations, meant for long distance communication. The essential

† See, for example, Casselman *et al.*, *Proc. I.R.E.*, May 1959, pp. 829–39.

‡ The literature before 1934 abounds with papers on propagation of very long waves. A good review is given by Tremellen in *Marconi Review*, 4th quarter, 1950, pp. 153–67.

Here follows a list of recent papers: Caradoc Williams, *Proc. I.E.E.*, *3*, March 1951, pp. 81–103; Straker, *Proc. I.E.E.*, part B, May 1955, pp. 396–400; Pierce, *Proc. I.R.E.*, May 1955, pp. 584–8; Shmoys, *Proc. I.R.E.*, Feb. 1956, pp. 163–70; Rust, *Marconi Review*, No. 120, 1st quarter 1956, pp. 47–52; Horner, *Proc. I.E.E.*, part B, March 1957, pp. 73–80; Horner, *Onde Électrique*, May 1957, pp. 535–8; Budden, *Proc. I.R.E.*, June 1957, pp. 772–4; Wait, *Proc. I.R.E.*, June 1957, pp. 739–40, 756–60, 760–7, 768–71; Berlose, *Proc. I.R.E.*, May 1959, pp. 661–80.

More generally, on propagation of "very long waves" and even waves of still lower frequency ("Extremely low frequencies") see: W.T. Blackband, *Propagation of Radio Waves at Frequencies below 300 kc/s*, Agardograph 74, Pergamon, 1964; Albrecht, *Proc. I.E.E.E.*, Nov. 1964, pp. 1377–8 and *C.C.I.R. Report No.* 265, 1963.

problem, which radio engineers, about 1920–5, hoped to succeed in solving, was to determine the necessary power and optimum wavelength to cover a given range, say, 2000 and 3000 miles, for example. Using the Austin–Cohen formula (Fig. 8.1), we immediately find the following values for the field strength (per kilowatt radiated):

TABLE 8.1

Wavelength (m)		5000	10,000	20,000
Field (in μV/m) at distance	2000 miles	10	21	34
	3000 miles	2·5	7	15

However, it is a question of dominating atmospheric noise whose level varies with frequency, time of day and season (§ 7.5, etc.). The following are reasonable experimental figures (in temperate latitudes) for a narrow band (telegraphy):

TABLE 8.2

Noise field strength (in μV/m)				
night		110	200	300
day	summer	50	120	250
	winter	5	10	20

It can be seen that for a transatlantic service (d = 3000 miles) a radiated power of 1 kW would just about ensure a guaranteed service by day in winter. Even a radiated power of 100 kW (corresponding, in view of the poor performance of aerials at very long waves, to a much higher power for the transmitter itself, say 200 or 300 kW) would still not always be sufficient. As to the choice of wavelength, it is uncertain, the noise increasing almost in the same proportion as the signal field strength when the wavelength increases.

Actually, the situation is better than this, because, for one thing, as stated above, the field deduced from the Austin–Cohen formula is clearly exceeded at night—often by a factor of 3 or 4.

Therefore, for a 100 kW station, it would clearly exceed atmospheric noise. Furthermore, the use of directional aerials (frames, Beverage aerials) improves the signal–noise ratio. However, the margin is not very great and the service is subject to interruptions. Telephonic transmission, which requires a signal level about 6 times higher, appears to be impossible.

This has in fact been confirmed by experience: all the big transatlantic telegraph posts have, in fact, worked with powers of this order and contact is sometimes difficult at certain times of day. As to wavelength, after trying 23,000 m, a return to 15,000 m was made. For the first transatlantic radiotelephone link, this was even reduced to about 5000 m; but, in spite of every artifice, notably the use of "single sideband", the results were disappointing since the reception stations lay in latitudes where the interference levels were as bad as those in Table 8.2. It was necessary to move them several hundred miles to the north in order to win back the decibels necessary for a satisfactory signal–noise ratio during the commercially useful parts of the day.

(Besides, no one need be under any delusions about the merit of this verification, because it was these very results that helped to determine the Austin–Cohen and noise-level formulae; in reconstructing them, therefore, we ought not to find any disagreement, whereas the slightest uncertainty in the propagation formula would lead to fantastic powers.)

The situation is a little better for a range of 2000 miles, which corresponds to the longest guidance ranges required for an aircraft or ship in the middle of the ocean. Further, since the speed of manipulation of guidance signals (simple identification) can be reduced, the bandwidth of the receiver can be decreased in proportion, for example, down to 20 c/s, which provides a further gain of about 10 dB. Under these conditions a transmitter with a radiated power of 10 kW should ensure a very good safety margin.

The gain resulting from a reduction in bandwidth can in, certain cases, be exploited even further. In fact, there is a proposal†

† C.C.I.R. Los Angeles, 1959, Doc. 195.

to use a 20 kc/s transmitter as a "frequency standard", stable to 1 part in nearly 10^9 (i.e. to 2×10^{-5} c/s). With a special receiver with instantaneous stability of the same order, it is then possible to integrate phase comparisons over a time interval of the order of $T = 15$ min, i.e. to have the same selectivity as an apparent bandwidth of $1/T = 1/(60\times15) \simeq 10^{-3}$ c/s. The effective atmospheric noise voltage is thus reduced by a fraction $\sqrt{10^{-3}/200} = 1/450$ from what it would be with a "telegraphy" bandwidth of 200 c/s, i.e. the "noise field strength" in the preceding table would fall (daytime, summer) to about $185/450 \simeq 0{\cdot}4$ μV/m. Now, the formulae of Pierce, Wait, Zinke (Fig. 8.2) show that with a radiated power of 1 kW, a useful field strength of several microvolts per metre would be obtained at up to 12,000 miles (i.e. *all over the world*): a world-wide "standard frequency" service is thus possible under these conditions. One could even without excessive outlay, radiate 10–20 kW,† which would give a respectable safety factor against poor propagation and increase in noise in equatorial and tropical zones.

8.1.5. Velocity and Phase during Propagation

Some guidance systems use very precise measurements of transit time of waves (or, in other words, phase) between fixed stations and the moving station. Their accuracy is thus dependent on the velocity of propagation by which the time is multiplied to calculate the distance. According to recent very careful measurements,‡ the velocity would be, in principle, 299,792·5±0·4 km/sec; but some authors maintain‖ that along the ground or in the atmosphere the actual velocity would be slightly less: 292,713 to 292,750 km/sec; it appears, in any case, that if a mean value of 292,740 or 292,750 is used, it is fairly certain that the error is not more than 1 part in 10,000.

† Not forgetting that, due to the poor performance of aerials at very long wavelengths, this corresponds to a much higher h.f. power in the transmitter, say 100 kW.

‡ International Radio-Scientific Union, 1957, resolution no. 6.

‖ Smith-Rose, *Proc. I.R.E.*, **38**, 1, Jan. 1950, pp. 16–20.

However, it must be remembered that this applies to direct, regular propagation, far from any obstacle; any change in the nature of the ground, the effect of any nearby obstacles, produces a local phase shift, i.e. an error in the transit time which may be considerable (sometimes a function of the path followed).†

This is all the more so, of course, in the case of a "space" wave reflected from the ionosphere; it can result in a lengthening of the path by several tens of kilometres (at large distances); this is one of the limitations in the use of the "Loran" guidance system; an attempt can be made to correct the difference but a certain error must remain.

Mme Stoyko‡ has summarized a large number of accurate measurements of horary signals in the following formula: between two points d (thousands of kilometres) apart on the surface of the Earth, propagation times correspond (within about a ten-thousandth of a second) to an average velocity of

$$v_d = \left[290 - \frac{139{\cdot}41}{d+2{\cdot}90}\right] \times 10^3 \text{ km/sec.}$$

8.1.6. Propagation along the lines of force of the Earth's magnetic field

Fairly recent work has shown that myriametric and longer waves (frequencies from a few kilocycles per second) are guided in the magnetosphere along the lines of force of the Earth's magnetic field—between conjugate magnetic North and South points—with a very small amount of absorption.

On arrival at the level of the ionosphere, reflection occurs but a fraction of the energy can pass through the ionosphere with a 10 or 20 dB attenuation and then return to earth.

By means of a double traversal of the ionosphere, a long-wave transmitter can therefore be received at a terrestrial station located at the conjugate magnetic point; and, using a single

† See, for example, Schneider, *J. Brit. I.R.E.*, March 1952, p. 181; Sanderson, *J. Brit. I.R.E.*, March 1952, p. 195; Hufford, *Proc. I.R.E.*, June 1950, pp. 614–18.

‡ *Ann. Telecomm.*, May 1937, pp. 39–40.

traversal, it is even easier for signals to be received by a satellite at high altitude intersecting the corresponding line of magnetic force in a magnetic meridian plane of the Earth.

This type of propagation, originating in a storm discharge, can give rise to "whistlers", as described in § 7.6.

8.2. Propagation of long and medium waves (2000–200 m)

8.2.1. Field strength and its variations. Dispersion

The general principles are the same but as the wavelength decreases:

(a) the range of the *direct ray* (ground wave) decreases and much more rapidly on land than sea;

(b) the *indirect ray*, reflected by the E layer,† becomes more variable in amplitude and phase; the absorption through the D layer decreases considerably at night.

On the whole, the effect of the indirect ray becomes more important and more irregular.

The conclusions can be summed up as follows:

The direct wave is preponderant in summer, around noon, as long as its value exceeds 10 μV/m *for* 1 kW *radiated.* The field strength can therefore be calculated from the graphs of § 4.4.2. These curves were originally given for the "day time" field strength but it can be seen that "day" has a very restricted meaning; radiowise, it is only "daytime" at noon in summer.

In all other cases, the indirect ray is appreciable or even preponderant.

At large distances it is practically the *only* ray, with amplitude variations due to irregularities in ionospheric reflection and possible interference from multiple reflections at neighbouring points. (An attempt is made to correct for these variations or fading by means of automatic gain control in receivers.)

† Just because the E layer does not produce any echoes of short waves in vertical probing, it must not be concluded that its electron density is zero; it is still great enough to reflect medium and long waves at oblique incidence.

But at shorter distances, there is always an intermediate zone where the direct and indirect rays are of the same order of magnitude, which produces particularly annoying interference; with the shortest waves in this range (200–250 m), the useful zone may be reduced—60–120 miles (this defect can be alleviated by constructing "antifading" transmitter aerials in which the radiation is tilted down towards the horizontal plane and is very weak in a cone of elevation greater than 45°).

It is important to study statistically the amplitude variations of the indirect ray, that is, the probability that it will or will not reach certain levels; the value of the service provided, the risk of interference, the safety margins required, the efficiency of certain corrective devices (automatic gain controls, "diversity" receivers, etc.) can all be deduced from this information.

We saw in § 1.2.1 how these amplitude distribution probabilities are represented by certain simple laws (normal logarithmic, Rayleigh, Gauss); we saw, in particular, the usefulness of certain particular values: the *quasi-minimum* (q.m.), which is exceeded 95% of the time and can be taken as the criterion for guaranteeing an acceptable service; and the *Quasi-Maximum* (Q.M.), which is exceeded only 5% of the time and can be regarded as the tolerable limit of interference.

Figure 1.4 gave a few examples of experimental results:

(a) A certain number of points (crosses) corresponding to values of Q.M. and q.m. observed in the course of readings taken at Paris on various European broadcasting stations:

Langenberg $\lambda = 455$ m;

Brussels $\lambda = 483$ m;

Strasbourg $\lambda = 349$ m.

These readings were taken (in 1935–6 by one of the authors) at night and lasted about 1 hr;

(b) the results of some American statistics (curve 2): 191 hr of recording (night) on station WLW (700 kc/s) at $d = 350$ miles (September 1938);

(c) the same results plotted in a different way: the means of the horary values of field strengths of various probabilities.

(These two curves were taken from the American proposals to the C.C.I.R. at Stockholm, 1948, p. 346, Fig. 4.)

These indications suffice to give some idea of the dispersion of the results: depending on the frequency, distance, period and method of plotting the readings, the ratio of Q.M. to q.m. varies between about 5 and 50, i.e. 14–34 dB, at *night*. By *day* it is clearly smaller.

We can deduce from this the order of magnitude of the *margin of safety* which would be appropriate against these variations if we calculate the protection of a signal against a source of interference by using the "horary median" values. Table 8.3† summarizes the values of this "safety margin" for the principal types of traffic and the estimated probability that this margin is sufficient:

TABLE 8.3 *"Safety margins" against fading*

Type of service	Margin to be taken (in dB) on the "monthly median value of the horary median intensity"		Probability of protection
	For interference or noise of fixed amplitude	For interference subject to fading and fluctuations	
	dB	dB	%
A_1 telegraphy (c.w.)			
aural	21	17	90
	25	20	89
automatic ("diversity")	32	27	99·99
A_2 telegraphy (modulated)	17	13	90
aural	20	17	98
Automatic telegraphy by frequency shift (F_1)	32	27	99·99
Commercial telephony	from 17 to 21	11–17	70–90
Broadcasting	21	17	

† From Recommendation 164, C.C.I.R. Los Angeles, 1959, where a few more details will be found. For aural telegraphy, the figures refer to speeds of 8 and 24 bauds; for automatic telegraphy, they apply to a 50 baud teleprinter.

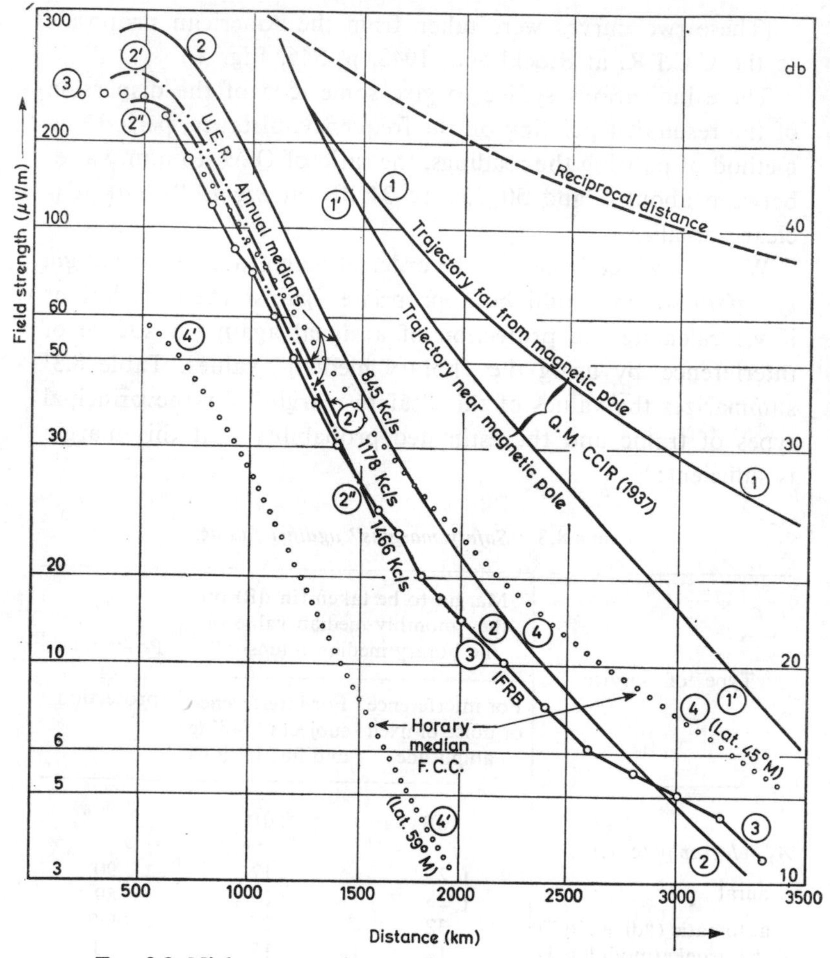

FIG. 8.3. Night propagation, medium waves (for 1 kW radiated)

Having thus fixed (approximately) the range of variation of the field strength, it is sufficient to give one of the parameters (Q.M. or median, for example) as a function of distance. Numerous measurements with this objective have been made by all interested bodies for more than 20 years; unfortunately, the results are not in very good agreement; one can hardly begin to unravel the complex effects: the rôle of time of day, frequency, trajectory in question (notably with relation to the magnetic pole), solar activity and even the phases of the Moon.

Figure 8.3 summarizes various official curves, the details being given in Table 8.4:

(a) curves 1 and 1′ from the C.C.I.R. 1937, giving the Quasi-Maximum, at all frequencies from 150 to 1500 c/s; they showed for the first time the supplementary attenuation near the magnetic pole;

(b) curves 2, 2′, 2″, deduced from more recent statistics of the Union Européenne de Radiodiffusion, giving the annual medians at three frequencies 845, 1178 and 1466 kc/s (reduced to a trajectory whose mid-point is at a geomagnetic latitude of 60°N and taken during a period of very weak solar activity; during periods of strong solar activity, predict a decrease of 2–4 dB);

(e) curve 3, used for calculating possible interference by the International Frequency Regulation Bureau (I.F.R.B.), in the band 285–1605 kc/s;

(d) curves 4, 4′, limits for geomagnetic latitudes (of the mid-point) or 45° and 59°, of the "horary median values exceeded on half of the nights during the 2nd hour after sunset". Source: Federal Communication Commission, U.S.A.

It can be seen that curves 2, 2′, 2″, 3 and 4 are quite close together and would also be in good agreement with values deduced from 1′ by reducing the ordinates by about 6 dB to convert from the Quasi-Maximum to the median. But it should be noted that curves 2, 2′, 2″ are reduced to a geomagnetic latitude of 60° and this should make them coincide with 4′ rather than 4, which applies at 45°.

In any case, several of these curves have been given as provisional and the measurements are still going on. C.C.I.R. Report No. 264 (1963) gives a method for calculating the night field strength in the European zone but issues the warning that "these forecasts are not necessarily usable in other zones" and that "preliminary comparisons" in other parts of the world "have disclosed the existence of divergencies".

We therefore do not wish to stress this method. Although there are these discrepancies, the data predict night ranges much

TABLE 8.4

Curve no.	Date	Origin	Frequency band (kc/s)	Parameter represented (per 1 kW radiated)	Latitude of middle of trajectory	Solar activity
1, 1′	1937	C.C.I.R.	150–1500	Quasi-Maximum (10)%	Far from mag. pole / Near mag. pole	?
2, 2′, 2″	1958	U.E.R. (C.C.I.R. 1959 Doc. 54)	845 / 1178 / 1466	Annual medians	Reduced to $l = 60°$N (geomagnetic)	Very Weak
3	1958	I.F.R.B. (norm. A.6)	285–1605	Median (?)	?	?
4, 4′	1946	F.C.C. (C.C.I.R. 1959 Doc. 58)	540–1600	"Horary medians exceeded on half of the nights[a] during the 2nd hour after sunset"	geogr. 36°, geomagn. 45° / geogr. 50°, geomagn. 59°	Weak (1944)

[a] To obtain the values exceeded on only 10% of the nights, or, on the contrary, exceeded regularly on 90% of the nights, add the following corrections to the ordinates of the curves:

	at 600 ml.	at 2500 ml.
for 10% of nights	+4 dB	+4 to +10 dB
for 90% of nights	−6 to −19 dB	−12 to −19 dB

greater than daytime ones: this is certainly confirmed by all radio listeners. In particular, 100 kW transmitters can produce, even at 3000 miles, fields of the order of 20–100 μV/m, which are easily detectable and explain certain interference phenomena at very great distances (between northern European and Canadian broadcasting stations, for example).

Given the restricted meaning applied to the word "day" when discussing the "direct" field, there are always numerous cases which can be classed as neither day nor night: for instance, summer mornings and evenings and all day in winter. The value of the field strength is then intermediate between the values of the direct and indirect field strengths, which obviously leaves a large gap of uncertainty in view of the difference between these two values (especially at the short-wave end of the band and at great distances over poor ground). This uncertainty has yet to be removed.

The importance and irregularity of the indirect ray in this band pose serious problems for navigators as far as the techniques of *radio-guidance* are concerned.

Goniometric techniques, based on the measurement of angles, may be affected by errors due to horizontal or elliptic polarization of the ray reflected from the ionosphere, hence the necessity to abandon the frame aerial in favour of more delicate and more cumbersome "Adcock" differential aerials.

Other techniques use the measurement of *distances* to fixed transmitters, the distance being evaluated by the transit time of short pulses (Loran) or phase change in continuous waves (Decca); but it is obvious that this distance, reckoned at ground level (direct ray), is grossly in error if it is calculated, in fact, from the indirect ray reflected from a height of 100 km. The use of correction tables is possible if it is certain that the indirect ray is involved and if the height of the reflecting layer is known sufficiently well. In all other cases, the range or the safety margin is decreased.†

† See for example, Caradoc Williams, *Proc. I.E.E.*, III, March 1951, pp. 81–103; Naismith *et al.*, *Wir. Eng.*, Sept. 1951, pp. 271–7; C.C.I.R., Geneva, 1951, Doc. 141 (impressive bibliography); Weekes *et al.*, *Proc. I.E.E.*, III, March 1952, pp. 99–105.

8.2.2. Obstacles, interference, etc.

The influence of *obstacles* also becomes more important in this band. In mountainous areas, the field strength is extremely variable as a result of "screening" and "reflection" effects; it is usually weak in transverse valleys or behind mountains and sometimes stronger in longitudinal valleys acting as "wave guides".† But many other peculiarities are observed. Among other things, electric power lines, timber or metallic structures, reinforced concrete buildings, hangars, etc., produce considerable absorption and important local perturbations on the polarization and phase of the field; the result is that, in towns, the attenuation is much more rapid than in the open country: the exponential coefficient in the Austin–Cohen formula (§ 8.1.1) would be multiplied by 10 or more; moreover, the techniques of location by measurement of angles, and even of phase, can cause considerable local errors; oblique passage from land to sea, elevated coastlines or cliffs, cause *appreciable goniometric errors;* on ships and aircraft, the presence of superstructures, fuselage, wings, etc., also introduce errors which can be considerable and very difficult (or even impossible, sometimes) to correct.

Finally, the role of atmospherics is still important: although, at the most favourable times and places, they scarcely limit the practical sensitivity (order of 1 μV/m in telegraphy, 10 in telephony), on the other hand, they are often much more troublesome, especially at night.

We can also include artificial interference, especially in towns or on board ships, aircraft, cars, etc., so that it is prudent in such cases to predict at least 20 μV/m in telegraphy; for high quality reception, broadcasting stations demand at the moment 5–10 mV/m and prefer even more.

† All these reflections from natural features of the ground also increase the energy directed upwards, i.e. the "indirect" wave; the efficiency of "antifading" aerials is found to decrease (as in Switzerland: Glinz, *Tech. Mitt. P.T.T.*, June 1950, pp. 224–8).

8.2.3. First example: broadcasting

The best-known example of the use of "medium" waves is obviously sound broadcasting.

It is distinguished from the others by the abnormally high level of field strength required for reception, since the users (who are numerous and politically powerful) demand very high quality reproduction with economical apparatus, simple and easily installed aerials, even in the midst of the inevitable artificial interference found in built-up areas.

As stated above, this leads to a requirement of 5–10 mV/m field strength in towns and the order of 1 mV/m in the country.

Furthermore, as the service must be maintained at all times, including midday in summer, we have to use the curves of "daytime propagation", relative to the ground wave, as a basis for calculating the necessary powers.

Now, we have only to refer to Fig. 4.13 (land) to obtain the following values of field strength per kilowatt radiated:

Table 8.5

	Field strength (mV/m)				
	Distance	200	300	500	650 miles
Field strength for wavelength	1500 m	0·6	0·16	0·05	0·02
	500 m	0·08	0·012	0·001	
	200 m	0·003			

The problem is therefore soluble for a "national" service on the scale of European countries, i.e. for an "acceptable operational radius" of about 300 miles, with wavelengths of the order of 1500 m and a power of the order of 1000 kW radiated (giving about 5 mV/m). In fact, some transmitters do go as high as 400–500 kW and even 1000 kW.† But these exceptional figures

† See *Tele. Techn.*, Jan 1953, pp. 50–1 and *Proc. I.R.E.*, Aug. 1954, pp. 1222–35.

are signs of a self-assertion and competition among various stations rather than a sound technical basis. The estimation of the minimum field strength necessary in towns is very uncertain. Even with 100 mV/m, interference can exist; and yet, 2 or 3 mV/m may sometimes be enough for good reception. The degree of satisfaction in the listener therefore increases quite slowly with the power of the transmitter, whereas the net cost increases extremely fast when the power exceeds 100 or 200 kW. Post-war regulations have therefore restricted the majority of stations to this order of magnitude without any marked ill-effect (inasmuch as mutual interference is often as bad as background interference).

However, this calculation shows up the great superiority of the 1500–2000 m band over the 200–600 m band. In the latter, the attenuation of the ground wave is much more rapid and, as a result, to guarantee a permanent service at 300 miles, a quite unreasonable amount of power would be necessary.

The same levels are therefore used and an attempt is made to improve the service:

(a) by setting up a network of stations dense enough for every listener to be within about 60–70 miles of one of them;

(b) by increasing the number of small local transmitters serving towns badly situated with respect to the main network;

(c) by combating artificial interference;

(d) lastly, by using the "indirect ray" which is present most of the time (at night all the year round and even by day in winter) to reinforce the direct ray calculated at ground level.

It must be noted, however, that this indirect ray is sometimes so little attenuated that not only does it increase the range of each station, considered in isolation, but also it increases, to a large extent, the interference between stations of the same or nearly the same frequency, even at considerable distances. This is an everyday experience with listeners. As has been pointed out already in connection with Fig. 8.3 (night field strength, C.C.I.R. figures), such interference has been observed not only between countries, but sometimes between continents, which introduces a pecular complication into the task of the international committees charged with frequency allocations.

8.2.4. Second example: distress calls at sea

Let us take as our second example the problem of the maximum range of a "distress call" sent out by a ship at 500 kc/s (600 m).

At this frequency, a recent survey† shows that the effective height of ships' aerials is generally (in 95% of cases) greater than 25 ft and their efficiency is therefore greater than 5·5%. The normal marine transmitter, about 250 W, therefore radiates 14 W (at least).

From Fig. 7.3 the level of atmospherics at this frequency, by day, in temperate latitudes, could be as high as 10 μV/m. But it is higher in tropical zones and, for this reason, it is generally assumed that 50–80 μV/m is necessary. Let us take 60.

For the (reduced) power of 0·014 kW, the range is read off the graphs for a field of $60/\sqrt{0{\cdot}014} = 500\ \mu\text{V/m}$.

For the surface wave, at sea Fig. 4.12b, this range is about 160 nautical miles.

For the indirect ionospheric wave, without absorption and at night (Fig. 4.14) it goes up to 430 nautical miles.

In fact the actual range will almost always lie between these two figures for there will be a small contribution from the indirect wave. Moreover, the efficiency figure assumed for marine aerials is a minimum; it would normally be 2 or 3 times as great and the range of the ground wave would theoretically be 190–220 nautical miles.

On the other hand, with an "emergency" transmitter of 40 W output, it could fall to 90 nautical miles.

Finally, it will be noted that at slightly shorter wavelengths, down to about 150 m (2 Mc/s), the range would be of the same order, because the attenuation of the surface wave, at sea, would be hardly any faster since the efficiency of the transmitting aerial would be better and the noise level lower. This is why intermediate waves are used by small boats.

† *R.C.A. Review*, Sept. 1953, p. 305.

8.3. Propagation of intermediate waves (60–200 m)

In this band, the trends resulting from the decrease in wavelength continue in the same direction:

(a) The direct ray is more quickly attenuated, to the extent that the corresponding range is of hardly any interest except at sea.

(b) The indirect ray gets stronger and stronger (being absorbed less by the *D* layer).

A new feature: ionospheric reflection can occur either at the *E* layer or, after passing through it, at the *F* layer, so that medium distances are quite evenly covered by one means or the other, especially *at sea.*

Furthermore, since these waves are transmitted with excellent efficiencies, even by small aerials, they are of interest for all categories of boats and hence are frequently used.

There are hardly any special formulae or published work relating to these wavelengths. They are truly "intermediate" from all points of view. Depending on circumstances, they can be extrapolated from either the medium wave class (preceding section) or the short wave class (next section).

Goniometry is particularly difficult in the vicinity of obstacles.

Although polarization is, in principle, always vertical, good results can be obtained with aerials of any orientation as long as they are elevated (§ 4.5).

8.4. Propagation of short (decametric) waves (10–60 m)

In this band, the surface wave is very quickly absorbed and can only be used at short distances: it is sometimes used in links between portable sets to take advantage of the small size of aerial required; but it is inefficient compared with the possibilities of the indirect ray.

By contrast, in this band the "indirect" ray (sometimes called the "sky wave") comes into its own. In fact it is often possible to find frequencies for which oblique radiation can pass through the *D* and *E* regions, with slight absorption, and then be reflected from the *F* region without appreciable loss of energy. This radiation therefore returns with great intensity at points as far away as 2500 miles; it can be reflected again from the ground and, by several "hops" in succession, can accomplish one or more circuits of the Earth.

In this type of radiocommunication, the determination of the power necessary is no longer the designer's chief objective. Of course, an increase in power is always advantageous in facilitating reception in the face of interference, fading and noise. But if the frequency is chosen well enough for the absorption over the whole trajectory to be very small, a negligible power is sufficient to ensure reception in the antipodes; this fact was discovered by the amateurs, rather by chance, around 1923–5, and mobile posts (aircraft, ships, etc). are trying to use it more scientifically every day.† On the other hand, if the frequency is badly chosen, the radiation passes through the *F* region and gets lost in space, or suffers such an attenuation in passing through the *D* and *E* regions that it gets nowhere, even with large numbers of kilowatts.

The essential problem, therefore, is to determine the *most favourable wavelength:* and since it depends on the state of the ionosphere and the latter, as we have seen, varies with time of day, season, year and position on the globe, the problem is extremely difficult. It is only in the last few years—and chiefly because of efforts demanded by a world war—that satisfactory methods have at last been evolved. "Ionospheric forecasting" stations had been built between 1940 and 1945, by both the Allies and the Germans.

We shall mention, particularly, the forecasting method of the American Central Radio Propagation Laboratory (C.R.P.L.) and the one used in France by the Division de Prévisions Ionosphériques (D.P.I.) of the Centre National d'Etudes des Télécommu-

† One must not lose sight of this point in discussing the merits of such-and-such a transmitter system or aerial, etc. The fact that it was picked up at a given instant in the antipodes proves little or nothing as to its relative value.

nications (C.N.E.T.)[†] which was inherited from the German method, with the addition of punched card computer techniques and various improvements.

The details of the internal workings of these services are not well known and are difficult to analyse; however, the general principles follow naturally from what we have seen above, and the results are presented in periodical bulletins, in various easily usable forms.

The essential features are summarized below.

The problem is subdivided into three parts:

(a) To find the maximum usable frequency (MUF), i.e. the maximum frequency *reflected* by the ionosphere, taking into account the angle of incidence corresponding to the range (at higher frequencies, the greater part of the energy passes through the layer and does not return to the Earth, so that the link becomes practically impossible whatever the power of the transmitter);

(b) To find the *lowest usable frequency* (LUF), i.e. that frequency below which the *absorption* in the lower layers of the ionosphere (which, as we have seen, increases with wavelength) is too great for the power of the transmitter to supply the necessary energy to the receiver; it is then a question of a quantitative determination of the field strength, taking account of the power and efficiency of the transmitter, the directional properties of the aerials, the sensitivity of the receiver and interference in the area in which it is situated:

(c) To assess the *most favourable frequency* between these two limits (optimum frequency for traffic, OFT), taking into account uncertainties in the prediction, approximations in the calculation, and possible irregularities in the ionosphere (sporadic E layer, magnetic storms, etc.).

The principle of the prediction methods is as follows: the state of the ionosphere depends for the most part on the Sun, its position, which varies with time of day and time of year but which can be predicted with precision, and its activity. There should therefore be a characteristic index of this activity with a suffi-

† Formerly the Section de Prévisions Ionosphériques Militaire (S.P.I.M.) and later the Section de Prévisions Ionosphériques Nationale (S.P.I.N.).

ciently regular variation to enable the past curve to be extrapolated several months ahead (in practice, about 6–12 months). On the basis of experimentally obtained correlation rules, the solar activity forecast is then transformed into an ionization forecast, or even directly into a critical frequencies forecast. The maximum usable frequencies for the oblique trajectory considered are then obtained. As to the forecast for the lowest usable frequency, this is derived more or less directly from that of the ionospheric absorption which is itself a function of solar activity.

The index of solar activity used at present in the forecasts is a factor proportional to the number of sunspots visible on the surface of the Sun, and called the "Wolf number" or "Zurich number". It is defined by the relation

$$R = k(N+10G), \tag{8.5}$$

where N and G denote respectively the number of isolated spots and the number of groups of spots visible during the period under consideration on the solar disc, and k is a coefficient which depends on the observer and the instrument used. k is chosen in such a way that the results obtained in the various observations are coherent. The range of variation of R extends from 0 to over 200.

To eliminate seasonal effects, the value used for a given month is a mean $\bar{R}$ taken over a period of 13 months bracketing the month considered, i.e. 6 months before and 6 months after. It would appear from some recent work that other indices would be more satisfactory. Indices formed from other solar characteristics have been proposed, such as the intensity of spectral rays (those of hydrogen or calcium) or the intensity of the solar radio emission at metric or centimetric wavelengths. Another proposal is to choose an index of solar activity based directly on ionospheric characteristics, for example an index which reflects the state of the F_2 region of the ionosphere, the region most sensitive to variations in solar activity. This is all under consideration, as well as the replacement of the present empirical method of forecasting, which consists in extrapolating the curve of past variations by a more scientific method using modern autocorrelation techniques.

Once the solar activity for the month under consideration has been forecast, then for every hour, following the principles indicated above and using observations passed from more than a hundred radio probing stations scattered all over the world, a world map of forecasts of critical frequencies can be drawn.

In fact, at a given point on the Earth's surface it is possible to establish a precise correlation (almost linear) between the Wolf number and the median value (i.e. the value exceeded for 50% of the time in the course of a given period) of the critical frequencies of the E and F regions. This correlation law can be determined for the various ionospheric probing stations and for each month of the year and each hour of the day (for example, for 5 p.m. in the month of February), by referring to the direct measurements of critical frequency made during previous periods.

For points of the same latitude and at the same local time, the Sun occupies the same position in the sky; it would therefore seem that the ionization and the critical frequencies ought to be practically the same and that the number of maps could be reduced from 24 (one per hour) to one giving the distribution of the critical frequencies as a function of latitude and local time.

But, in fact, at least insofar as the F layer is concerned, the ionization and critical frequencies also vary with longitude (in particular under the influence of the earth's magnetic field). For links which depend on reflection from the F layer (actually, the F_2 layer), it is only in limited areas that a single forecast map, in terms of local time and latitude, will suffice without too great an error. On a world-wide scale, it is necessary to produce maps which are valid for a given time of day, as a function of both latitude and longitude.

8.4.1. Determination of the maximum usable frequency (MUF)

It has been seen that the possibility of reflection depends uniquely on the ionization level of the layer and the angle of incidence.

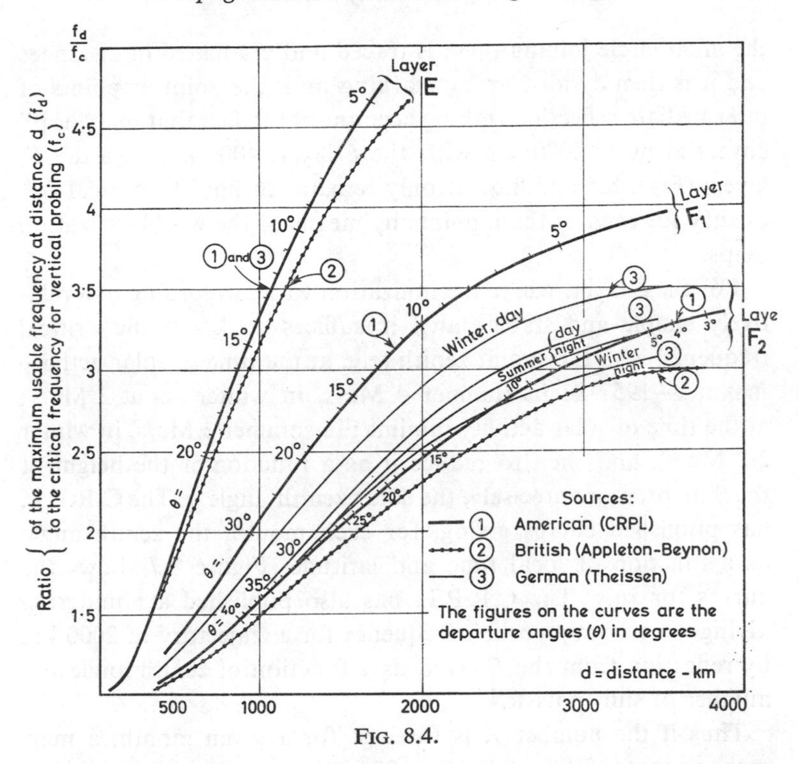

FIG. 8.4.

There is no difficulty as far as incidence is concerned: it is assumed that the height of the layer is more or less known—which is true, especially for the *E* layer—the curves of Fig. 8.4, deduced from the formulae of § 6.2.3 and from experiment, give immediately, as a function of the distance *d*, the factor f_d/f_c by which the critical frequency for vertical probing is to be multiplied. (These curves also give the departure angle θ.)

The two stations are therefore plotted on a map,† the arc of

† Naturally, any projection can be used for this and that of Mercator was actually used at first. But in it the poles recede to infinity which is inconvenient in radio propagation in which they play a great part. The Americans have therefore substituted the special projection (Fig. 8.5) (latitudes plotted linearly) on which arcs of great circles (full lines) and distances in thousands of kilometres (broken lines) can be plotted with the help of Fig. 8.6.

To use this chart, take a sheet of tracing paper and trace the equator on it. Then superimpose it on Fig. 8.5 and plot on it the stations. Next superimpose the great circle joining them is traced and graduated in distances and it is then divided up by marking on it the point or points of intermediate reflection (taking account of the fact that one "hop" covers at most 2000 km with the E layer, 4000 km with the F_2 layer) (Figs. 8.5 and 8.6). It only remains to find the critical frequency for each of these points by means of the world ionization maps.

For the E layer, where the ionization varies according to a relatively simple and stable law,† it suffices to know the critical frequency with the sun at zenith (e.g. at the time of solar activity maxima—1957–8: in summer 4 Mc/s, in winter about 3 Mc/s; at the time of solar activity minima: in summer 3 Mc/s, in winter 2·2 Mc/s), and then to reduce it as a function of the height of the Sun, or, more precisely, the solar zenith angle χ. The C.R.P.L. has published curves giving, for each month, the zenith angle as a function of local time and latitude. Figure 8.7 shows the curves for July. The C.R.P.L. has also published a nomogram giving the maximum usable frequency for a trajectory of 2000 km by reflection from the E layer, as a function of zenith angle and number of sun spots $\bar{R}$.‡

Thus if the number $\bar{R}$ is forecast for a given month, a map such as that of Fig. 8.8 ($R = 20$, July) can be drawn, giving in megacycles per second the maximum usable frequencies for a trajectory of 2000 km by reflection from the E layer as a function of latitude and local time (with respect to the meridian of the point of reflection, which is the midpoint of the trajectory).

For the F_2 layer the principle is the same but the variations are much more irregular. Since the height of the layer is variable, two series of maps are necessary, one for critical frequencies (or MUF for a zero trajectory) and the other for the MUF corre-

---

it on Fig. 8.6 and slide it along the equator until the two stations lie on the same full line: this is the arc of the great circle; trace it and graduate it in distances with the broken curves.

† The critical frequency varies with the zenith angle χ of the Sun, proportionally to $\cos^n \chi$, where the constant n is roughly equal to 0·3.

‡ The height of the E layer being fixed, maximum usable frequencies for a trajectory of 2000 km can be obtained from critical frequencies by multiplying by a well-defined coefficient: 4·78.

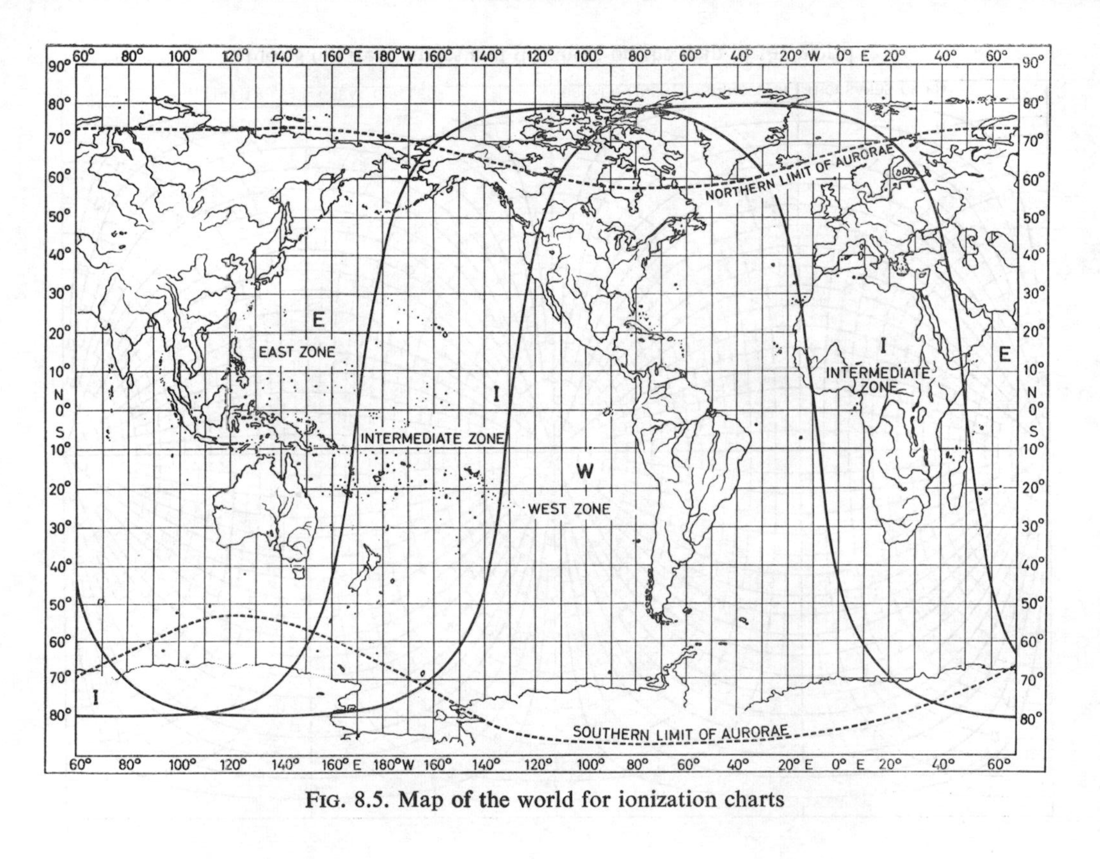

FIG. 8.5. Map of the world for ionization charts

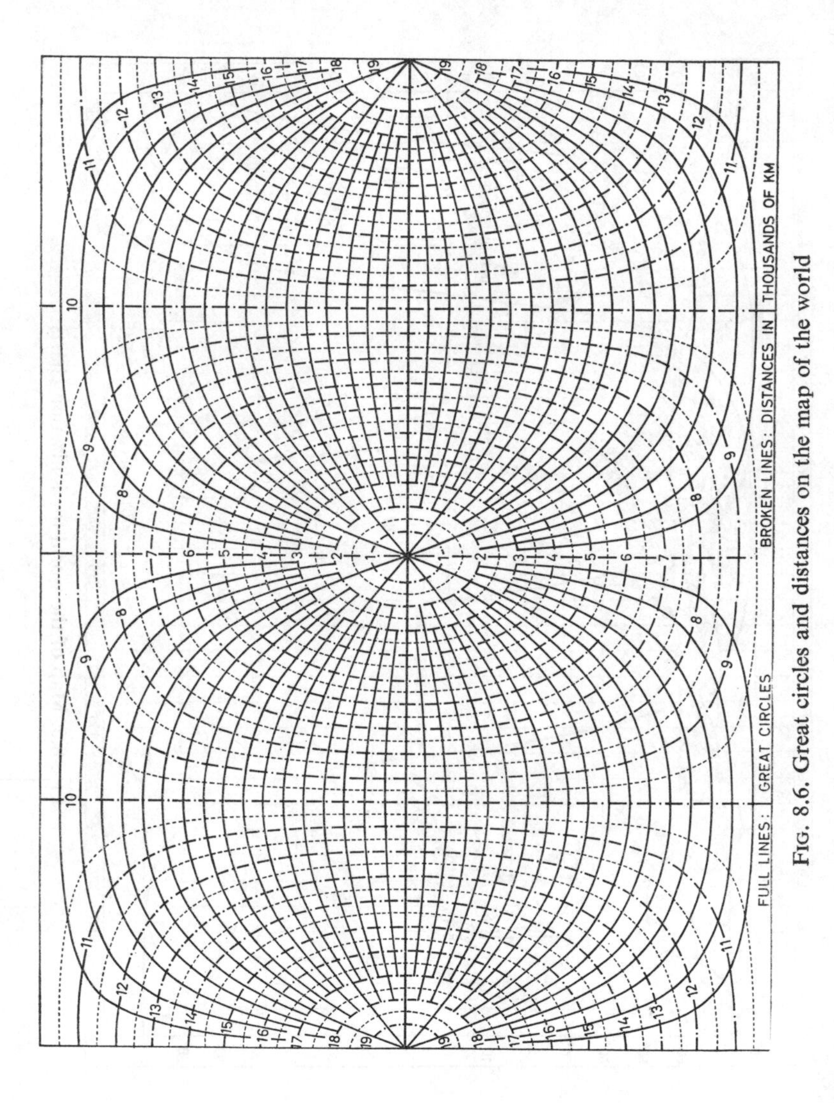

FIG. 8.6. Great circles and distances on the map of the world

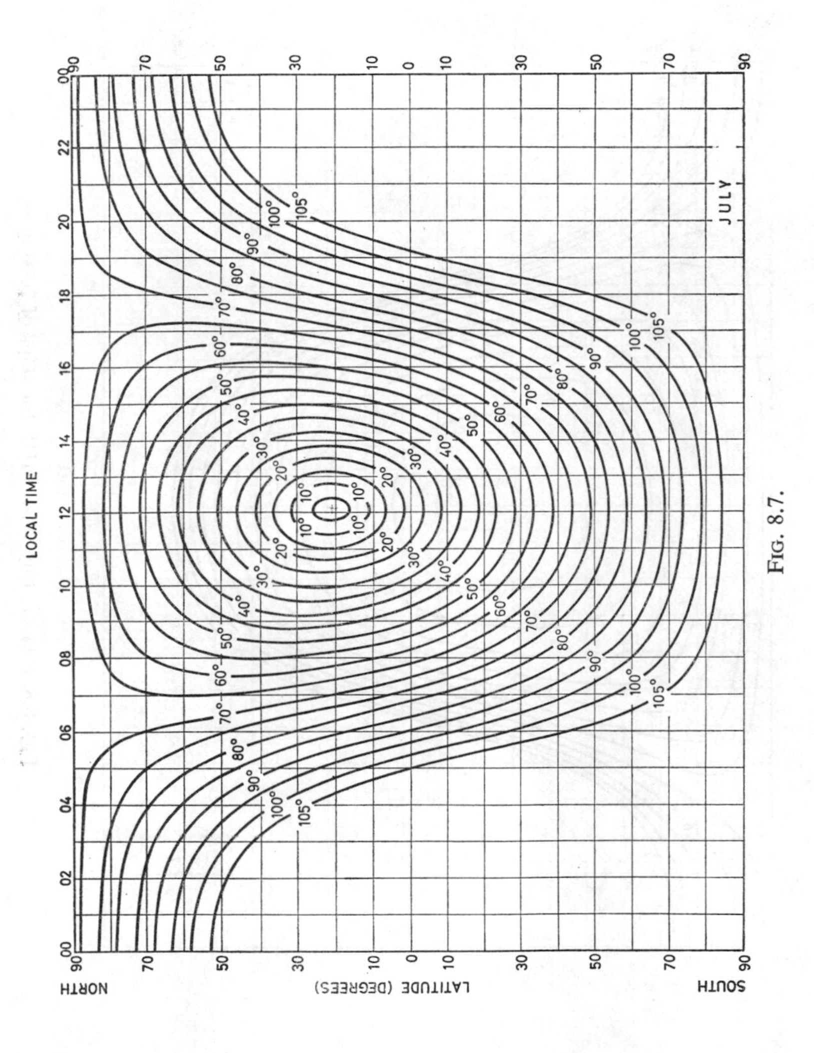
LOCAL TIME
00
02
04
06
08
10
12
14
16
18
20
22
00
LATITUDE (DEGREES)
NORTH
SOUTH
90
70
50
30
10
0
10
30
50
70
90
JULY
10°
20°
30°
40°
50°
60°
70°
80°
90°
100°
105°

FIG. 8.7.

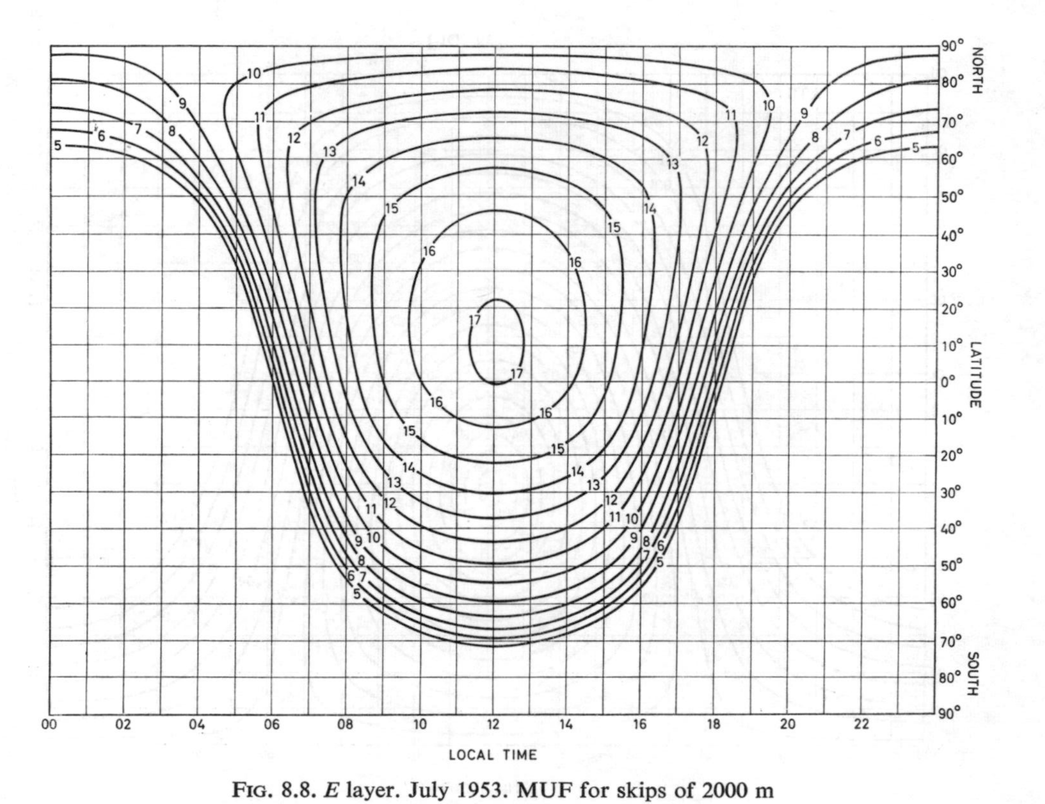

FIG. 8.8. *E* layer. July 1953. MUF for skips of 2000 m

sponding to the maximum trajectory of 4000 km; and up to June 1963, the C.R.P.L. did publish, for each of these ranges, three maps corresponding to three different zones of the world: East, West and Intermediate, as shown in Fig. 8.5; the East and West zones covered 120° sectors on the axis of the geomagnetic poles (the former including Asia and Russia and the latter America); the Intermediate zone consisted of two 60° sectors, one covering Europe and Africa and the other the Pacific; the maps for each zone were prepared from measurements carried out in the probing stations of that zone only. Figures 8.9 and 8.10 reproduce, by way of example, the two maps for the Intermediate zone for July 1953, one for the critical frequency (MUF, 0 km) and the other for the MUF for a "hop" of 4000 km.

Since June 1963, for the reasons indicated above (end of § 8.4), the C.R.P.L. prefers to publish maps for MUF zero and MUF 4000 for each of the even hours, in universal time, from midnight to 22.00, as a function of longitude and latitude; data from all probing stations are used, with interpolation by computer, which has permitted an increase in the number of maps (12 for each range instead of 3) and, therefore, in their accuracy. By way of example, Figs. 8.14 and 8.15 show two extracts from the forecasts for June 1964 for 6 a.m. U.T.

For trajectories of length less than 2000 km (1200 miles) off E or 4000 km (2500 miles) off F_2, one can determine, with the help of diagrams supplied by the C.R.P.L. to users of its forecasts, the MUF's from the 2000 km (E) MUF's or the zero kilometre (F_2) MUF's and the 4000 km (F_2) MUF's. One can also, more simply, make use of the curves of Fig. 8.4.

If there are several hops in succession, the calculation is done for each of the points of reflection and the smallest MUF thus found is taken as the maximum frequency for the trajectory (since reflection must take place at every point).†

† In the American method, if there are more than two hops, only the two extremes are considered; it turns out in practice that the wave seldom fails on the intermediate hops. In the D.P.I. method, on the other hand, all the intermediate hops are considered, at least, as long as the distance is less than 12,000 km. This difference in technique sometimes results in appreciable discrepancies in the forecasts of MUF's.

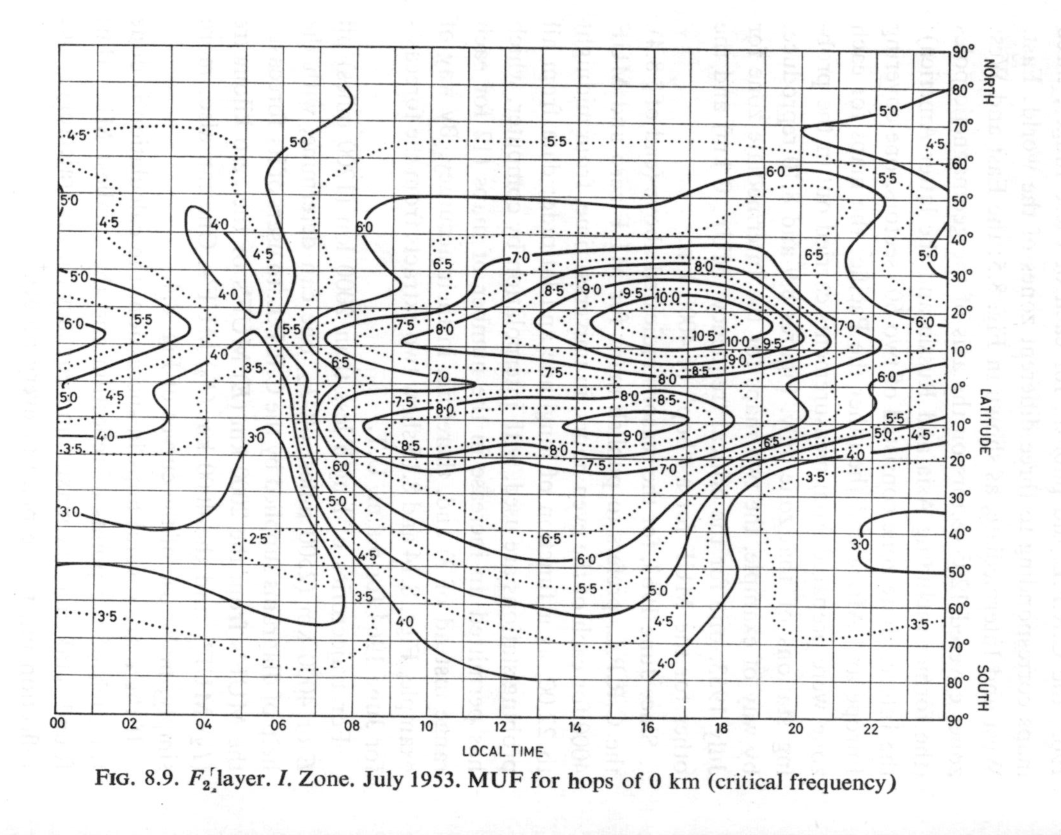

FIG. 8.9. F_2 layer. *I*. Zone. July 1953. MUF for hops of 0 km (critical frequency)

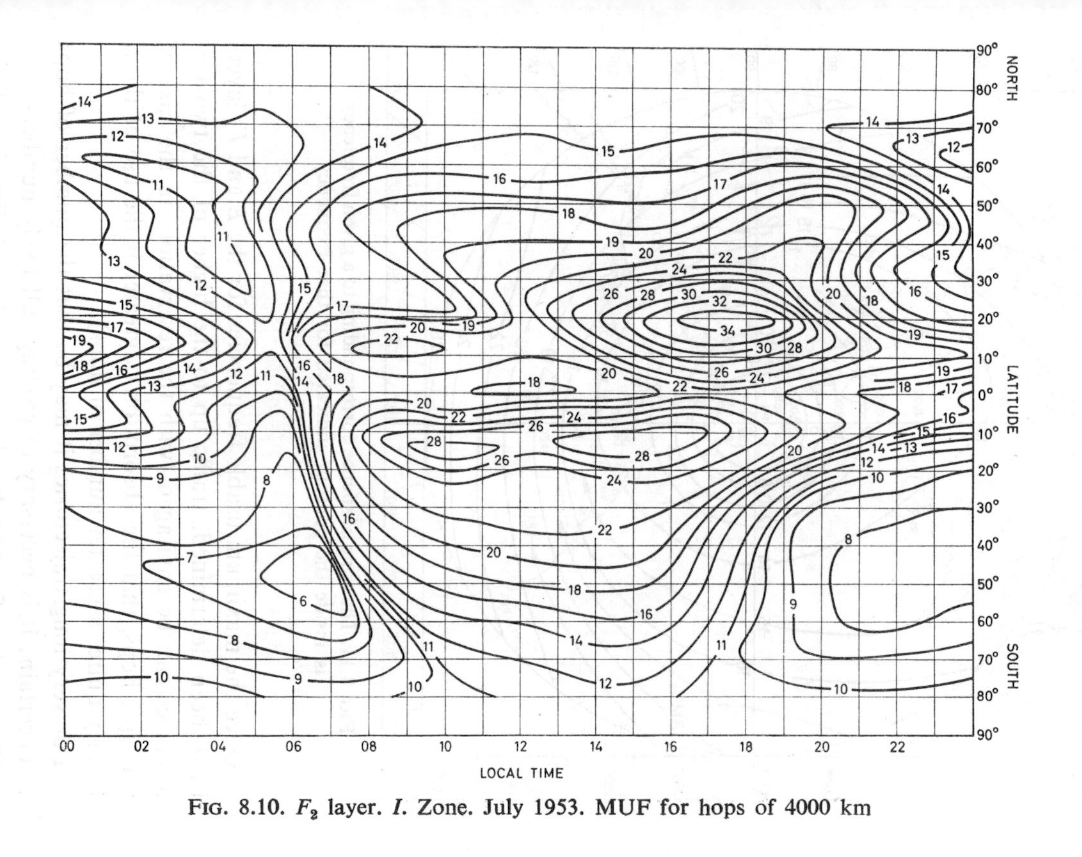

FIG. 8.10. F_2 layer. *I.* Zone. July 1953. MUF for hops of 4000 km

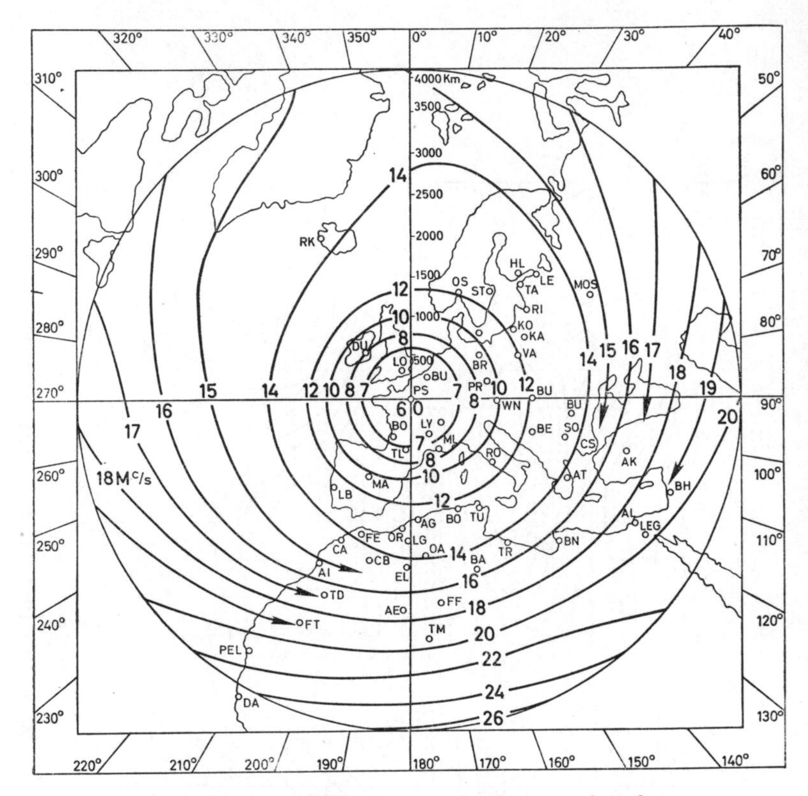

FIG. 8.11. The zone of silence corresponding to a given frequency is inside the contour indicated for that frequency

Once the maximum usable frequencies for the E and F_2 layers have been determined, traffic up to the higher of the two frequencies can be envisaged (with the reservation, in the case of multiple hops, that the absorption is greater if the use of the E layer requires a greater number of hops).

For very long trajectories it may be that the number of hops is uncertain; it is necessary to envisage all likely numbers.

Once such a forecast has been made for all directions from a given centre (for example, Paris), the loci of the minimum ranges corresponding to each frequency can be plotted on the map. The "zones of silence" where this frequency cannot be received are thus obtained. Figure 8.11 represents such a map, constructed by the C.R.P.L. method, for July 1953, at 4 p.m.†

It can be clearly seen that at this time, the Paris–Algiers link can only work at frequencies below 13 Mc/s, the Paris–Dakar at frequencies below 24 Mc/s.

The map in Fig. 8.12 is an extension of the above for distances greater than 4000 km (2500 miles); on this planisphere (on the cylindrical projection already mentioned), the transmission zone with Paris at a maximum frequency is the area inside the contour line on which that frequency is indicated; it can be clearly seen that South America can be served with frequencies below 15 Mc/s.

Finally, if we are content to observe the variation of the maximum usable frequency, as a function of the time, between two fixed points (or in a fixed zone of small extent), we can confine ourselves to adapting the graphs of Fig. 6.12: the time from midnight to midnight is always plotted as abscissa (stating precisely at what point they are taken if the longitude changes over the trajectory), and as ordinate, instead of the critical frequency f_c, the MUF is plotted (in Mc/s).

Figure 8.13 shows four graphs of this type,‡ borrowed from the D.P.I. forecasts for July 1953; the top curve represents the MUF (we shall see the meaning of the others later). Three of them relate to the "European" zone for ranges of 500, 1000 and 1500 km; the increase in the MUF with distance, due to the more oblique incidence on the ionosphere, is clearly shown. The fourth relates to the Paris–Madagascar trajectory (about 8000 km) and will be discussed later.

† These maps take account of reflection by the *E* and *F* layers, but not the sporadic *E* layer.

‡ We regret that the frequency scale is not *logarithmic*; several advantages would accrue (constant precision, ease of correction), as P. David pointed out in 1940 (Naval Radio Officers' Course).

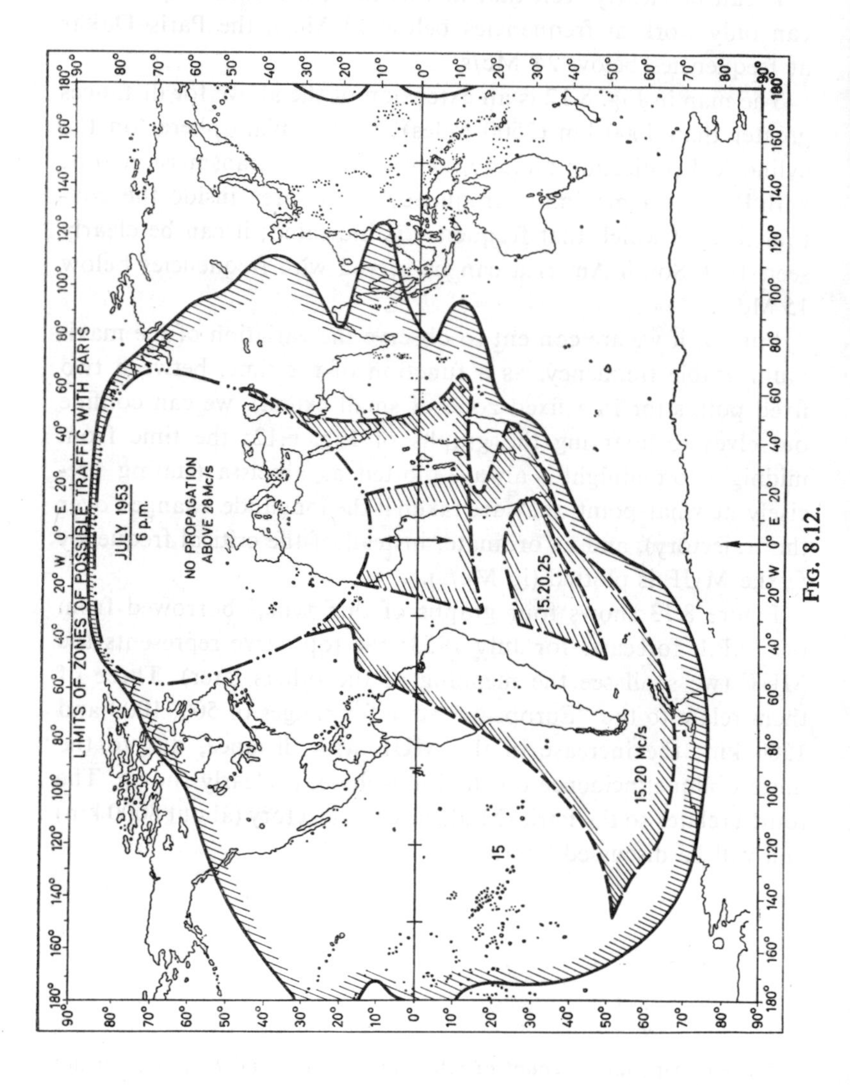
LIMITS OF ZONES OF POSSIBLE TRAFFIC WITH PARIS
JULY 1953
6 p.m.
NO PROPAGATION
ABOVE 28 Mc/s
15.20.25
15.20 Mc/s
15
20° W 0° E 20°

FIG. 8.12.

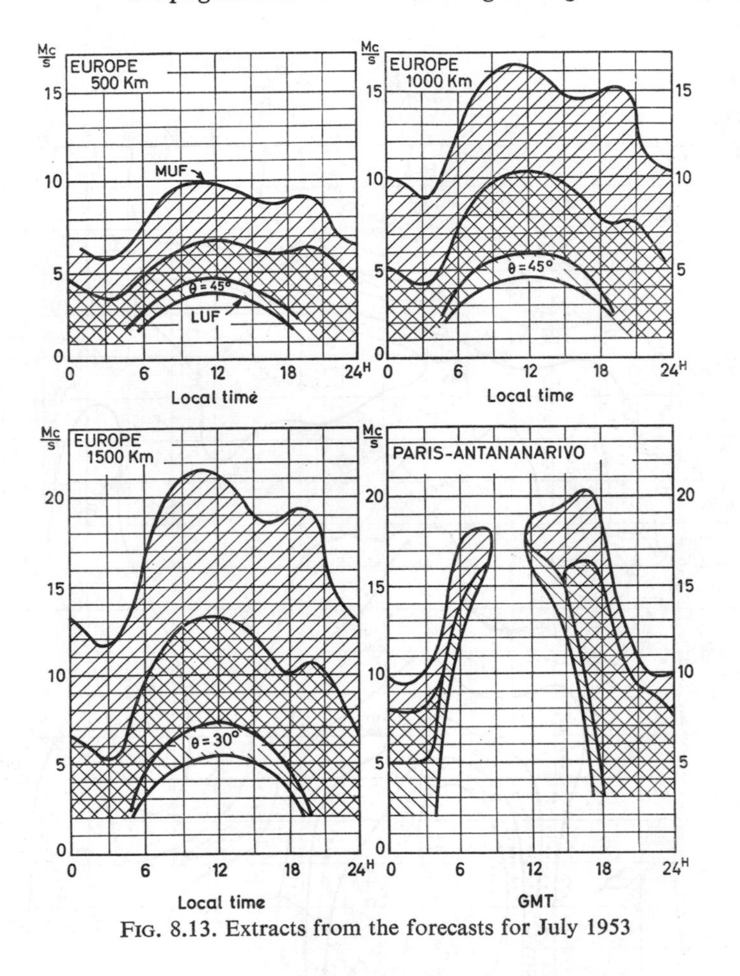

FIG. 8.13. Extracts from the forecasts for July 1953

8.4.2. DETERMINATION OF THE FIELD INTENSITY AND LOWEST USABLE FREQUENCY (LUF)

All frequencies below the MUF, being reflected by the ionosphere, are *a priori* usable for transmission purposes. But, as the absorption increases with wavelength (§ 2.5), there must be a minimum frequency beyond which the received field strength remaining after absorption is too low. It depends on various factors which must be taken into consideration:

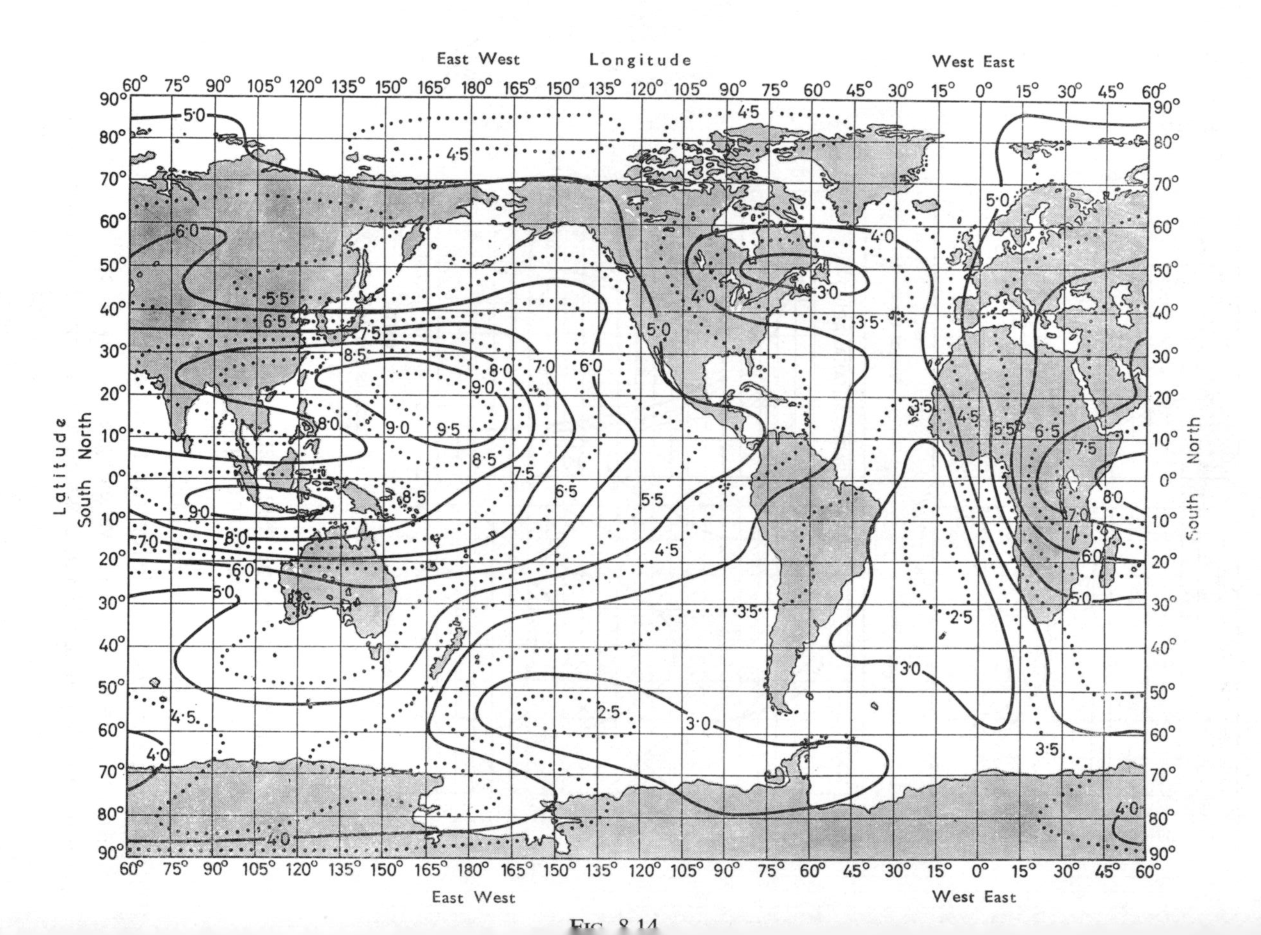

Fig. 8.14

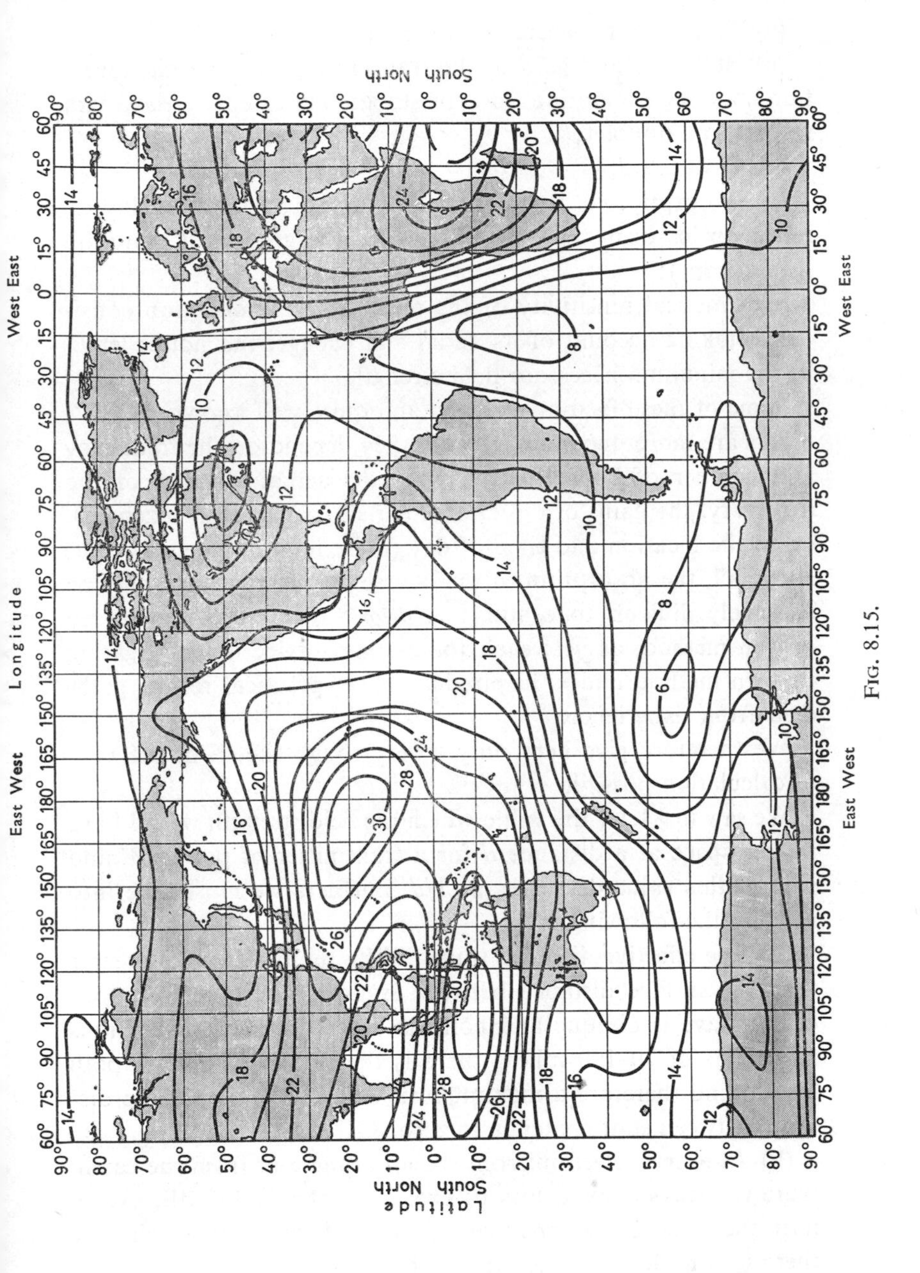

FIG. 8.15.

(a) Transmitter power: W (kilowatts);

(b) Efficiency and gain of the transmitting aerial in the direction under consideration (in azimuth and elevation): g, say, with respect to a doublet;

(c) Distance d;

(d) Absorption e^{-A} on passing through the lower layers, reflection losses in the ionosphere (and the ground, if there are several hops);

(e) Practical sensitivity of the receiver in the direction considered, taking account of its aerial and background noise: let E_r be the minimum necessary field strength.

Some of these factors are fixed and quite well known, e.g. W; others are more uncertain because they depend on the frequency and number of hops (for example, the actual distance of the trajectory, the gains of directional aerials, particularly lozenges), or on the location and time of day (noise level in receiver). But, above all, the absorption in the course of the trajectory may be extremely difficult to evaluate *a priori*; there seems to be no proven method, as yet; each forecasting service is investigating its own method and is inspired largely by practical results gathered from experience.

We therefore give here only a very rough sketch of the type of calculation possible.

We saw in § 1.1.1 the value which the field strength would have in free space at a distance d for a transmitter of power W and gain g; this value decreases as $1/d$; but here we must take into account the following facts.

(a) The effective distance traversed by the radiation is greater than the surface distance between stations since the trajectory of the wave is oblique and consists of one or more excursions to heights of 70–190 miles; there is therefore a minimum path length (and therefore a maximum field strength) even if the distance between stations tends to zero.

(b) At every reflection from the ionosphere or from the earth, there is always a small loss of energy: probably 1–3 dB, except near the critical frequency of the ionized layer, in which case there is considerable "selective" absorption.

(c) The Earth's magnetic field produces a splitting of the ray and one of the components is, in general, lost (§ 6.3.6), giving a further source of attenuation.

(d) On the other hand, the concave curvature of the ionosphere

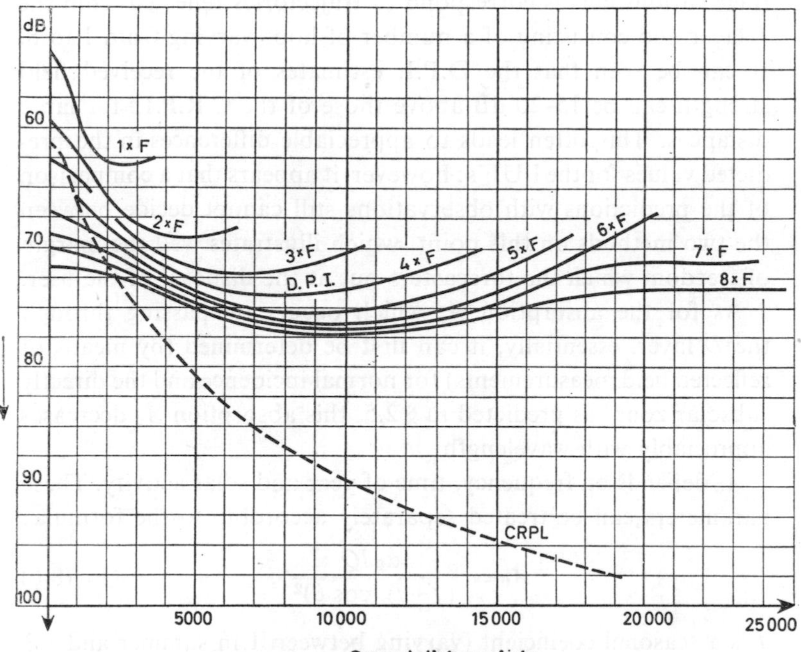

FIG. 8.16. Variation of field strength with distance, in the absence of ionospheric absorption, by the C.R.P.L. and D.P.I. methods. The ratio of the field strength which would exist in free space at 1 km to the received field strength at a given distance is plotted against the distance

can produce "focusing" effects, capable of concentrating the energy at certain distances and increasing the field strength.

Figure 8.16 shows the variation in the field strength as a function of distance, in the absence of ionospheric absorption (night field strength), according to the forecasting services of the C.R.P.L. (broken line) and the D.P.I. (full lines).

The C.R.P.L., on the basis of experimental data, assumes, for distances d greater than 3000 km (2000 miles) a $d^{-1.4}$ decay law and for distances below 3000 km an r^{-1} law, where r is the length of the radio trajectory reflected from the ionosphere. The D.P.I., on the other hand, bases its calculation on a more theoretical estimate of the "focussing" effects which reinforce the field: their various curves correspond to trajectories reflected from the F layer and consisting of a number of hops varying from 1 to 8. It can be seen that the D.P.I. estimates of the received field strength can be 15–20 dB above those of the C.R.P.L. for large distances. This often leads to appreciable differences in the predicted values for the LUF's: however, it appears that a comparison of the predictions with observations still cannot decide between the two methods on this point, which illustrates well the margin of freedom which the forecasters put at the disposal of the user.

As for the absorption A, which occurs on passing through the D layer, essentially, it can first be determined (by means of reflected field measurements) for normal incidence and the directly subsolar zone; as predicted in § 2.5, this absorption A_0 decreases appreciably with wavelength.

A_0 depends on frequency, time of year and solar activity. These parameters can be treated separately according to the formula:

$$A_0 \simeq \frac{\alpha_0 JQ}{(f+f_H \cos\theta)^2}; \tag{8.6}$$

J is a seasonal coefficient (varying between 1 in summer and 1·3 in winter for the C.R.P.L. and about the same for the D.P.I.); Q is a coefficient of solar activity (equal to $1+0{\cdot}005\bar{R}$ for the C.R.P.L., and $1+0{\cdot}0035\bar{R}$ for the D.P.I.); f is the frequency, f_H the gyrofrequency (order of 1·5 Mc/s,† see § 6.3.6) and θ the angle between the direction of propagation and the Earth's magnetic field.

To a first approximation, the C.R.P.L. adopted an $f^{-1{\cdot}92}$ law of variation of A_0 in terms of frequency, as being sufficiently close to the theoretical variation as $(f+f_H \cos\theta)^{-2}$.

Two corrections are then applied to A_0:

(a) A *decrease* as one moves away from the subsolar region, to account for the decrease in ionization level; the decrease

† 0·7 Mc/s at the geomagnetic equator, 1·6 at the geomagnetic poles.

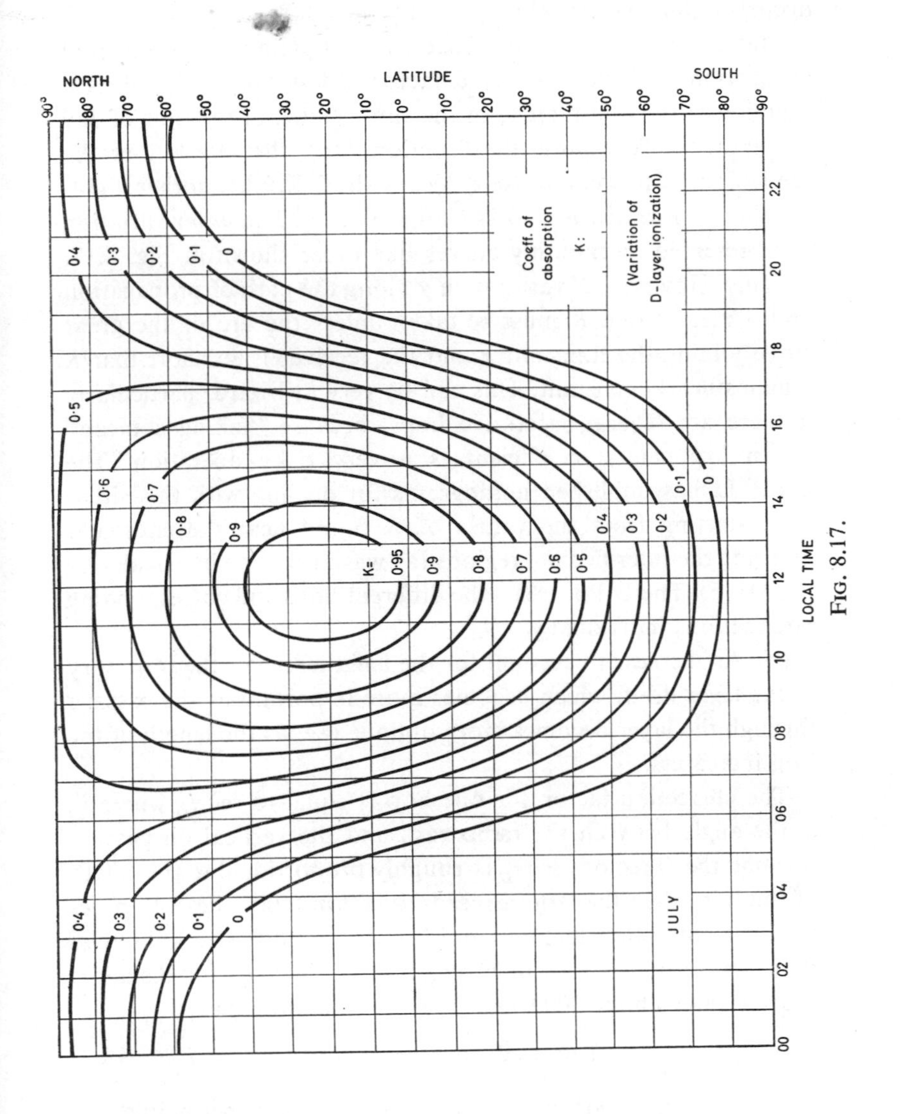
NORTH
LATITUDE
SOUTH
90°
80°
70°
60°
50°
40°
30°
20°
10°
0°
10°
20°
30°
40°
50°
60°
70°
80°
90°
Coeff. of absorption
K:
(Variation of D-layer ionization)
K=
0·95
0·9
0·8
0·7
0·6
0·5
0·4
0·3
0·2
0·1
0
JULY
LOCAL TIME
00
02
04
06
08
10
12
14
16
18
20
22

FIG. 8.17.

practically vanishes about midnight when the D layer seems to disappear almost entirely.

This correction can be obtained by multiplying the exponential absorption coefficient A_0 by a factor K less than unity. K is a function of the zenith angle of the sun (χ), i.e. of the time of day and latitude. To express the function $K(\chi)$, the D.P.I. uses the semiempirical formula $K=\cos^{3/4}\chi$ and the C.R.P.L. another slightly different formula $K = 0{\cdot}142+0{\cdot}858 \cos \chi$. The latter variation is represented by monthly curves like those shown in Fig. 8.17, for July. However, K varies with χ along the path of propagation and a mean value $\bar{K}$ must be taken unless the arc of the great circle joining the transmitter and the receiver is so short that K remains nearly constant. This can be very awkward, particularly if there are several successive hops, because "taking a mean" of an exponential coefficient is a very risky operation. The C.R.P.L. has published nomograms for dealing with this situation, starting from the values of K found near the ends and ignoring the intermediate regions (as was also done in calculating the MUF). The D.P.I. treats the different hops and corresponding attenuations individually.

(b) An *increase* to account for the lengthening of the trajectory in the layer itself, which becomes more important as the passage through the layer becomes more oblique (i.e. as the length of the hop increases).

The correction factor for one hop is equal to sec i_D where i_D is the angle between the radio rays and the vertical on passing through the D region; sec i_D is roughly proportional to the length of the hop when the latter exceeds 1000 km (600 miles). In practice, the correction factor is unity for short distance links (less than 400 km); it varies in proportion to distance for distances greater than about 3000 km. We can therefore put:

$$A = A_0 K \sec i_D = S_0 J Q K d, \tag{8.7}$$

where S_0 corresponds to the absorption coefficient per unit length of trajectory for a Wolf number of zero ($Q = 1$), in a subsolar region ($\bar{K} = 1$) and during summer ($J = 1$).

Figure 8.18 shows the variation of S_0 in terms of frequency,

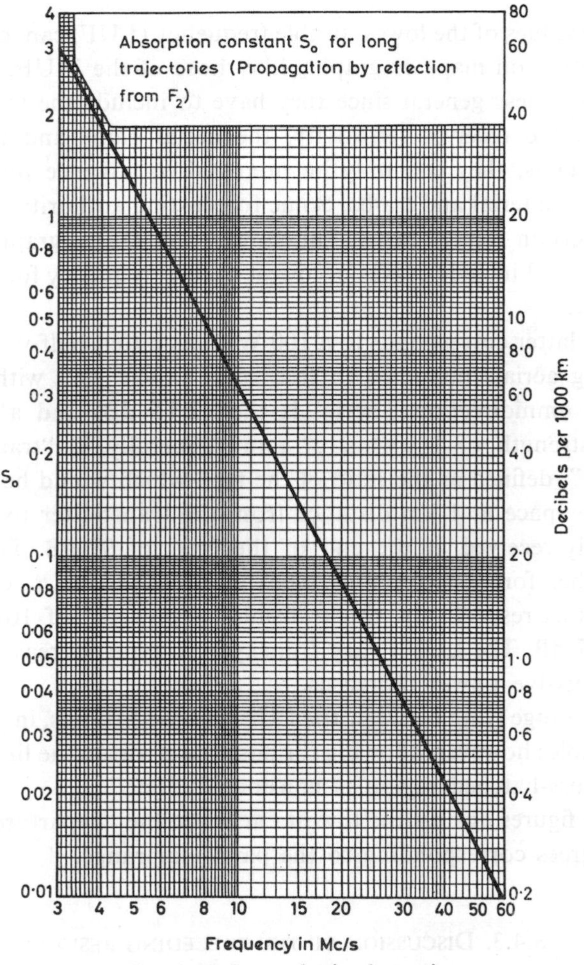

FIG. 8.18. Ionospheric absorption

according to the C.R.P.L. estimates. For distances lying between 400 and 3000 km, the C.R.P.L. has plotted curves giving the variation in the received field strength as a function of frequency over trajectories varying in length from 400 km in 400 km steps and each of the possible modes of propagation (one or two hops with reflection from *E* or *F*). D.P.I., on the other hand, again treats the various hops corresponding to each possible mode of propagation separately, whatever the length of the trajectory.

The values of the lowest usable frequency (LUF) can, of course, be plotted on maps or graphs, like those of the MUF; but they are much less general since they have to include the transmitter power, the receiver sensitivity, the aerial gains and the noise level. Thus, they can only be prepared in advance for a given type of material and traffic. Some forecasting authorities do envisage certain typical cases. This is why the D.P.I. graphs shown in Fig. 8.13 include curves of lowest usable frequency for a certain case *A*.

The latter corresponds to a 100 W transmitter, half-wave transmitting aerials (assumed in free space: their gain with respect to an omnidirectional aerial is then 2·15 dB) and a received field strength of 1 μV/m, or more generally, a "transmitting power", defined as the ratio of the field which would be received in free space at 1 km distance from the transmitter to the field actually received at the end of the link, of 97 dB. The D.P.I. publishes forecasts for four types of link—*Z*, *A*, *B*, *C*—corresponding, respectively, to "transmissive powers" of 107, 97, 87 and 77 dB. The LUF's vary from type to type, increasing as the transmissive power decreases.

The range lying between the LUF and the MUF is, in principle, all usable; however, allowing for possible errors in the limits, only the cross-hatched region is considered "safe".

The figures against the lowest curve are the "departure angles" in degrees corresponding to the particular range.

8.4.3. Discussion of the preceding results

It is quite evident that the preceding methods for forecasting the preferred wavelength suffer from several causes of inaccuracy. The problem, therefore, is to know what degree of confidence we can place in them. It is an awkward problem; but it appears to be possible to solve it by means of the following indications:

Let us first examine the case of "normal" propagation conditions, in the absence of "accidental perturbations, storms, etc.". Everything is then based on the prediction of the critical frequencies of the *steady E* and *F* layers. What is the accuracy?

For the relatively stable E layer, experience shows that the instantaneous values rarely fluctuate by more than $\pm 5\%$—and never by more than 10%—of the predicted value.

For the F_2 layer, the spread of the instantaneous values is clearly greater (Fig. 8.19): spreads of ± 20–25% are not uncommon. But between the *monthly average* of observed values and the curve *predicted three months ahead*, the agreement is usually better than 5%.

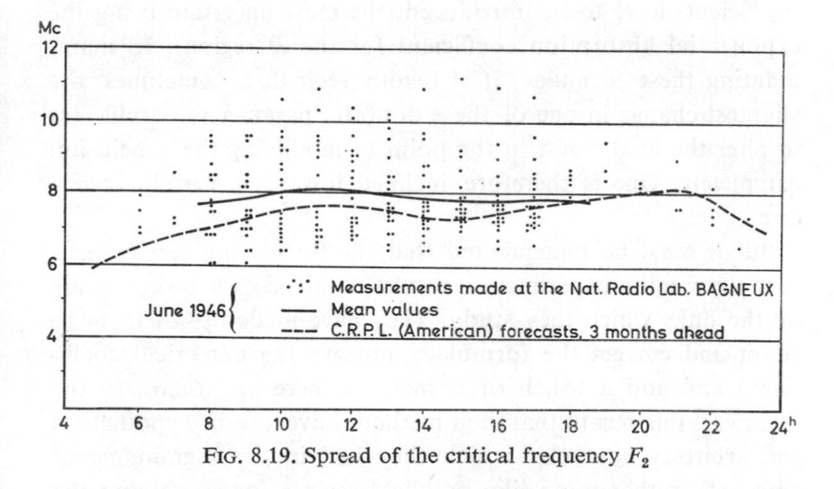

FIG. 8.19. Spread of the critical frequency F_2

The monthly spread of the values of the critical frequencies and, therefore, of the MUF's, relative to propagation trajectories reflected from the F_2 layer being accounted for, the monthly values of MUF forecast may be taken as median values not exceeded on 50% of the days in the month considered. But if a quasi-permanent traffic is desired, it will be wise to operate well above the median MUF. This is why the C.R.P.L. usually takes as the optimum frequency for traffic, for trajectories reflected from F_2, a frequency equal to 0·85 of the median MUF: this OFT is theoretically exceeded on only 10% of the days in the month. For trajectories reflected from E on the other hand, the C.R.P.L. makes no distinction between the median MUF and the optimum frequency for traffic. The D.P.I. also takes account of the monthly spread of the MUF's and the LUF's, calculating for each of them the values corresponding to probabilities of 90 and 30%.

Secondly, if we assume that the values of the critical frequencies (at normal incidence) for the E and F regions are exact, what errors do we introduce in calculating the maximum and lowest usable frequencies at oblique incidence for a given range?

As we have seen, some of our formulae are only approximate; they do not take everything into account; or else almost unknown coefficients have to be introduced (the most uncertain being the exponential absorption coefficient for the D region). In manipulating these formulae, it is readily seen that, sometimes, the slightest change in one of these doubtful parameters is sufficient to alter the final result to the point of modifying the conclusion completely. One is therefore inclined towards a certain scepticism.

But it must be remembered that the forecasting services can now draw upon experimental results obtained over several years on the links which they study. They have made use of them to adapt and correct the formulae, improve the numerical coefficients and add a touch of refinement, here and there, to the theories;† this means that their methods have lost the hypothetical and arbitrary character which they had in the beginning and now look rather more like practical recipes for combining the experience of the users with the researches of the geophysicists, astronomers and specialists in all categories.

This explains how, starting sometimes with very different formulae, but adding to them corrections and safety margins dictated by daily observation, the various services end up with results which agree with each other and, *on the average*, satisfy the users.‡

† We have already seen an example in the way in which the "American" method confines its calculations to what happens at the two ends of the trajectory, neglecting the intermediate hops, a simple-minded approach justified by the argument: "It works", to which there is no reply.

‡ On the subject of forecasting methods, their comparison and the value of the results, consult: Rawer, *Bulletin du S.P.I.M.*, particularly R.5, R.6 and R.7; *Die Ionosphäre* and *Wir. Eng.*, Nov. 1952, pp. 287–301; Notes prélim. Lab. Nat. Radio., nos. 150 and 166; Report 160 of the C.C.I.R.,

In particular, since the curves of MUF and LUF are calculated with a certain amount of care, the limits which they indicate are generally correct and there is even an appreciable probability of success for a radio link some way beyond them. If their ratio is greater than or equal to 1·4, then there exists between them a zone in which transmission looks as if it should be *very safe*, under normal ionospheric conditions. This is the zone indicated by cross-hatching in the graphs of Fig. 8.13.

Of course, there is still room for improvement.

For example, we have only considered the E and F_2 regions. But reflection is also possible from the F_1 region when it exists, and it can be taken into account, although its influence is small.

We have also seen (§ 6.3.2) that there is a "sporadic E" region, formed by heavily ionized clouds in the E region. The current tendency is to attribute a certain importance to it; the American bulletin of the C.R.P.L. gives two maps for its distribution, one of the "median critical frequency", which can be as high as 7–8 Mc/s in the subsolar region, and one of the "probability that the maximum usable frequency for a range of 2000 km reaches 15 Mc/s"; this probability is rarely less than 20% and often more than 80%; thus the result, quite often, is an appreciable increase in the MUF calculated from E or F_2 alone. Some authors† hold that "the sporadic E layer governs propagation over quite long periods, especially in summer".

However, there is still not sufficient data on the ionization of this layer.

Finally, we must remember that all the above is concerned exclusively with propagation under "normal" conditions (forecast 3 months ahead).

All these forecasts, therefore, go by the board in cases of

Los Angeles, 1959; Beynon, *Proc. Phys. Soc.*, July 1947, pp. 521–35; Circular no. 462 of the Nat. Bureau of Standards; Richard, *J. Télécomm.*, Aug. 1950, pp. 338–70; Niguet, *Revue des Transmissions*, May–June 1953, pp. 49–63; Halley, Lepechinsky and Mouchez, *Ann. Télécomm.*, Sept.–Oct. 1958, pp. 254–64.

† Doc. 224 (U.S.A.) to the C.C.I.R. London, 1953.

"abnormal" propagation, for all the types of perturbations listed in §§ 6.4.1–6.4.4: magnetic and ionospheric storms, sudden black-outs, etc.

These perturbations can completely modify the density and position of the layers and cause regular links to disappear, either by lowering the critical frequency (reduced ionization) or by increasing the absorption (increased ionization).

Ionospheric storms are particularly marked in the polar regions (auroras) where "normal" conditions of propagation may obtain for less than 50% of the time. But they can happen anywhere.

In the history of radio, several cases have been recorded in which radio contact was practically impossible over a more or less large part of the world for periods of several hours, and, in exceptional cases, several days (for example: October, 1926, interruption all over the world; from 8 to 15 June 1928, very marked attenuation; numerous ionospheric storms in August 1950, etc.).

One can obviously attempt to re-establish contact on some other frequency; but it does not seem to have been possible yet to establish the precise technique for this.†

Since users lack the ability to find a sure remedy for these accidents, it would be of some interest to them to be warned of them a little in advance. This is difficult, for most of them spring from unforeseeable solar perturbations (sunspots, flares, etc.). However, if some of these perturbations can be detected (with a "coronograph") the moment they appear at the edge of the Sun, one may get some idea of the effect which they will produce when they reach the middle of the disc, 6 days later.‡ This fact was used by the Germans in their wartime forecasting service.

In particular, ionospheric storms, due to the action of particles ejected by the Sun (§ 6.4.2), often occur 24–48 hr after sunspots pass the central meridian of the Sun. This appreciable transit time of the particles enables an attempt to be made at forecasting

† Maire indicated, for this case, the possibilities of frequencies above 30 Mc/s (*Annales Radioélec.* July 1951, pp. 197–203).

‡ It will be remembered that the Sun rotates on its own axis about once every 25 days.

storms. But this requires a knowledge as to whether the corresponding sunspots are really "active"; certain indications exist: reinforcement of solar noise at metric wavelengths (Chapter 7); a decrease in the amount of cosmic radiation received at ground level as the particles pass through interplanetary space; magnetic activity, etc. These "harbingers" of ionospheric storms enable them to be anticipated several hours in advance, with some degree of success. "Perturbation warnings" can thus be broadcast for the use of operators. But the percentage of successes is still limited (60–70%) and experiments are continuing.

All this shows clearly that ionospheric predictions still leave much to be desired. It would, moreover, be especially interesting for important stations to know at each moment and with certainty, in terms of frequency, the parts of the world with which contact is possible. To this end, one can use "back-scatter" probing apparatus, the principle of which was described in § 6.2.4. If one operates at a fixed frequency with a rotating aerial, one can measure the limits of the zone covered by reflection at this frequency (Fig. 8.20). More elaborate apparatus (with swept frequency) is being developed in the United States and Britain, and it is conceivable that in the near future the big stations will be equipped with them.†

In spite of the above-mentioned reservations, it can be claimed that the efforts made since the last war have considerably improved the reliability of radiocommunication at decametric wavelengths. The forecasts are, *on the average*, very exact and are extremely helpful to the user in adapting himself to diurnal, seasonal and yearly variations. They cannot, however, give *a guarantee* for any given time on a given day, either because of (normal) dispersion of the ionization of the layers or because of "abnormal disturbances, impossible to predict 3 months in advance"; still, for a certain number of the latter, "urgent short-term warnings" may inform users about probable trouble.

† We would mention here the suggestion that measurement apparatus using pulses enables two stations to determine for themselves their most favourable frequency and when and how to change it (Covill, *5th National Symp. Global Comm.* Chicago, 1961, pp. 65–8).

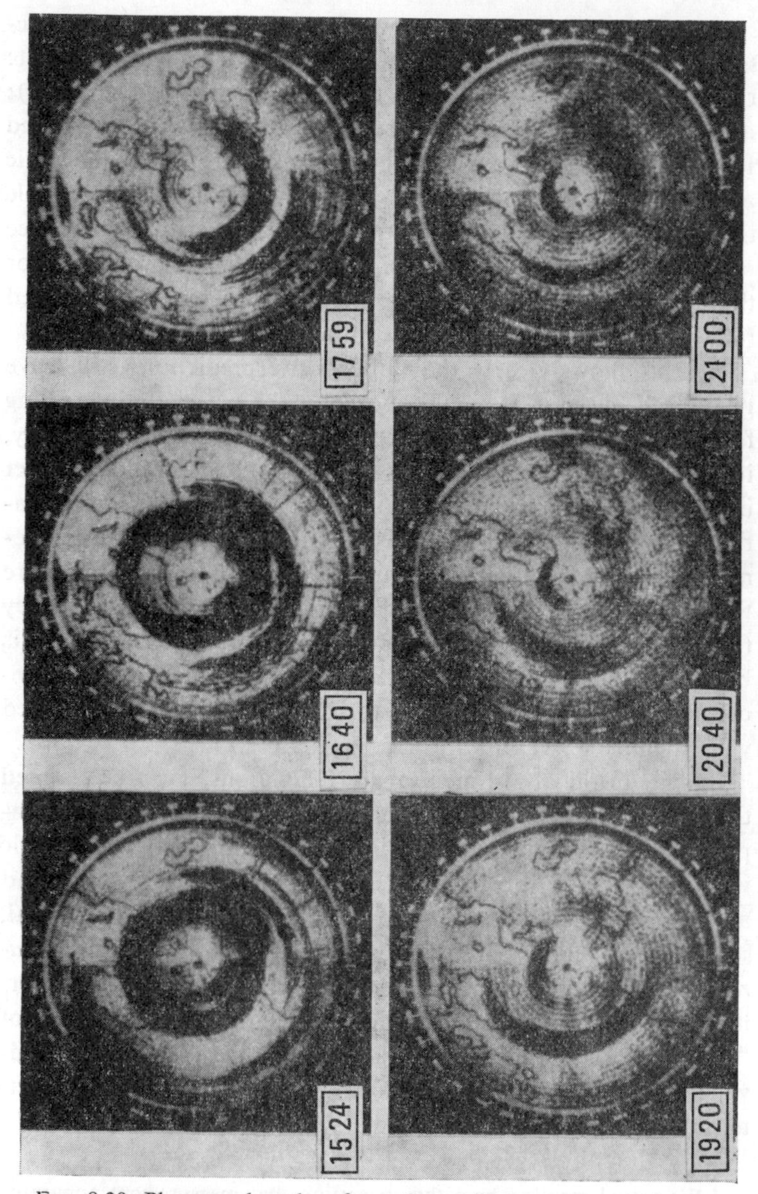

FIG. 8.20. Photographs taken from the oscilloscope screen of a set operating at 17 Mc/s (about 17 m wavelength), at six successive times on the same day. The radius of the area covered is about 5000 km around London (from Shearman and Martin, *Wireless Engineer*, August 1956)

In any case, it is very likely that the margin of certainty will increase further in the future as observations and theoretical analysis give us a better understanding of the variations in the ionosphere.

8.4.4. Examples

As the case of decametric waves is rather complicated, we believe that it is helpful to illustrate these considerations by a few examples.

We shall start from the curves and maps given above, both to demonstrate their value and to give some idea of the way in which the forecasting services work. But this is a purely pedagogic objective; in general, the user will not have to go to so much effort, the operation being too lengthy and intricate for him; it is essentially a job for the forecasting specialist who has at his disposal the results of old probings, all the recent data on solar activity, the results from existing links, etc., plus special training and a gift for interpreting data.

In principle, therefore, the user is assumed to possess the *periodical bulletins* of one of the forecasting services and to be content to extract the data, already prepared and best adapted to his own case.

The D.P.I. publishes[†] monthly forecasts, which are complete and ready for use, for:

(a) links of fixed length, in steps from 250 to 3000 km, in certain areas (Europe, Mediterranean, North Africa, northern Europe, West and Central Africa, Madagascar, Indo-China);

(b) point-to-point links for a large number of fixed trajectories;

(c) realizable links starting from Paris and Dakar on a certain number of fixed frequencies.

The C.R.P.L. monthly forecasts consist of a number of world maps which require the user to perform a series of fairly simple

†The reader can refer to this subject in the *Instructions for the Use of Forecasts for the Ionospheric Propagation of Radio Waves*, edited by the Centre National d'Etudes des Télécommunications (C.N.E.T.).

operations in order to determine the MUF's,† but which permit him also to make predictions for trajectories anywhere in the world, and in particular for trajectories which are not covered by the D.P.I. forecasts.

The C.R.P.L. has also announced a very complicated method for a monthly forecast of LUF's, introducing the predicted value of the Wolf number for the month considered.‡

We shall now give a few examples.

8.4.5. First example: broadcasting from France to Algeria

Consider a radiotelephony service from a transmitter of useful power 25 kW and a lozenge aerial of gain 16 dB, situated at the centre of France; reception in Algeria, at an average distance of 1200 km (750 miles) on any aerial, requires in summer a field strength of 100 μV/m (night and morning) up to 350 μV/m from 2 p.m. to 8 p.m. due to local interference.

What are the favourable wavelengths for July 1953?

For the *E* region, the maximum usable frequency for July 1953 and a trajectory of 2000 km with reflection in the subsolar zone is, according to Fig. 8.8, 17 Mc/s. The critical frequency corresponding to the *E* region will be $17/4{\cdot}78 \simeq 3{\cdot}5$ Mc/s.

As reflection occurs at a latitude of about 40°, the *E*-layer ionization map (Fig. 8.7) shows that it is necessary to reduce this figure as a function of time in the following ratio:

time	noon	10 a.m., 2 p.m.	8 a.m., 4.30 p.m.	6 a.m., 6 p.m.	4.30 a.m., 8 p.m.
factor	0·95	0·92	0·80	0·60	0·30

and during the night, to an even smaller figure.

† Instructions for using these forecasts have been published (*Handbook for C.R.P.L.*, "Ionospheric Predictions Based on Numerical Methods of Mapping", edited by the National Bureau of Standards, U.S.A.).

‡ A detailed description of the method of forecasting LUF's will be found in Circular No. 462 of the National Bureau of Standards.

The values thus obtained are plotted as curve 1, Fig. 8.21 and we take this as the MUF for the *E*-region (no increase in the safety margin seems necessary in view of the stability of this region).

Let us now pass to the F_2 layer. The C.R.P.L. ionization map for the given month (Fig. 8.9) gives us the values of the maximum usable frequency at $d = 4000$ km (2500 miles); for latitude 40°,

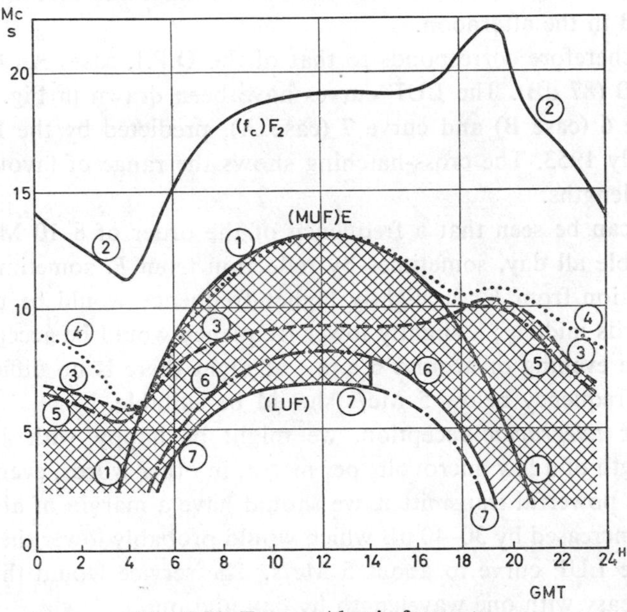

FIG. 8.21. France–Algeria forecasts

we obtain curve 2, in terms of time. To obtain the maximum usable frequency at $d = 1200$ km, it is necessary to multiply by the coefficient $(f_d)_{1200}/(f_d)_{4000}$ taken from Fig. 8.4, say about $1{\cdot}70/3{\cdot}30 = 0{\cdot}52$.

A small margin of safety might be added, say 10%, to take account of horary dispersion and in this way curve 3 was obtained, giving the MUF for the F_2 layer.

Naturally, the higher of the two MUF's for the *E* and F_2 regions constitutes the limit (heavy line on the diagram).

By way of comparison, curve 4 shows the MUF forecast by careful application of the C.R.P.L. method for Paris–Algiers and curve 5 the "upper limit of the zone of safe frequencies" according to the D.P.I. for the same path.†

Let us pass to the other limit: the lower limit due to ***absorption.*** In free space, the "transmissive power" of the link for a 25 kW transmitter, a transmitting aerial gain of 16 dB and a field strength at the receiver of 100 μV/m, night and morning, and 350 μV/m in the afternoon, would be about 97 dB, night and morning and 87 dB in the afternoon.

It therefore corresponds to that of the D.P.I. cases A (97 dB) and B (87 dB). The LUF curves have been drawn in Fig. 8.21: curve 6 (case B) and curve 7 (case A), predicted by the D.P.I. in July 1953. The cross-hatching shows the range of favourable wavelengths.

It can be seen that a frequency of the order of 8–10 Mc/s is possible all day, sometimes by reflection from E, sometimes by reflection from F_2; a slightly higher frequency would be usable towards midday; a slightly lower frequency would be acceptable in the evening or around 6 a.m.; at night there is no difficulty: any frequency below 5 Mc/s should be suitable.

For *telegraphic* reception, we might be happy with a field strength of a few microvolts per metre; in other words, even with a less powerful transmitter, we should have a margin of absorption increased by 30–40 dB which would probably lower the peak of the LUF curve to about 5 Mc/s. The service would then be very easy with one wavelength by day and one by night.

The only thing to watch is that (according to the figures against the curves in Fig. 8.4) the departure angle of the transmitted beam above the horizon is quite different, depending on whether reflection is from the E layer at a height of 120 km (75 miles) (it is then about 12°) or from the F_2 layer at 250–400 km (150–250 miles) (it rises to about 35°). It will be necessary to take account of this in the choice of directional aerial.

All the above is valid for July 1953. For any other time it would be necessary to begin the operation again with the appropriate maps; the difference would be appreciable from one month

† On the D.P.I. graph the curve for the MUF rises to 20 Mc/s at noon, probably by possible consideration of the sporadic E region.

to the next. In winter, the useful frequencies will be decreased; on the other hand, as the solar activity increased from 1953 to 1958, the useful frequencies also increased up to 1958 and returned to about the same values in 1964.

8.4.6. Second example: Paris–Madagascar aircraft

We now come to a more complicated case in which the distance varies rapidly. An aircraft leaves Paris at 12 noon (G.M.T) in July 1953 at a speed of 500 k.p.h., bound for Madagascar. It uses a transmitter of radiated power $W = 0{\cdot}01$ kW; the transmission is received in Paris by c.w. telegraphy on a well-oriented lozenge aerial whose mean gain is greater than 10 dB. A forecast is required for the wavelength to be employed for a direct link with Paris.

Preliminary calculations

The great circle arc *PM* from Paris to Madagascar is plotted on the map, Fig. 8.5 (by means of the graph 8.6), and marked off in distances. It is seen (Fig. 8.22) that the maximum range s about 7500 km and that the difference in longitude is up to 45°, i.e. a time difference of 3 hr. The first columns of Table 8.6 can now be deduced (knowing the speed).

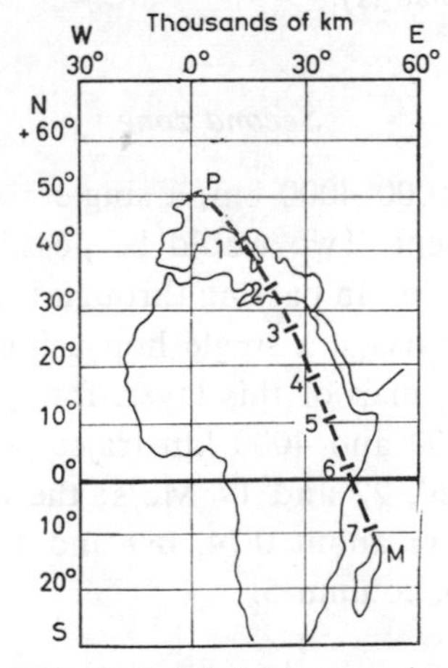

Fig. 8.22. Paris–Madagascar trajectory

Determination of the optimum frequency—first zone

In a first zone, up to about 1500 km, reflection can take place (by day) either from the E layer or from the F layer, at points of latitude 45° to 40°N and local time 1 to 4 p.m. For the E layer, the map (Fig. 8.7) shows that the ionization level at these points is about 0·9–0·85 of the subsolar maximum, i.e. about 3·5 Mc/s (see preceding example).

However, the critical frequency f_c at normal incidence must be multiplied by the coefficient f_d/f_c appropriate to the obliquity, i.e. the distance; from Fig. 8.4 it goes from 2 for $d = 500$ km to 4·2 for $d = 1500$ km. We thus obtain the values of the $(\text{MUF})_E$ indicated in column 4 of Table 8.6.

For the F_2 layer, the map (Fig. 8.10) gives, in the reflection region, a critical frequency $f_d \simeq 19$ Mc/s for an obliquity corresponding to a link of 4000 km.

For links of only 500–1500 km, this figure must be reduced, according to Fig. 8.4 in the ratio of $1{\cdot}1/3{\cdot}4 = 0{\cdot}32$ to $2{\cdot}1/3{\cdot}4 = 0{\cdot}62$. We thus obtain the values of the $(\text{MUF})_F$ shown in column 5, Table 8.6.

The rest of Table 8.6 gives the values of the MUF and LUF deduced from the use of the C.R.P.L. (map of the zones of silence, Fig. 8.11) and D.P.I. methods (Fig. 8.13; our data correspond approximately to case A).

Second zone

At distances of 1000–4000 km, a single hop using the E layer is no longer sufficient. Two would be possible but at the price of excessive absorption in passing through the D layer. We therefore continue to envisage a single hop using the F_2 layer.

According to the map of this layer, for reflection at the mid-points of 2000, 3000 and 4000 km trajectories, the frequencies $(f_d)_{4000}$ are about 20, 22 and 17 Mc/s; the obliquity correction factors (Fig. 8.4) are about 0·74, 0·9 and 1, giving the MUF's shown in Table 8.6, column 5.

TABLE 8.6. *Predictions for Paris–Madagascar flight* (July 1953)

Distance (km)	Time (Paris)	Local time at point of ionospheric reflection	Calculated here		C.R.P.L. predictions			D.P.I. predictions		
			MUF *E* layer (Mc/s)	MUF *F* layer (Mc/s)	MUF (Mc/s)	LUF (Mc/s)	Optimum (Mc/s)	MUF (Mc/s)	LUF (Mc/s)	Optimum (Mc/s)
1	2	3	4	5	6	7	8	9	10	11
500	13	13	6·1	6·1	6·5			9	3·5	5–6·5
1000	14	14.30	7·5	8·7	10			14	4	6–9
1500	15	16	12	11·6	13			18	5	7–12
2000	16	17		15	16			20	5	6–15
3000	18	19.30		20	19			22	<2	2–18
4000	20	21.30		17	17					
						(Jibuti)			(Jibuti)	
5000	22	0		9–12	12	4	3–10	14	<4	5–10
6000	0	2.30								
						(Antananarivo)			(Antananarivo)	
7500	3	6			10–12		6–8	9	<4	5–8

Third zone

From 4000 to 7500 km, two hops using the F_2 layer are necessary. The obliquity for each hop therefore decreases, as does the factor f_d/f_e; the map of the F_2 layer shows that both reflections occur in regions where the limit frequency for 4000 km hops varies from 11 to 15 Mc/s; for a 2500 km hop it would be reduced to about 9–12 Mc/s.

For the Paris–Antananarivo link, the results of the D.P.I. and C.R.P.L. predictions indicate "safe" frequencies of 6–8 Mc/s up to about 4 a.m. but, from about daybreak, the link becomes precarious (the LUF rising to coincide with the MUF) or even impossible (see Fig. 8.13, bottom right). The prediction would therefore have to be completely revised and would become more pessimistic if the aircraft set off in the evening to arrive in the morning.

8.4.7. Third example: very long range links

When we come to study very long range links, especially between points with a wide difference in longitude—for example, Paris–Saigon, Dakar–Tahiti, etc.—i.e. requiring a large number of successive hops in areas with various degrees of sunlight, we encounter more and more frequently the difficulty of the previous example (Paris–Madagascar); in daylight: the MUF curve is lowered because parts of the trajectory are in "darkness"; the LUF curve rises because of the "daylight" parts. The link requires more and more numerous and precise changes of frequency and finally becomes impossible at certain times.

If the trajectory passes near the pole, it is necessary to predict an additional absorption as well.

These conditions lead to the conclusion (already mentioned in § 6.2.3) that the shorter arc of the circle joining the stations is not always the best possible trajectory and that it may be advantageous to use the other part of the great circle, for example, to cover 30,000 km at night instead of 10,000 by day. This conclusion is confirmed by experience and it often happens that operators turn their reflectors round so as to send the energy in the direction opposite to that of the correspondent.

Figure 8.23 shows a D.P.I. prediction relevant to this case: on the Dakar–Tahiti trajectory (about 15,500 km, longitude diffe-

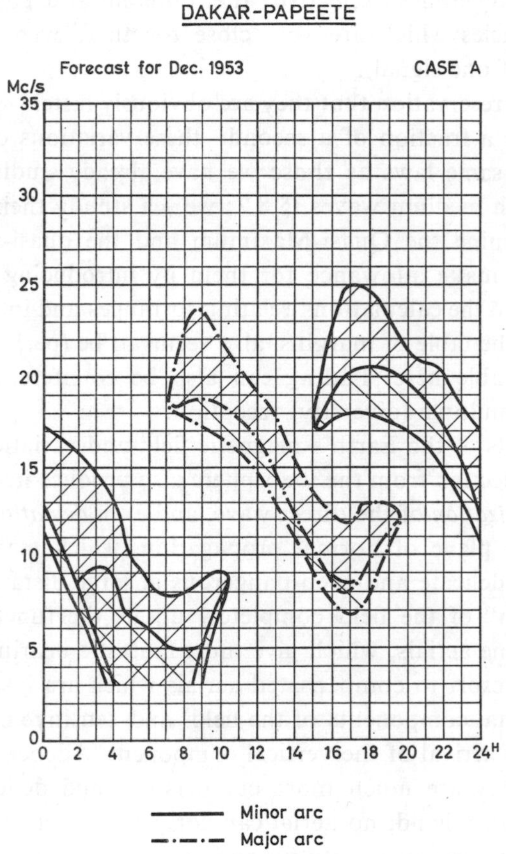

DAKAR-PAPEETE

Forecast for Dec. 1953 CASE A

Minor arc
Major arc

FIG. 8.23.

rence 9 hr) the direct link (*minor* arc) becomes impossible from 10 a.m. to 4 p.m. G.M.T. when the Sun illuminates one part of the trajectory whilst the rest is in darkness. But it is then possible to communicate using the *major* arc (dashed curve).

8.4.8. Fading at decametre wavelengths

All the above applies to the median values of the field strength. But interference with the various components reflected from nearby points of the ionosphere introduces rapid variations or "fading", sometimes "selective" (i.e. different at a given instant for frequencies which are very close together, even inside the spectrum of one signal).

With the reservation that they are obviously faster—sometimes taking only a fraction of a second—these variations obey more or less the same laws as those we have already studied in connection with medium waves (§ 8.2; we)ca stnudy their distribution, determine the Quasi-Maximum and the quasi-minimum, and finally make allowance for them by introducing a "safety margin" into the calculations relating to ranges and interference; Table 8.3, the table of margins, also seems to be (perhaps especially) applicable here. Fading can also be countered, to some extent, by multiple (diversity) reception.

The effects of the Earth's magnetic field and variations in the point of reflection from the ionosphere also produce irregularities in the *polarization* of the direct wave, and even *deviations* outside the vertical plane of normal propagation, which makes *goniometry* very delicate and even dangerous at large distances. "Depolarization" of the field completely upsets the functioning of normal frame aerials, which give most fantastic bearings.

We can resort to compensated aerials which are insensitive to the horizontal components of the field† and sensitive only to the direction of arrival of the vertical component (Adcock and similar); but they are much more cumbersome and delicate.

On the other hand, no aerial can correct for accidental deviations of the ray outside its normal trajectory. Such deviations can amount to several degrees in the zone of good reception at

†The horizontal components of the field do not vanish in the vicinity of the ground because at these frequencies the ground is no longer a perfect conductor (see § 2.2). Since aerials are slightly elevated, moreover, horizontal polarization is usable and nearly equivalent to vertical polarization (§ 4.5.1.): this is the principle of lozenge aerials and also explains why any shape fo aerial whatever can be used for reception.

large distances but are worse still in and near the "zone of silence" where the ray can be returned by diffraction and diffusion from the normal or "sporadic" E layer, as if it came from a point situated thousands of kilometres from the transmitter.

This is a serious handicap in the utilization of short waves for aerial navigation and shows why the use of long or medium waves (which are reflected far more uniformly by the E layer) retains its supporters.

The influence of *obstacles* near the receiver is also formidable and, on ships for example, necessitates putting S.W. goniometers at the mastheads, with all sorts of precautions.

8.5. Propagation of metric and shorter waves

In this section we recapitulate briefly:

(a) the role of the ionosphere;
(b) that of the ground and the troposphere;
(c) that of obstacles;

and we shall take as an example the calculation of the electromagnetic detection of an obstacle.

But first, let us remember that in this range, the polarization is almost irrelevant: although vertical polarization is theoretically better near the ground, horizontal polarization becomes equivalent as soon as aerials are slightly elevated. One sometimes chooses between the two for secondary reasons, such as slightly unequal sensitivity to certain types of interference or to re-radiation from certain obstacles.

Notice also that very short waves, which are not reflected by the ionosphere (except occasionally in the case of metric waves) and which lend themselves well to the use of highly directional aerials (particularly at decimetre and centimetre wavelengths), are obviously ideal for direct line-of-sight links (such as television) as well as for communication between the Earth and space-craft orbiting above the ionosphere (§ 8.7). The propagation conditions then approach those of "free space"; however, they may be appreciably modified by absorption and refraction in the various atmospheric layers. We shall first make the necessary reservations on the subject.

8.5.1. Role of the ionosphere

The boundary between "decametric" and "metric" waves is obviously not very well defined as far as their propagation is concerned.

However, as we have already seen, *normal* ionospheric layers are rarely sufficiently ionized to reflect frequencies higher than 30 Mc/s (even when transmitted tangentially to the ground); thus, when we come to "metric" waves, we find we have a new problem to cope with: that of guaranteeing a *regular long-distance service* with a link of the same type as those currently in use at decametric waves and, in particular, with transmitter powers of the same order.

This does not mean to say that ionospheric propagation over large distances is quite impossible at metric waves; as we have already pointed out, the ionization level of the F_2 layer can accidentally exceed the normal (especially in November–December, in periods of maximum solar activity); the "sporadic E" layer can reflect obliquely frequencies up to 50, 60 or sometimes even 80 Mc/s for a considerable proportion of the time. "Diffusion" by these layers is also possible. This explains:

(a) the reception of European television signals (at wavelengths of about 6–7 m) in the U.S.A. in certain winters;

(b) the interception of London police signals (on wavelengths of the same order) by German listening posts during the 1939–45 war;

(c) mutual interference between television or f.m. stations at unexpected distances.

But although these phenomena are common at certain times, they do not have the permanent uniformity ordinarily required by radio services.

They are, in any case, difficult to predict, in our present state of knowledge.

We shall not, therefore, discuss them any further.†

On the other hand, we have seen that quasi-permanent links are possible over trajectories of the order of 1000–2000 km (600–1200 miles) at wavelengths of the order of 7–10 m, provided very high transmitter powers and aerial gains are used ("ionospheric diffusion" links, described in § 6.3.5). However, these links are subject to perturbations, and sometimes interruptions, due to interference caused by distant stations, even less powerful ones as a consequence of abnormal ionospheric reflections. The risks of interference are particularly likely during years of strong solar activity; they decrease as the frequency rises (for example, from 30 to 40 Mc/s), but the useful "diffusion" signal decreases at the same time, which complicates the choice of optimum frequency.

We have also mentioned, in § 6.3.7, the possibility of establishing communications by reflection from meteor trails. At wavelengths of 6–8 m the field from meteoric reflection appears as a train of pulses with amplitudes greater than the mean level corresponding to the "ionospheric diffusion" field and with very variable durations and recurrence frequencies: the durations are normally of the order of a few hundredths to a few tenths of a second and the mean rate is of the order of a few pulses per minute, although occasionally pulses of several seconds (or even, exceptionally, several tens of seconds) are observed, and sometimes several minutes or even tens of minutes go by between two successive pulses.

It has been suggested that, if it was possible to design a system which only transmitted during pulses, a small power (of the order of a kilowatt instead of several tens of kilowatts for ionospheric diffusion) and fairly low gain aerials would suffice to set up a link. The Canadian Army has built such a system, in radiotelegraphy, known under the code name of "Janet". The wavelength is of the order of 7·5 m. Since transmission is possible, on average, for only 5–10% of the time, the signal is recorded on magnetic tape which is played back, at suitable times, at a speed of about 20 times normal so as to preserve the average speed of the tele-

† See, for example, Morgan, *Proc. I.R.E.*, May 1953, pp. 583–7.

graphic signal. If it is desired to transmit from a transmitter T towards a receiver R, modulation is applied to T only during the periods when the pulses are being received at R. In order to indicate when these occur, a return link transmitter, t, placed near R, transmits a c.w. signal on a wavelength close to that of T; t is received on a receiver r, placed near T. The arrival of a pulse switches on the modulation of T. The switching time is of the order of a few hundredths of a second. It would appear that practically permanent transmission would be ensured, with a percentage error in telegraphy of the order of only one in a thousand. A power of 1–3 kW and aerials with a gain of the order of 12–15 dB are sufficient. As we saw in § 6.3.7, the regions of maximum reflection may be situated outside the vertical plane containing the transmitter and receiver, which means swinging the aerial beams by a few degrees in azimuth with respect to this axis. One great feature of these links is their relative secrecy; in fact, pulses received outside a limited region around the receiving point very rarely correspond to periods when modulation is applied. Plans are already being discussed for transmitting other than telegraphy: attempts to transmit pictures, multiplex telegraphy and even telephony have been made. Attempts have also been made at higher frequencies, up to about 80 or 100 Mc/s.

8.5.2. Role of the ground and the troposphere

As we saw in Chapters 4 and 5, the nature of the surface is almost irrelevant, except in so far as it is irregular. The essential factor is the height of the stations above the surface, or, in other words, the ratio of the distance to be covered to the "optical range".

We indicated the necessary subdivision of the problem according to the values of this ratio: the zone of "free propagation", "interference" zone, "diffraction" zone, separated by an "intermediate" zone; and we gave examples of the calculation of the field strength in these various cases.

We have also indicated the role of the troposphere in "normal

TABLE 8.7. *Propagation of very short waves*

Distance d (km); height h_1, h_2 (m); optical range d_0 (km) $= 4(\sqrt{h_1}+\sqrt{h_2})$

Case	Rôle of the ground	Rôle of the troposphere	Results	Examples
Large heights $h_1 \gg d$ $h_2 \gg d$	None. Only direct wave appreciable (§ 4.2)	Absorption or refraction small as long as $\lambda > 3$ cm	Field strength good and stable	Links between aircraft
$d < d_0$ "line of sight"	"Fresnel zone" reflection (§ 4.3.2) and *interference* with direct ray (§ 4.3.1)	Ditto	"Lobe" diagram, periodic variations with d, h_1 or h_2, maxima doubled, minima dropping possibly to 1/10 of the maxima	Air–ground links, Hertzian beams, television, etc. (§ 4.3)
$d \approx d_0$ optical range	*Intermediate* zone. Interpolate between the previous case and the following (§ 4.6)	Increase in the apparent Earth ray (variable with time)	Rapid fall in field strength but without discontinuity, variations with time	Ditto (§ 4.6)
$d_0 < d < < 2d_0$	"Shadow" or "diffraction" zone. Calculate the field strength at ground level, or at the critical height, and then multiply by a "height gain" (§ 4.5)	Ditto	Ditto	Ditto (§ 4.5.3)
$d > 2d_0$ (up to about 1000 km and beyond)	Negligible diffracted field	1. At certain periods: *superrefraction* 2. At any period: *tropospheric diffusion*	Formation of "ducts", abnormal ranges Weak permanent field, but usable for transmission	Radar (§ 4.4.2) Interference (5.3.2) "Trans-horizon" links (§ 5.3.3)

refraction" (§ 5.2), "superrefraction" (§ 5.3.2), "diffusion" or "partial reflection" (§ 5.3.3) and finally "absorption" (§ 5.4).

There is therefore no point in repeating these details here. Instead, we shall summarize, by means of a general table (see Table 8.7), the various cases encountered, the formulae to use and some examples.

Bibliography

1. C.C.I.R. *Recommendation 370* and *Report 240*, 1963.
2. Direct "optical" links or links just beyond the horizon:

BULLINGTON, *Proc. I.R.E.*, Oct. 1947, pp. 1121–36.
BULLINGTON, *Proc. I.R.E.*, Jan. 1950, pp. 27–32.
GUENTHER, *Proc. I.R.E.*, Sept. 1951, pp. 1027–34.
NORTON and ALLEN, *J. Brit. I.R.E.*, March 1951, pp. 93–100.
NORTON and RICE, *Proc. I.R.E.*, April 1952, pp. 470–4.
VOGE, *Onde Électrique*, June 1954, pp. 491–8.
BATTESTI and BOITHIAS, *Ann. Télécomm.*, July–Aug., 1964, pp. 173–87. *C.C.I.R. Report 228*, 1963.
ARNAUD, *Onde Électrique*, Nov. 1964, pp. 1099–106 (Television).

3. "Trans-horizon" links:

MELLEN, MORROW, POTÈ, RADFORD and WIESNER, *Proc. I.R.E.*, Oct. 1955, pp. 1269–81.
NORTON, *I.R.E. Trans. Comm. Syst.*, March 1956, pp. 39–49.
WIESNER, *Onde Électrique*, May 1957, pp. 456–61.
YEH, *Commun. and Electronics*, Nov. 1958, pp. 707–16.
VOGE and DU CASTEL, *Echo des Recherches*, 1959, No. 35, pp. 22–3.
DU CASTEL, *Onde Électrique*, Jan. 1960, pp. 9–18.
X., *Proc. I.R.E.*, Jan. 1960, pp. 30–44.
CHAVANCE, *et al.*, *Revue technique C.F.T.H.*, June 1961, No. 34.
BARBERA *et al.*, *Revue télécomm. I.T.T.*, **37**, 3, 1962, pp. 80–8.

8.5.3. ROLE OF OBSTACLES

The role of obstacles is important and complex. Here again a distinction has to be made.

Let us consider first one isolated obstacle, near one of the stations, for example the receiver. It produces a re-radiation which is superposed on the primary radiation from the transmitter.

If the receiving aerial is sufficiently directional and the arrival directions of the primary and secondary fields are different enough, it may be that the influence of the secondary field is small or negligible: this is usually the case for centimetre waves and this is what makes "Hertzian beams", for example, fairly independent of the siting of neighbouring objects.

However, if the receiving aerial cannot discriminate between the directions of arrival of the two fields, their effects are additive with a certain phase difference: the resultant e.m.f. may therefore be increased or decreased and the slightest relative displacement of the aerial and the obstacle or the slightest variation in frequency can cause the resultant to pass through marked maxima and minima. This is what happens in television receivers at metric waves, where many peculiarities have been noted: the picture is good or bad depending on whether certain nearby conduction objects (or the aerial) are placed this way or that (whether a venetian blind is up or down, etc.). But this effect, although usually objectionable, can sometimes be put to good use as a "screen" against interference.†

These irregularities may be "selective", i.e. critical functions of frequency, to the point of showing up in the actual spectrum of a signal and introducing considerable distortion into the transmissions by "frequency modulation".

An obstacle interposed between the transmitter and the receiver can produce very different and paradoxical effects.

In general, it reduces the field strength (see § 4.8); however, if it lies in the path of the ray reflected by the ground, and especially in the "Fresnel zone" surrounding the point of reflection, it may attenuate the reflected ray more than the direct ray, which (in the lower lobe of the diagram or in the neighbourhood of the minima) eventually leads to an *increase* in the received field. It has often been proposed, therefore, and sometimes realized with success, either to mask the said "Fresnel zone" by an "absorbent cover" or a small vertical screen, or, if the ground is irregular in that region, to move the stations about (by trial and error) until the unwanted reflection passes through a minimum.‡

† Morrow *et al., Proc. I.E.E.E.* June 1963, pp. 955–6.

‡ See, for example, Bussey, *Proc. I.R.E.*, Dec. 1950, p. 1453; Bussey, *Proc. I.R.E.*, June 1951, p. 718; Gough, *Marconi Review*, Oct.–Dec. 1949, pp. 121–39.

On the other hand, even if the obstacle is in the path of the *direct* ray, it cannot be said with certainty that it will always produce an attenuation; we saw, in § 4.8.3 (Fig. 4.29, especially) that this is not always true: by suppressing the second Fresnel zone it can, on the contrary, produce a reinforcement.

This fortunate result can be even more marked if we compare the effect of the screen with that which would be produced, in its absence by the curvature of the Earth; for stations very near the ground, it was seen that this curvature created an absorbent "mask" all the way along the trajectory; this can result in an attenuation greater than that due to diffraction by a "knife-edge" screen placed at the mid-point.

This point is easy to verify: for two stations 10 km apart, at an elevation of 1 m at a wavelength of 1 m, there is no optical path; the field strength is obtained by taking the ground level value (Fig. 4.13, over land): 2 μV/m for 1 kW radiated, and then multiplying by the "height factor" (Fig. 4.21, dry ground, vertical polarization): approximately 3; total $2 \times 3 \times 3 = 18$ μV/m.

Let us now suppose that a thin screen (a hill, cliff, etc.) of height 10 m is interposed at the mid-point of the trajectory and let us apply Sacco's formula (Fig. 4.30); we have

$$d_1 = 5, \quad d_2 = 5, \quad h = 0{\cdot}1, \quad \lambda = 0{\cdot}001 = z_1 = z_2;$$

the second term in the brackets is negligible, we find $v = -2{\cdot}8$ and, from Fig. 4.29, an attenuation of 0·08 with respect to the free field (which is 30 mV/m for 1 kW radiated), giving a diffracted field strength of 2·4 mV/m, much greater than field strength after following the Earth's contour.

This is an extreme, simplified case,† but the reinforcement produced in a metre or decimetre wave link, by an intervening obstacle, such as a hill or a mountain range, has been observed in practice.‡

† See also §§ 4.8.2 and 4.8.3 for the case of a spherical hill.

‡ See, for example, Dickson *et al.*, *Proc. I.R.E.*, Aug. 1953, pp. 967–9. Reference is made therein to Doc. 190 (Japan) of the C.C.I.R. London 1953. There is also the case where a "passive relay", consisting of a receiving aerial feeding a transmitting aerial, is placed at the top of a hill; this is equivalent to retransmission.

For example, over a 260 km link at a frequency of 38 Mc/s, with a 2600 m mask at the mid-point of the trajectory, the received level (38 dB below the level for propagation in free space) was 73 dB above the theoretical level of propagation by diffraction round a spherical Earth in a "standard" atmosphere, and more than 20 dB above the mean field strength from "tropospheric diffusion" over links of comparable length. The stability of the received signal with time was excellent as well. Similar observations have been made at decimetric wavelengths.

This fortunate effect—at first sight paradoxical—of intervening obstacles playing an important part in certain links has been the subject of numerous experiments; the results are not always in agreement with the theoretical predictions of § 4.8.3, which is not surprising since real obstacles are neither infinitely thin nor spherical nor cylindrical, as was assumed.

For example,[†] Lacy (*Annales Telecommunications*, May 1957, pp. 66–70 and Institute of Radio Engineers, National Convention, 1957, vol. I, p. 32) summarized the results of his experiments on forty sites in California in an array which seems to fit the following formula:

$$L\,(\text{dB}) = 10 \log d + 20 \log h + 30 \log F + 8,$$

where L is the total transmission loss, h is the height of the obstacle (in metres), d the distance between the stations (in kilometres), F the frequency (in megacycles per second); it is as if an "obstacle attenuation" proportional to h^2F/d were added to the "normal free space loss" proportional to d^2F^2.

It is difficult to estimate the accuracy of this formula whose merit is obviously its great simplicity.

However, it can only be applied when there is on the trajectory of a link an obstacle whose diffracting edge is approximately perpendicular to the trajectory and is unique and visible from the two ends of the link; the received field seems to fall considerably when we are dealing with certain more complex obstacles

† See also Bullington, Radio Propagation Fundamentals, *Bell Syst. Tech. J.*, May 1957, p. 687; and Jowett, *Proc. I.R.E.*, part B, Mar. 1960, pp. 141–9.

consisting of several diffraction edges close together (mountainous country with several approximately parallel ridges, for example).

When the two stations are masked from each other by a screen but there exists, either on the crest or at the side, a high point visible directly from both, an artifice which is sometimes employed is to place at this point a *passive reflector* (plane or parabolic mirror) to convey a fraction of the radiation from the transmitter to the receiver (§ 4.8.6). Naturally, the feasible dimensions of this reflector soon limit its efficiency but the method has been employed in certain cases.†

Let us finish by considering the case in which there is, in the vicinity of one of the terminal stations of a link, not one obstacle but a *large number* of obstacles of various shapes and sizes (trees, houses, metallic superstructures, cars, etc.)—such as would normally be found in built-up areas for broadcasting, police radio, etc. The effects of all the secondary re-radiations then combine in a disordered manner to give unpredictable fluctuations; their amplitude ranges from 12–20 dB along a road to 40 dB in a town. In addition, obviously, as all the energy dissipated by the currents induced in these obstacles themselves is supplied by the primary field, the latter becomes, on average, considerably attenuated; as we saw in § 4.8.5, a reduction of 10–35 dB can be expected; it can only be prevented by elevation of the aerials into the perfectly clear region, above all the obstacles.‡

So far, we have only spoken of the amplitude of the field; but, of course, its polarization and apparent direction of arrival can also be changed by reradiation from obstacles.

Consequently, *goniometry* becomes unreliable.

If we use "differential" goniometers, with frame or Adcock aerials, which work on the principle of the extinction of the received wave, the presence of a spurious component, however weak, can lead to gross errors: this is the case for the majority

† See Rider, *Marconi Review*, 2nd quarter 1953, pp. 96–106; Colavito *et al.*, *Proc. I.E.E.E.*, Nov. 1963, pp. 1423–30.

‡ See, for example, Goldsmith *et al.*, *Proc. I.R.E.*, May 1949, pp. 556–63; Dreyer, *Electronics*, Oct. 1949, pp. 82–5; Bertaux *et al.*, Note prélim. Lab. Nat. Radio., no. 141, March 1950; Epstein, *Proc. I.R.E.*, May 1953, pp. 595–611.

of metric wave goniometers, for which it is necessary to make the same reservations as for decametric wave instruments.

However, when we come to decimetre and centimetre waves, we generally use "maximum" aerials (or "equality" aerials, which comes to the same thing), in which the useful component of the field, in phase over a large surface, is selected (paraboloids, cornets, horns, etc.). Under these conditions, the presence of an irregularity or a very localized deformation, as a result of neighbouring obstacles, is much less troublesome. The "mask" and "deviation" effects decrease; objects placed behind the aerial have no effect except through the "secondary lobes" of the diagram. This is why radars can be pointed correctly at their targets, even on board warships with cumbersome superstructures.

8.5.4. "The radar equation"

This is the name given to the relationship which gives the field reflected by an obstacle and received by electromagnetic detection.

The relationship is very simple as long as we confine ourselves to *free space;* it comes out immediately from the basic formulae for radiation in free space (§ 1.1.3) and a knowledge of the "equivalent cross-section" of the obstacle (§ 4.8.6).

Consider a transmitter of power W (watts) using an aerial whose energy gain with respect to an "isotropic radiator" is G. In the preferred direction, it is as if the radiated power were GW and the energy density per unit area (m²) at distance d (m) is

$$\delta\left(\frac{\mathrm{W}}{\mathrm{m}^2}\right) = G\frac{W}{4\pi d^2}.$$

Let us assume that there is in this zone an obstacle whose "equivalent" cross-section is σ, i.e. which, by definition, captures the energy passing through the area σ and re-radiates it isotropically; it will, therefore, behave like a "secondary radiator" of power:

$$w_2\,(\mathrm{W}) = \frac{\sigma\,(\mathrm{m}^2)}{4\pi d^2\,(\mathrm{m})} \times G\cdot W\,(\mathrm{W}),$$

and the same reasoning, in the opposite direction, shows that a receiving aerial of cross-section $S'(\text{m}^2)$ will capture an energy:

$$w'\,(\text{W}) = \frac{\sigma S' G W}{16\pi^2 d^4};$$

the range corresponding to a given sensitivity w' of the receiver is, therefore,

$$d = \sqrt[4]{\frac{\sigma S' G W}{16\pi^2 w'}}.$$

It can be seen that it is proportional to the *fourth* root of the factors σ, S', G, $1/w'$, and, especially, the power W.

This formula can be expressed in various forms, depending on whether the *gains*, or the *cross-sections* of the aerials are given, and the "equivalent cross-section" σ has several variants.

In particular, we saw in § 1.1.1 that the *maximum* theoretical gain of an aerial of cross-section S, with respect to an isotropic aerial, is

$$G = \frac{4\pi S}{\lambda^2},$$

but that in practice, it is difficult to achieve half this figure.

If, therefore, the transmitting aerial has a cross-section S, we can replace G in the preceding formula by (2 to 4) $\pi S/\lambda^2$; and conversely, for the cross-section S' of the receiving aerial, we can substitute its gain:

$$G' = (2 \text{ to } 4)\,\pi \frac{S'}{\lambda^2},$$

which gives the two variants:

$$d = \sqrt[4]{\frac{\sigma S S' W}{(4 \text{ to } 8)\,\pi \lambda^2 w'}} = \sqrt[4]{\frac{\sigma G G' \lambda^2 W}{(32 \text{ to } 64)\,\pi^3 w'}}.$$

In general, for mobile radars, one is limited by the cross-section of the common transmitting–receiving aerial; therefore, $S = S'$ is given, and the first variant shows that it is advantageous to *reduce the wavelength* (at least, as long as the decrease in the transmitted power W and the increase in the power necessary at the receiver w'—due to the noise factor—do not cancel this advantage).

Since this also increases the sharpness of the beam and therefore the accuracy of aiming, it is easily understood why the evolution of radar has, in fact, tended towards smaller λ.

But in practice, the gain does not increase indefinitely with S; imperfections in the reflectors limit G (to about 10,000, for example); once this value has been obtained, the second variant of the formula shows that it would be better to increase λ and the cross-section of the receiving aerial. Large ground radars still use relatively long waves, for example metric waves.

Numerical application is simple; take, for example,

$W = 100$ kW,

$S = S' = 4$ m^2,

$\lambda = 0{\cdot}1$ m,

$w' = 10^{-13}$ W (normal sensitivity of a good receiver).

We saw in § 4.8.6 that the equivalent cross-section σ of an aircraft can easily be 10 m^2.

We therefore have:

$$d = \sqrt[4]{\frac{10 \times 16 \times 10^5}{(4 \text{ to } 8)\,\pi \times 10^{-2} \times 10^{-13}}} = 160\text{–}190 \text{ km},$$

which is the order of magnitude of the range obtained for powerful radars.

The above applies to propagation *in free space*, i.e. for example, the detection of aircraft or rockets at high altitude.

For low-altitude aircraft, and even more so for ground or floating targets, the wave reflected by the surface must be taken into account, giving the factor $2 \sin (2\pi h_1 h_2/\lambda d)$ from § 4.3.1, which reduces to $4\pi h_1 h_2/\lambda d$ in the bottom lobe.

This factor multiplies both the outgoing and incoming fields; it therefore appears squared in the field strength and to the fourth power in the received energy, which becomes:

$$w' = \left[2 \sin\left(\frac{2\pi h_1 h_2}{\lambda d}\right)\right]^4 \times \frac{\sigma S' G W}{16\pi^2 d^4}.$$

The sine factor naturally introduces lobes with maxima and minima into the vertical diagram, as in § 4.3.1 and consequently considerable irregularities in range into the "coverage" zone of the radar. Given that it is important to know the angles at which an aircraft can, because of this, escape detection, a "coverage" diagram is often constructed for each radar (i.e. for given h_1, λ, S', G, W). This diagram consists of lines of "constant field strength" (or constant w'), with distance d in abscissa and height h_2 of aircraft in ordinate (on an exaggerated scale). The Earth's curvature is usually also plotted.

Figure 8.24 shows such a diagram for a transmitter situated at height h_1 = 110 ft above sea level, for three wavelengths

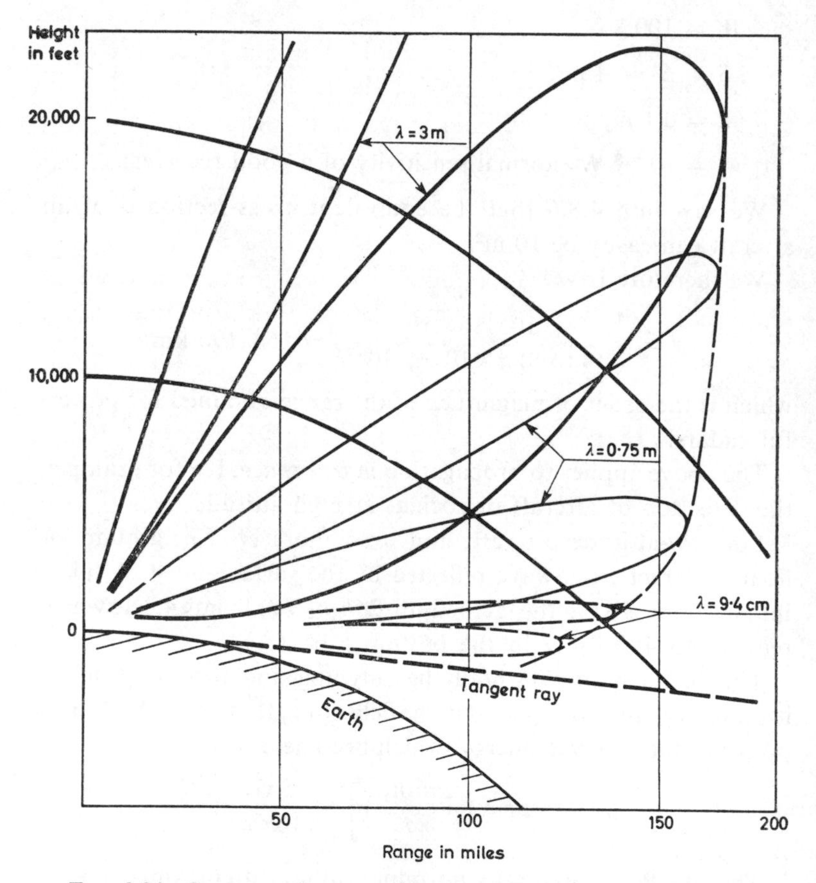

FIG. 8.24. Coverage diagram with h_1 = 110 ft (curves −43 dB relative to 1 km). Wavelengths: 3 m, 0·75 m, 9·4 cm

$\lambda = 3$ m, $\lambda = 0{\cdot}75$ m and $\lambda = 9{\cdot}4$ cm (only the lowest two lobes of the interference zone are shown for each wavelength, for a field 43 dB below the free field at $d = 1$ km).

Figure 8.25 shows a similar diagram for a transmitter situated at $h_2 = 9$ m above sea level and $\lambda = 10$ cm (vertical polarization). The levels of the curves are in decibels with respect to the free field at 1 km.

The range at the minima is of the order of a tenth of the range at the maxima.

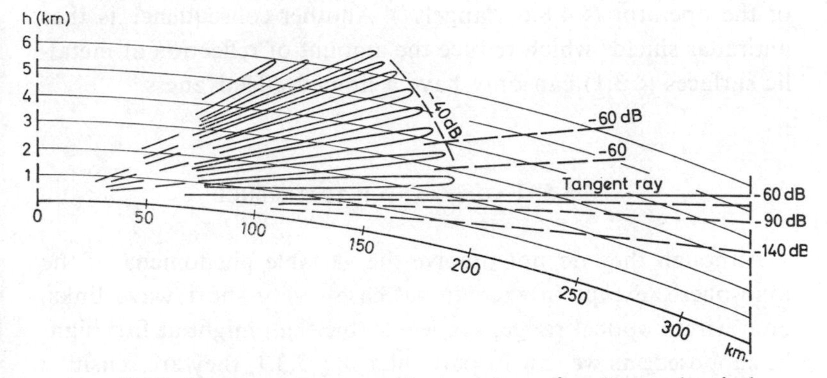

FIG. 8.25. Coverage diagram with $h_1 = 9$ m, $\lambda = 10$ cm (vertical polarization over sea). Field strength curves given in decibels below the field at 1 km

At limiting range corresponding to minimum altitude, the sine becomes equal to the angle and the formula becomes

$$d = \sqrt{\frac{h_1 h_2}{\lambda}} \cdot \sqrt[8]{16\pi^2 \sigma S' G \frac{W}{w'}}\,.$$

The range only increases as the eighth root of the power; but it increases as the square root of the heights and of $1/\lambda$, and hence, it is advantageous to increase h_1 and decrease λ more and more as one wishes to detect lower and lower obstacles; this is why navigation or surveillance radars at surface level are often at wavelengths of $\lambda = 3$ cm and placed relatively high in the masts.

Finally, we should naturally have to take account of random superrefraction at sea, as indicated in § 5.3.2.

The above applies to the detection of relatively large obstacles: aircraft, ships, etc., whose "equivalent cross-section" is of the order of several square metres (or more). The formula shows that the range decreases quite slowly with equivalent cross-section σ; in other words, obstacles which are much smaller or less reflective are still detectable at considerable distances: thus, mortar shells, birds and sometimes even insects, pedestrians, clouds, discontinuities in the atmosphere, appear—systematically or accidentally—on radar screens, sometimes catching the imagination of the operator (§ 4.8.6, "angels"). Another consequence is that antiradar shields which reduce the amount of reflection at metallic surfaces (§ 3.1) can only have a limited effectiveness.

8.5.5. Fading and interference

Although they do not involve the variable phenomena of the ionosphere, except in exceptional cases, very short wave links, even within optical range, are less stable than might at first sight be supposed: as we saw in particular in § 5.3.1, they are sensitive to variations in atmospheric refraction and absorption, to variations in ground reflections, to interference from all the components reflected from obstacles, following multiple trajectories whose lengths are not fixed.

In the section mentioned above, we gave one or two indications as to the techniques one can adopt to reduce the dispersion of fadings and attenuate their effects, especially, by means of "diversity". They are in current usage in Hertzian beams, using relays in line of sight, to ensure transmission of a large number of telephone channels or television programmes. The techniques of diversity in space, frequency or arrival angle are also in general use in "trans-horizon" Hertzian beams and links using "ionospheric diffusion" (§ 5.3.3).

The type of interference which affects the long wave bands is quite harmless at very short waves. Atmospherics are almost non-existent. Artificial interference becomes less intense, except that due to motor vehicle ignition systems, which is very strong at metric waves and seriously affects television in towns; but it is practically negligible at decimetre and centimetre waves.

The sensitivity is ultimately limited either by the intrinsic background noise of the receiver (hence the importance of the "noise factor" or its "noise temperature"), the parametric amplifier or the maser, or by extra-terrestrial interference, galactic noise or the thermal noise of the atmosphere (see §§ 7.2, 7.3 and 8.7).

8.6. Submarine and subterranean propagation

We studied in § 2.6 the propagation and attenuation of a plane wave propagating in a conductive medium and in the sea, in particular.

These formulae apply to all subterranean and submarine propagation as if the waves were propagated vertically downwards after crossing the surface of separation without discontinuity. Thus the field strength at a depth z is obtained by taking the field strength at the surface (whatever its origin: direct wave, indirect wave, etc.) and applying to it a reduction factor $e^{-\alpha z}$, where α is given by the formula in § 2.6 (the "skin effect" formula) in terms of the conductivity σ and the magnetic permeability μ.

Figure 8.26 shows this attenuation for sea water ($\sigma = 4$ mho/m). It can be seen that reception is only possible, for submerged submarines, a few metres below the surface and at very long waves.

The problem of penetration into the earth is sometimes posed (underground shelters, mine shafts, etc.). As the conductivity in this case is 400–4000 times smaller, the penetration is the same for frequencies 400–4000 times higher, i.e. for example, at 100 kc/s the attenuation would be 1 neper at a depth of 15–50 m (45–160 ft). Reception of medium waves is thus possible at a certain depth. It must be noted, however, that in a large number of cases, subterranean passages have metallic conductors running along them (rails, electric cables, etc.) which pick up the field outside and facilitate its propagation inside the passage.

This very rapid attenuation, however, appears to cast doubts on the feasibility of subterranean transmitting aerials and communication between deep shelters.

Certainly, such conditions are unfavourable. However, we should not be too pessimistic. For one thing, the conductivity of most soils is very much less than that of the sea—so, therefore, is the attenuation. Secondly, recent work† has shown that if we consider two stations, buried at depths Z_1 and Z_2, at a horizontal distance of D much greater than Z_1 and Z_2, the propaga-

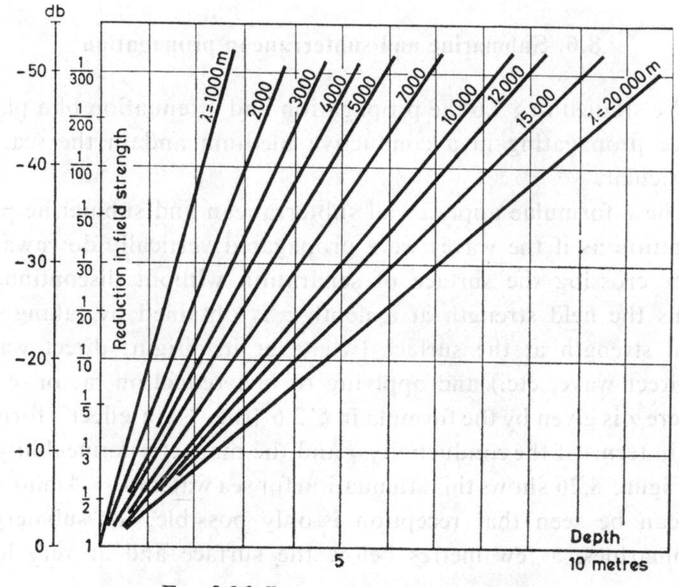

FIG. 8.26. Penetration below the sea

tion attenuation is not the same as would occur over a distance D in the soil: instead the wave rises to the surface, becomes refracted and then propagates along the surface and penetrates the ground again; the attenuation is thus merely that which results from the trajectory (Z_1+Z_2) in the soil, plus the trajec-

† See, for example, Lunden, *Radio-communications in the 100-miles Range using Buried Antennas*, Globe Com. IV, Washington, 1960; Earth current communications system, *Aviation Week*, Dec. 19, 1960, p. 31; Ghose, Subsurface communications for survival, *Electronics*, Feb. 3, 1961, pp. 43–5; Anderson *Proc. I.R.E.*, Mar. 1961, p. 645.

tory *D* on the surface. Ranges of the order of 60 miles are not therefore impossible.

Finally, we can no longer discount the possibility that certain subterranean layers, or perhaps the Earth's crust under the sea, give rise to appreciable wave-guide effects.†

8.7. Space telecommunications‡

Whether for actual communication, telemetry, telecommand, radiodetection or radionavigation, space radiocommunication takes place, in the large majority of cases, over direct-line-of-sight trajectories, sometimes of very great length (over 100 million miles for Mariner IV). One hopes to obtain free space propagation conditions, but, more often than not, the link between a ground station and a space craft passes through the troposphere and part, if not all, of the ionosphere where ionization is particularly high. We shall now study the most important consequences of these conditions.

8.7.1. Influence of the Troposphere

As we saw in § 5.4, the gases present in the air (oxygen and water vapour) and occasionally rain, storms and even fog produce appreciable absorption at centimetre and shorter wavelengths. The absorption due to the gases in the air increases as the altitude of the ground station decreases (Fig. 8.27) or the humidity of the atmosphere increases. It also increases as the inclination ϕ of the propagation trajectory above the horizon decreases (approximately as $1/\sin\phi$). The attenuation (in decibels) for a trajectory of inclination ϕ is equal to the attenuation corresponding to a vertical trajectory multiplied by 2 for $\phi = 30°$, 5·5 for $\phi = 10°$, 11 for $\phi = 5°$ or 80 for $\phi = 0°$.

† Wheeler, *J. Res. Nat. Bur. Stand.*, March–April 1961, **65**, 2, pp. 189–91.

‡ C.C.I.R. Report 205, Geneva, 1963.

The attenuation factor for rain or fog can be determined with the help of the curves of Fig. 5.18; but, in order to make a precise evaluation it would be necessary to possess statistical data on the horizontal and vertical extent of the rainy, cloudy or foggy zones.

Statistics valid for the south of England lead, for 1% of the time, to an attenuation of 0·25 dB by rain and 0·7 dB by clouds, for an elevation of 5° and a frequency of 4000 Mc/s.

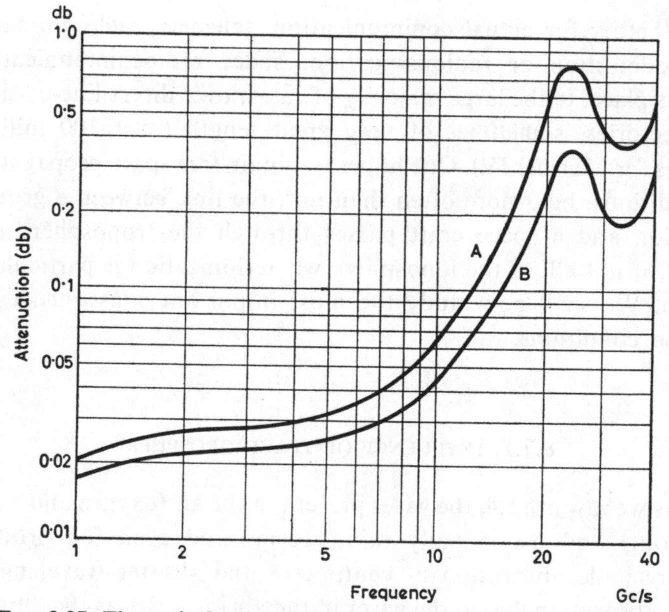

FIG. 8.27. Theoretical attenuation for one direction of propagation, vertical trajectory, area of average humidity (Washington, August). Curve *A*: sea level; curve *B*: at a height of 2 km above sea level

We saw in § 7.3 that tropospheric absorption increased the noise level (or temperature) elsewhere at very high frequencies.

Taking account of these phenomena we can settle on an average value of 20 Gc/s for the maximum usable frequency for ground-space craft communication. Under favourable conditions (high altitude station, dry climate), we might go as far as 50 Gc/s; in rainy climates or seasons, it is not suitable to exceed 10 Gc/s.

8.7.2. Influence of the ionosphere

As we saw in Chapter 6, the *E* and *F* layers can constitute an impermeable screen and the *D* layer can give rise to appreciable absorption at frequencies in the decametre range or lower.

In order to avoid these phenomena, it may be assumed that frequencies of 10–20 Mc/s constitute a lower limit for space communications under average conditions (average solar activity, vertical trajectory in the polar regions or oblique in the tropics). Under favourable conditions (vertical trajectory, night-time, in the polar regions and during a period of weak solar activity), the limit could be as low as 2 Mc/s. On the other hand, with an oblique trajectory, by day, in the tropics and during a period of intense solar activity—or, in general terms, in the presence of a severe ionospheric perturbation (§ 6.4)—a high lower limit, say 70 Mc/s might be necessary.

8.7.3. Choice of frequencies for space radiocommunications

The above considerations explain the choice of frequency bands reserved for space communications: they fall in the decimetric, upper centimetric and lower metric ranges. Bands lying at the ends of this "transparent" range (10, 15, 18, 20, 30, 40 Mc/s and 15, 31, 34 Gc/s) are allocated for space research as the absorption due to the ionosphere or the troposphere can then provide information about the characteristics of the medium traversed.

8.7.4. Other propagation phenomena

8.7.4.1. *Refraction and scintillation*

Since the atmosphere is not homogeneous (the temperature, pressure and humidity in the troposphere and the ionization density in the ionosphere vary with altitude), radio waves between the ground and a space craft are bent by refraction. Consequently, the rays reach the ground at a different angle of elevation ϕ from that which would obtain in a homogeneous atmosphere (straight rays), say ϕ_0. The order of magnitude of the deviation $\Delta\phi = \phi - \phi_0$ can thus be fixed. It depends on the inclination ϕ_0:

$\Delta\phi$	$\phi_0 = 1°$	$\phi_0 = 5°$	$\phi_0 = 10°$
Average troposphere	0·5°	0·2°	0·1°
Ionosphere, daytime, 400 Mc/s	0·2°	0·1°	0·03°

The deviation $\Delta\phi$ increases with humidity in the case of the troposphere. In the case of the ionosphere, it varies approximately as $1/f^2$, where f is the frequency, and the indicated values should be reduced by a factor of about 3 for use at night.

Atmospheric turbulence also introduces scintillation in the received levels and a rapid fluctuation in the direction of incidence, with an amplitude which is often of the order of a tenth of the deviation $\Delta\phi$, but sometimes much greater.

These refraction and scintillation errors—which are also seen in radar sets—can cause errors in the location of spacecraft. It is also possible to lose contact if very narrow beam aerials are programmed to point in the theoretical direction.

8.7.4.2. *The Doppler–Fizeau effect*

When a spacecraft transmits a signal to a ground station the received frequency differs from the transmitted frequency, the frequency shift being proportional to the ratio of the radial velocity v_r of the space-ship to the velocity of light in free space c:

$$\frac{\Delta f}{f} = \pm \frac{v_r}{c} \tag{8.8}$$

(negative if the spacecraft is receding from the ground station).

For a radial velocity of 10 km/s, the velocity required to escape from the Earth's pull, the relative frequency shift would thus be 3×10^{-5}.

Formula (8.8) is, however, rigorously valid only in the case of propagation in a vacuum (refractive index = 1). Since the propagation medium is, in reality, inhomogeneous ($n \neq 1$) and the inhomogeneities cause the rays to bend, the optical lengths of the propagation trajectories become modified and the Doppler–Fizeau shift is not exactly what it would be *in vacuo*, say Δf, but $\Delta f + \delta \Delta f$, where $\delta \Delta F$ denotes the deviation due to the inhomogeneities of the medium.

The order of magnitude of $\delta \Delta f$ for an average troposphere and ionosphere can be calculated, assuming $v_r = 10$ km/s.

f	100 Mc/s	1000 Mc/s
Δf ($v_r = 10$ km/s)	3 kc/s	30 kc/s
$\delta \Delta f$ (average troposphere, $\phi = 5°$)	1 c/s	10 c/s
$\delta \Delta f$ (average ionosphere, daytime, $\phi = 5°$)	60 c/s	6 c/s

$\delta \Delta f$ varies roughly as $1/\sin \phi$ where ϕ is the elevation of the propagation trajectory above the horizon. $\delta \Delta f/f$ is independent of frequency for the troposphere and varies as $1/f^2$ in the case of the ionosphere. Finally, the value of $\delta \Delta f$ corresponding to a night-time ionosphere would be about 3 times smaller than that indicated for the day-time ionosphere.

From a practical point of view, this phenomenon forces us to use receivers with a wide enough bandwidth to include the Doppler frequency shift or which have automatic frequency or phase control. If the frequency transmitted by a spacecraft is very stable and known precisely, measurement of the Doppler shift in a receiver on the ground permits the speed of the spacecraft, and thus its position with respect to the receiving station, to be determined: this is the principle of the "Transit" system of geodesy and radionavigation by satellite whereby, in the case of satellites with accurately known orbits, the converse operation can be performed, namely the location of a ground station (fixed or mobile, ships in particular). The precision is limited by tropospheric and ionospheric influences. A first estimate of the tropospheric error can be provided by a meteorological probe of the atmosphere. The ionospheric error can be eliminated if transmissions are made simultaneously on two frequencies, say 150 and 400 Mc/s and the Doppler shift measured, using the fact that, theoretically, $\delta\, \Delta f/f$ varies as $1/f^2$.

Note. In the case of vehicles travelling at speeds approaching the speed of light, the Doppler shift $\Delta f/f$ would be very great. But another phenomenon would then appear: the radiation diagram of the spacecraft aerials would become very deformed and concentrated forwards, in the direction of motion (as in the case of relativistic electron radiation). If the vehicle were moving away from Earth, communication with a ground station would become very difficult.

8.7.4.3. *The Faraday effect*

It is known that when a rectilinearly polarized plane wave is propagated in a dispersive medium (with refractive index a function of frequency) in the presence of a d.c. magnetic field, rotation of the direction of polarization takes place around the direction of the magnetic field. The effect is a maximum when propagation is longitudinal in the direction of the magnetic field.

The ionosphere (and, indeed, the exosphere and interplanetary space), which consists of an ionized plasma, is a dispersive medium. The earth's magnetic field (which is of the order of a fraction of an oersted at sea level and decreases as r^{-3} with distance r from the centre of the earth) therefore induces the Faraday effect in radiation passing through it.

The angle of rotation of the polarization direction varies directly as the ionization density N, the amplitude H_0 of the magnetic field (or, more accurately, its component in the propagation direction) and the length of the trajectory in the ionosphere and inversely as the square of the frequency.

For a satellite at a height of 1000 km and a direction of propagation coincident with that of the Earth's magnetic field, the following orders of magnitude can be calculated for the angle of rotation α of the polarization between the satellite and the ground:

	100 Mc/s	1000 Mc/s
Day-time, strong solar activity	9200°	92°
Night-time, weak solar activity	75°	0·75°

A rotation of 90° would result in a vertically polarized transmission from a spacecraft being received at ground level as a horizontally polarized signal. There is thus a risk of serious fading. This is often avoided by using circular polarization; diversity reception can also be used, with two receiving aerials, one for horizontal polarization and the other for vertical polarization.

8.7.5. Communication with spacecraft during re-entry into the atmosphere†

During re-entry, at high speed, into the atmosphere a spacecraft is subject to extreme deceleration accompanied by a considerable increase in temperature, and the formation, by thermal ionization, of a "shield" or thin layer of plasma surrounding the rocket and able to stretch out behind it like a sort of wake.

The ionization density N reaches quite high levels (order of 10^{12} electrons/cm^3) and frequencies below the critical frequency cannot penetrate the shield. The formulae connecting ionization density and critical frequency for the ionosphere still apply to a first approximation, although for precise calculations, it is necessary to bring in the inhomogeneities of the plasma and the frequency of the shocks to which the electrons are subjected, a frequency which appears to be relatively low compared with the critical frequency. The calculation of the ionization density in the plasma shield follows from a preliminary determination, itself quite intricate, of the aerodynamic and thermodynamic characteristics of the flux surrounding the spacecraft; these characteristics depend on the shape of rocket, its speed and its height. Figure 8.28 shows the theoretical speed–height curves for re-entry for the Mercury and Apollo space capsules.

† C.C.I.R. Report 222, Geneva, 1963.

The critical frequencies—below which communications are practically impossible—often lie between 1 and 10 Gc/s and sometimes even higher. In the case of the Mercury capsule the crew had the use of a radio-telephone transmitter on a frequency of the order of 200 Mc/s and a radar beacon on a frequency of about 6 Gc/s (*C*-band). The telephonic link was broken in the part of the trajectory lying below the curve marked "VHF" whilst radar detection at *C*-band was assured at all times, although with appreciable attenuation in the region where the "Mercury" curve and the "*C*-band" curve coalesce. For the "Apollo" capsules, cut-

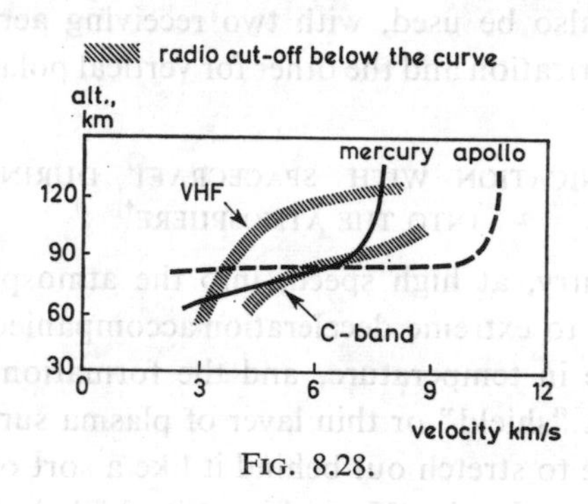

Fig. 8.28.

off would occur even at *C*-band and it would be necessary to have recourse to even higher frequencies. We then land in the range where atmospheric absorption is appreciable (§ 5.4) and full use would have to be made of the minimum absorption windows for oxygen and water vapour at frequencies of the order of 90 Gc/s or 140 Gc/s. But at these frequencies the slightest rain can cause interruption of communications.

The unfavourable influence of the plasma shield also makes itself felt in the functioning of the aerials by modifying their impedance and reducing their efficiency.

It has been suggested that the plasma shield might be injected with substances which have a strong affinity for electrons (so as to neutralize it) or that a magnetic field intense enough to produce a discontinuity in the shield might be created. Work on these suggestions is still going on.

INDEX

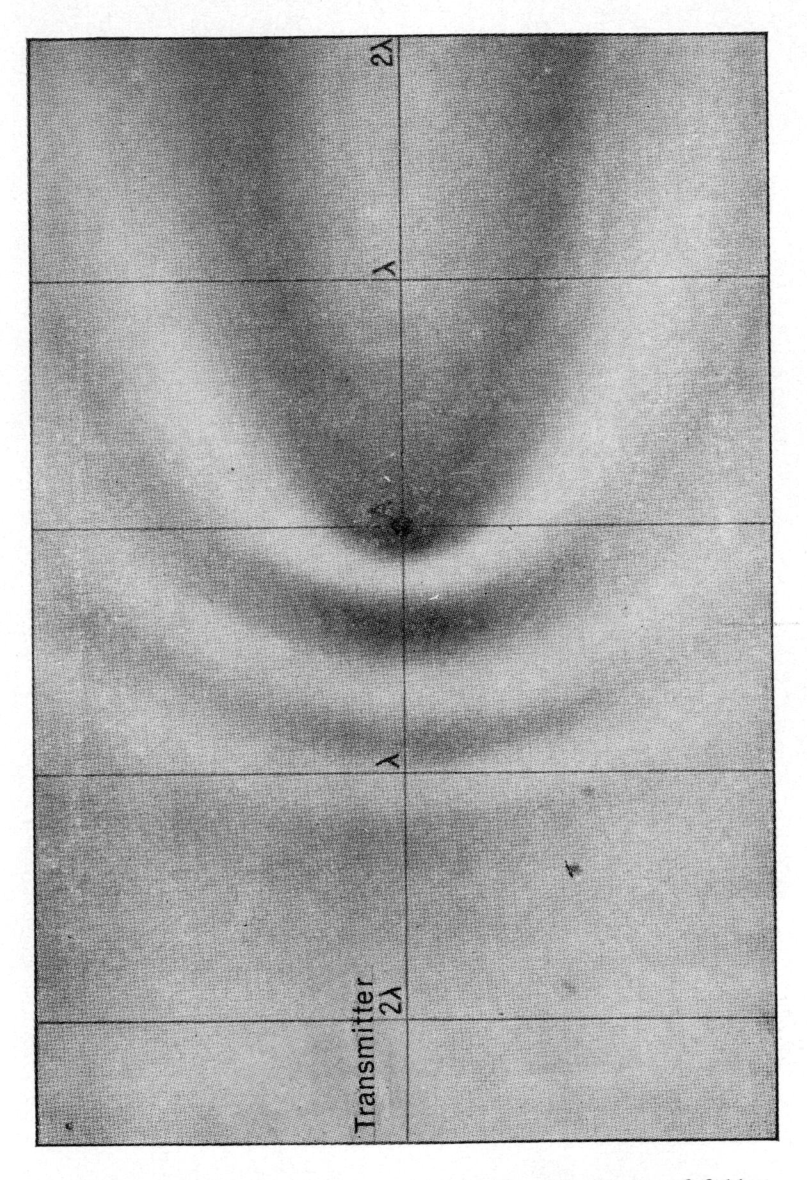

PLATE 1. ~~Hyperbolic~~ reinforcement and attenuation zones of field strength in the vicinity of an obstacle (tuned dipole)

PLATE 2. (Top) Automatic ionospheric probe equipment in the Ionospheric Forecasting Division (D.P.I.) of the Centre National d'Etudes des Télécommunications. (Bottom) Specimen photographic recording of probing results. Horizontal scale: frequency in megacycles per second (logarithmic scale, increasing to the left. Vertical scale: virtual height of the layer (points at 150 km intervals). At the right can be seen the *E* layer, for which the critical frequency is about 3·2 Mc/s; then comes a small intermediate layer *I*; the F_1 layer, with two distinct components, the "ordinary" and the "extraordinary"; and, on the left, the F_2 layer, also split, and above it an echo reflected from F_2 then the ground and then F_2 again (from a D.P.I. probe made at Lwivo, Congo, at 9 a.m. on 19 January 1953)

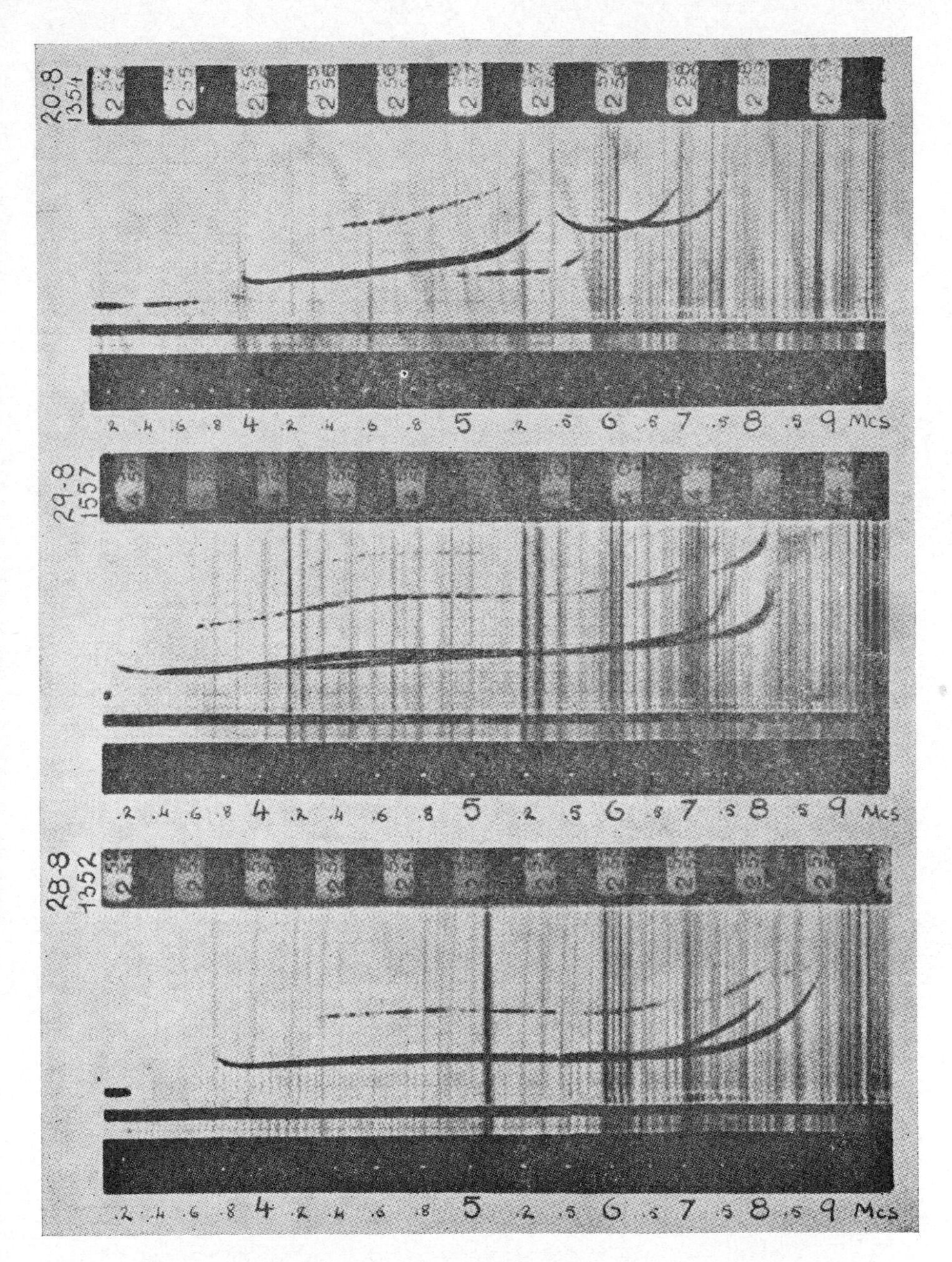

PLATE 3. Features of F_1 layer probes: *below*, the two components (ordinary and extraordinary) are coincident up to a frequency of 7 Mc/s; their critical frequencies then differ by about 0·7 Mc/s; in the other two illustrations (especially the upper one), the splitting is more marked. (Measurements made at Poitiers by M. Rivault, 1948.)

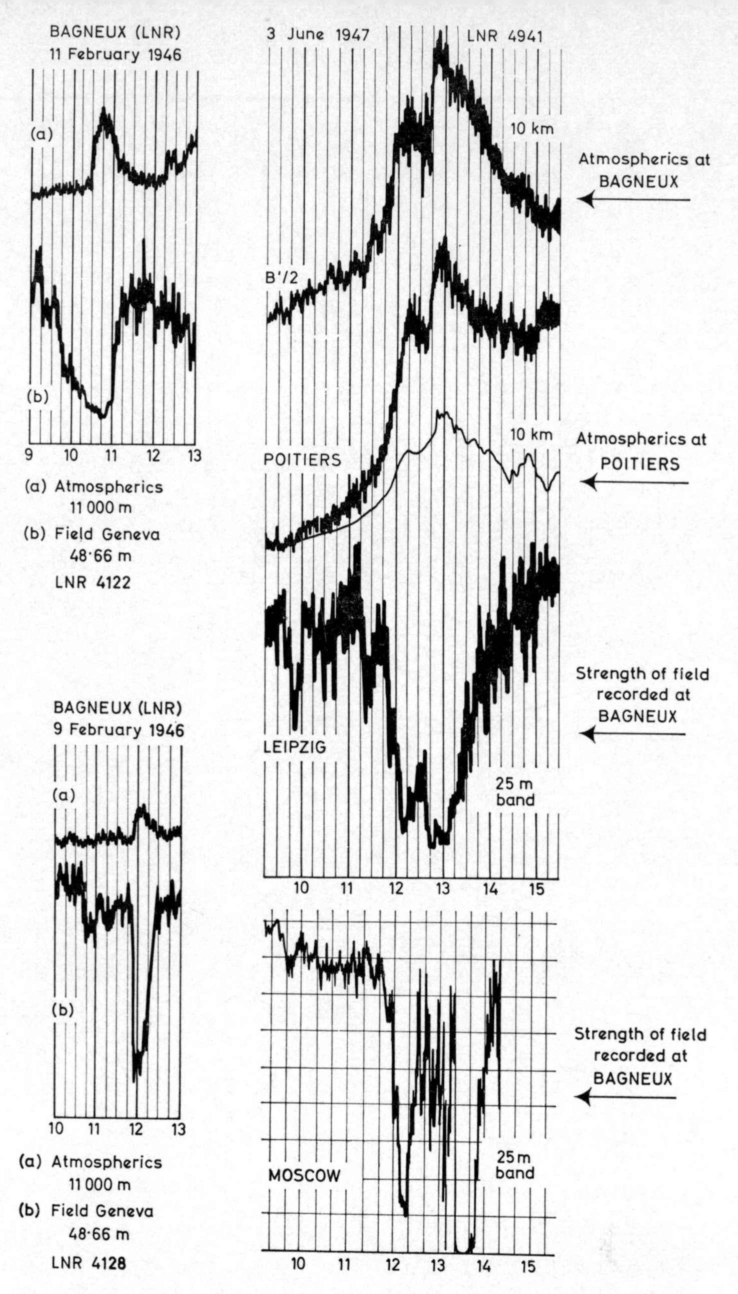

PLATE 4. Simultaneous measurements of atmospherics on long waves and the field due to distant signals on short waves, showing that the same ionospheric perturbation often manifests itself as an increase in the level of the former and a decrease in the latter. *Top left:* upper curve: atmospherics at 11 km; lower curve: field strength from Geneva on 48 m recorded at Bagneux, 11 February 1946. *Bottom left:* ditto, 9 February, 1946. *Right*: the upper two curves are the atmospherics on 10 km at Bagneux and Poitiers—the lower two curves are the field strengths from Leipzig on 25 m and Moscow on about 25 m, recorded at Bagneux, 3 June 1947. *(Laboratoire National de Radio-Électricité.)*